AF358648

Recent Advancements in Metallic Glasses

Recent Advancements in Metallic Glasses

Editor

Vitaly A. Khonik

MDPI • Basel • Beijing • Wuhan • Barcelona • Belgrade • Manchester • Tokyo • Cluj • Tianjin

Editor
Vitaly A. Khonik
Department of General Physics,
Voronezh State Pedagogical University
Russia

Editorial Office
MDPI
St. Alban-Anlage 66
4052 Basel, Switzerland

This is a reprint of articles from the Special Issue published online in the open access journal *Metals* (ISSN 2075-4701) (available at: https://www.mdpi.com/journal/metals/special_issues/metallic_glasses).

For citation purposes, cite each article independently as indicated on the article page online and as indicated below:

LastName, A.A.; LastName, B.B.; LastName, C.C. Article Title. *Journal Name* **Year**, *Volume Number*, Page Range.

ISBN 978-3-0365-0880-1 (Hbk)
ISBN 978-3-0365-0881-8 (PDF)

Contents

About the Editor

Vitaly A. Khonik born in 1955, city of Kemerovo, Russia. Candidate of Sciences (Physics and Mathematics) degree in Solid State Physics awarded in 1983, Doctor of Sciences (Physics and Mathematics) degree Solid State Physics awarded in 1992. Professor academic rank since 1994. Honorary Worker of higher professional education of the Russian Federation (2011). Published 220 papers on metallic glasses in international peer reviewed physics and materials science journals.

Editorial

Recent Advancements in Metallic Glasses

Vitaly A. Khonik

Department of General Physics, State Pedagogical University, Lenin St. 86, 394043 Voronezh, Russia;
v.a.khonik@vspu.ac.ru

Received: 5 November 2020; Accepted: 17 November 2020; Published: 19 November 2020

Investigations of the structure and properties of metallic glasses constitute a subject of unabated interest. The Web of Science database gives 1400 to 1500 papers on metallic glasses published per year from 2015 to 2019. The present Special Issue "Recent Advancements in Metallic Glasses" presents ten interesting works considering both general scientific and application issues related to metallic glasses.

Seleznev and Vinogradov [1] report a detailed study of shear bands occurring in several bulk metallic glasses deformed by compression and indentation. The high spatial resolution achieved by scanning white-light interferometry gives the evidence for interaction between slip bands, as well as provides information on different scenarios of their propagation. The elastic field around a tip of the slip band was found similar to the elastic field of a Volterra-type macro-dislocation, revealing edge-like and screw-like shear components. The authors conclude that the slip bands can be described in terms of Volterra's macro-dislocations and their behavior under the load can be rationalized accordingly.

Tian et al. [2] studied the structural and mechanical properties of hydrogen-charged metallic glass. They found that nanoindentation reveal a clear increase in the elastic modulus and hardness as well as in the load of the first pop-in with increasing hydrogen content. At the same time, the probability of a pop-in decreases, indicating that hydrogen hinders the onset of plastic instabilities while allowing local homogeneous deformation. The hydrogen-induced stiffening and hardening is rationalized by hydrogen stabilization of shear transformation zones, while the improved ductility is attributed to a change in the spatial correlation of these zones.

Gunderov and Astanin [3] present a review on the influence of high-pressure torsion on the structure and properties of amorphous alloys. The variation of structure due to the processing parameters (shear strain, processing temperature, imposed pressure) is considered. Amorphous alloys with a nanocluster structure and nanoscale inhomogeneities significantly differ in their physical and mechanical properties from conventional metallic glasses. High-pressure torsion processing constitutes an efficient way to produce nanocluster structures and improve the properties of metallic glasses.

Aronin and Abrosimova [4] present a brief overview of the structure and properties of amorphous–nanocrystalline metallic alloys. The transformation of a homogeneous amorphous phase into a heterogeneous phase, the dependence of the scale of inhomogeneities on the component composition, and external influences are considered. The mechanical properties in the structure in and around shear bands are discussed.

Permyakova and Glezer [5] present a detailed study of the preparation method and specific features of the changes in the structure and properties of amorphous–nanocrystalline composites formed by high-pressure torsion of melt-quenched iron- and cobalt-based amorphous ribbons and Cu–Nb crystalline nanolaminates.

Abrosimova et al. [6] studied the devitrification of $Zr_{55}Cu_{30}Al_{15}Ni_5$ bulk metallic glass upon annealing and high-pressure torsion deformation. They found the formation of different crystalline phases upon heating and deformation. At the first crystallization stage, a metastable phase with a hexagonal structure is formed. A further temperature rise results in the formation of stable crystalline phases. It was determined that nanocrystals upon deformation are formed primarily in the subsurface regions of samples.

Liu et al. [7] studied the mechanical relaxation behavior of La-based metallic glasses probed by dynamic mechanical analysis. They found that the intensity of the secondary β relaxation decreases after physical aging below the glass transition temperature, which is probably due to the reduction in the atomic mobility induced by physical aging.

Khonik and Kobelev [8] present a brief overview of the Interstitialcy theory as applied to different relaxation phenomena occurring in metallic glasses upon structural relaxation and crystallization. The basic hypotheses of this theory and their experimental verification are briefly considered. The main focus is given to the interpretation of recent experiments on the heat effects, volume changes, and their link with the shear modulus relaxation.

Makarov et al. [9] apply the Interstitialcy theory to show that the kinetics of endothermal and exothermal effects occurring in the supercooled liquid state and upon crystallization of metallic glasses can be well-reproduced using the temperature dependences of their shear moduli. They argue that the interrelation between the heat effects and shear modulus relaxation reflects the thermally activated evolution of an interstitial-type defect system.

The paper by Wang et al. [10] reports on a new type of Mg-based metallic glass, which demonstrates excellent corrosion resistance and favorable biocompatibility. In their study, the authors prepared amorphous/crystalline composite Mg–Rare Earth alloy sheets by a twin roll caster method and studied their corrosion resistance behavior. The samples were implanted into the femur of rats to study its prospect as biological transplantation material. The experiments showed that Mg–RE (La, Ce) sheets have good corrosion resistance and, as an implant material, induce new bone formation, which is why they can be considered as a promising implant material.

Conflicts of Interest: The author declares no conflict of interest.

References

1. Seleznev, M.; Vinogradov, A. Shear Bands Topology in the Deformed Bulk Metallic Glasses. *Metals* **2020**, *10*, 374. [CrossRef]
2. Tian, L.; Tönnies, D.; Hirsbrunner, M.; Sievert, T.; Shan, Z.; Volkert, C.A. Effect of Hydrogen Charging on Pop-in Behavior of a Zr-Based Metallic Glass. *Metals* **2020**, *10*, 22. [CrossRef]
3. Gunderov, D.; Astanin, V. Influence of HPT Deformation on the Structure and Properties of Amorphous Alloys. *Metals* **2020**, *10*, 415. [CrossRef]
4. Aronin, A.; Abrosimova, G. Specific Features of Structure Transformation and Properties of Amorphous-Nanocrystalline Alloys. *Metals* **2020**, *10*, 358. [CrossRef]
5. Permyakova, I.; Glezer, A. Amorphous-Nanocrystalline Composites Prepared by High-Pressure Torsion. *Metals* **2020**, *10*, 511. [CrossRef]
6. Abrosimova, G.; Gnesin, B.; Gunderov, D.; Drozdenko, A.; Matveev, D.; Mironchuk, B.; Pershina, E.; Sholin, I.; Aronin, A. Devitrification of $Zr_{55}Cu_{30}Al_{15}Ni_5$ Bulk Metallic Glass under Heating and HPT Deformation. *Metals* **2020**, *10*, 1329. [CrossRef]
7. Liu, M.; Qiao, J.; Hao, Q.; Chen, Y.; Yao, Y.; Crespo, D.; Pelletier, J.-M. Dynamic Mechanical Relaxation in LaCe-Based Metallic Glasses: Influence of the Chemical Composition. *Metals* **2019**, *9*, 1013. [CrossRef]
8. Khonik, V.; Kobelev, N. Metallic Glasses: A New Approach to the Understanding of the Defect Structure and Physical Properties. *Metals* **2019**, *9*, 605. [CrossRef]

9.   Makarov, A.; Afonin, G.; Mitrofanov, Y.; Kobelev, N.; Khonik, V. Heat Effects Occurring in the Supercooled Liquid State and Upon Crystallization of Metallic Glasses as a Result of Thermally Activated Evolution of Their Defect Systems. *Metals* **2020**, *10*, 417. [CrossRef]
10. Wang, H.; Ju, D.; Wang, H. Preparation and Characterization of Mg-RE Alloy Sheets and Formation of Amorphous/Crystalline Composites by Twin Roll Casting for Biomedical Implant Application. *Metals* **2019**, *9*, 1075. [CrossRef]

Publisher's Note: MDPI stays neutral with regard to jurisdictional claims in published maps and institutional affiliations.

Review

Influence of HPT Deformation on the Structure and Properties of Amorphous Alloys

Dmitry Gunderov [1,2,*] **and Vasily Astanin** [3]

[1] Institute of Physics of Advanced Materials, Ufa State Aviation Technical University, 12 K. Marx str., Ufa 450008, Russia

[2] Laboratory of Nanostructured Materials Physics, Institute of Molecule and Crystal Physics UFRC RAS, 71 pr. Oktyabrya, Ufa 450054, Russia

[3] Joint Research Center "Nanotech", Ufa State Aviation Technical University, 12 K. Marx str., Ufa 450008, Russia; v.astanin@gmail.com

* Correspondence: dimagun@mail.ru; Tel.: +7-927-635-3744

Received: 3 March 2020; Accepted: 20 March 2020; Published: 23 March 2020

Abstract: Recent studies showed that structural changes in amorphous alloys under high pressure torsion (HPT) are determined by their chemical composition and processing regimes. For example, HPT treatment of some amorphous alloys leads to their nanocrystallization; in other alloys, nanocrystallization was not observed, but structural transformations of the amorphous phase were revealed. HPT processing resulted in its modification by introducing interfaces due to the formation of shear bands. In this case, the alloys after HPT processing remained amorphous, but a cluster-type structure was formed. The origin of the observed changes in the structure and properties of amorphous alloys is associated with the chemical separation and evolution of free volume in the amorphous phase due to the formation of a high density of interfaces as a result of HPT processing. Amorphous metal alloys with a nanocluster structure and nanoscale inhomogeneities, representatives of which are nanoglasses, significantly differ in their physical and mechanical properties from conventional amorphous materials. The results presented in this review show that the severe plastic deformation (SPD) processing can be one of the efficient ways for producing a nanocluster structure and improving the properties of amorphous alloys.

Keywords: high-pressure torsion; severe plastic deformation; bulk metallic glass; transmitting electron microscopy; X-ray diffraction; differential scanning calorimetry; free volume

1. Introduction

Amorphous alloys are one of the most attractive areas of modern materials science [1–7]. Amorphous metals and alloys are usually produced by melt quenching at typical rates of $\geq 10^6$ K/s down to temperatures of 0.2–0.3 T_{melt} (T_{melt} is the melting temperature), at which diffusion and crystallization processes are suppressed [1–4]. The advanced functional properties (mechanical, magnetic, corrosion, etc.) of amorphous materials promoted active research in this field. In recent years, the fabrication of the so-called "bulk metallic glasses" (BMGs) [4,5] has been a developing area of research. BMG compositions are selected in such a way that an amorphous structure can be obtained even at cooling rates of 10^2 K/s, and this enables producing bulk amorphous samples with a diameter of up to several centimeters.

High tensile strength, large elastic strain limit, high hardness, low friction coefficient, good resistance to corrosion, and wear provide great potential for various commercial applications of amorphous alloys [6,7]. Amorphous alloys with unique soft magnetic properties are widely used [6,7]. The atomic structure of metallic glasses determines their mechanical properties. In particular, their tensile strength and yield stress are higher in comparison to their crystalline counterparts. For example, yield stress reaches 2 GPa for Cu-, Ti-, and Zr-based BMGs, 3 GPa for Ni-based, and 4 GPa for Fe-based alloys [8].

The combination of high strength, good corrosion resistance, and reduced elastic modulus provides a promising outlook for amorphous materials for medical equipment and implants, including biodegradable CaMg-based implants [9].

Unfortunately, amorphous materials exhibit extremely low tensile ductility, since their deformation occurs via the formation and propagation of shear bands (SBs), and they fracture catastrophically along the very first SB [1,2]. In this regard, numerous attempts have been made to find a way to improve the ductility via structural modifications of amorphous alloys [10–14]. The main idea of these routes is to form SBs in amorphous phases preliminarily or to obtain amorphous structures consisting of nanoclusters [13]. Preliminary deformation by compression, cold rolling, etc., enables increasing the ductility and other service properties of amorphous alloys [10–12] due to the nucleation of secondary SBs and branching of SBs. However, conventional thermomechanical treatment does not allow producing large strains in the case of brittle amorphous materials. A promising way to introduce high strain and, therefore, a high density of structural defects into amorphous solids is the use of severe plastic deformation. Severe plastic deformation (SPD) is a group of metal-working techniques involving very large strains under high pressure [15,16]. In crystalline materials, they result in the formation of "ultrafine" grains (UFG) ($d < 500$ nm) or nanocrystals (NC) ($d < 100$ nm) [15,16]. Among various SPD methods, high pressure torsion (HPT) is especially interesting as it can be readily applied for brittle and hard-to-deform materials [15,16].

There are many research groups actively involved in studying the effect of HPT processing on amorphous alloys. In particular, the following groups can be identified: Zhilyaev and Langdon [16,17]; Kovács et al. [18–20]; Boucharat and Wilde [21]; Pippan [22]; Zhu [23], Meng [24]; Glezerand Sundeev [25–27]; Edalati and Horita [28,29]; Aronin and Abrosimova [30–32]; Czeppe and Korznikova [33–35]; and other groups. It is worth noting that the effect of HPT processing on amorphous alloys was examined in the section of the book [16], and also quite recently (in 2019), an extensive overview was published on this topic [36]. However, the topic of the effect of HPT processing on amorphous alloys is very broad, and it contains many subtasks; the present review will be interesting for the audience as well. We should note that the author Valiev is a pioneer in the research areas on SPD processing, including HPT processing, of metallic materials for producing a nanocrystalline state [15,37,38], and a leading expert in this field. He and Gunderov, together with other co-authors, conducted one of the first studies on the effect of HPT processing on amorphous alloys [39–48], and since then, they have performed quite many studies in this area. In this connection, the present review includes, in the first place, the research results obtained by its authors.

In recent years, the influence of HPT processing on amorphous alloys has been investigated in numerous publications [41,47,49–60]. It has been shown that in amorphous alloys of certain chemical compositions, SPD processing leads to partial nanocrystallization [41,51–54]. In amorphous alloys of some other compositions, nanocrystallization during SPD is not observed; however, SPD processing leads to the creation of an extremely high density of SBs in the amorphous matrix and the formation of internal heterogeneity in the amorphous structure [24,50,57]. The produced materials significantly differ in their physical and mechanical properties from amorphous materials with a homogeneous structure. In particular, the formation of internal heterogeneities in the amorphous structure leads to changes in the magnetic, electrical, mechanical, and other properties of alloys. The aim of this review is to present and analyze the effect of SPD processing on the atomic structure of amorphous alloys, their properties, internal energy, and the relaxation processes during heating, as well as the crystallization kinetics of amorphous alloys during subsequent annealing.

2. Atomic Structure of Amorphous Alloys Subjected to SPD

HPT processing results in extremely high densities of shear bands in the amorphous matrix [50]. In general, the spacing between adjacent shear bands is found to be reduced with increasing plastic deformation. In [50], $Au_{49}Ag_{5.5}Pd_{2.3}Cu_{26.9}Si_{16.3}$ BMG samples were deformed using the HPT technique, under a quasi-hydrostatic pressure of 6 GPa with five revolutions ($n = 5$) [50]. A modulated structural feature with a large number of shear bands is visible in SEM and TEM micrographs for the cross-section of the HPT-treated samples (Figure 1a,b). The spacing between the shear bands is about 50–100 nm [50].

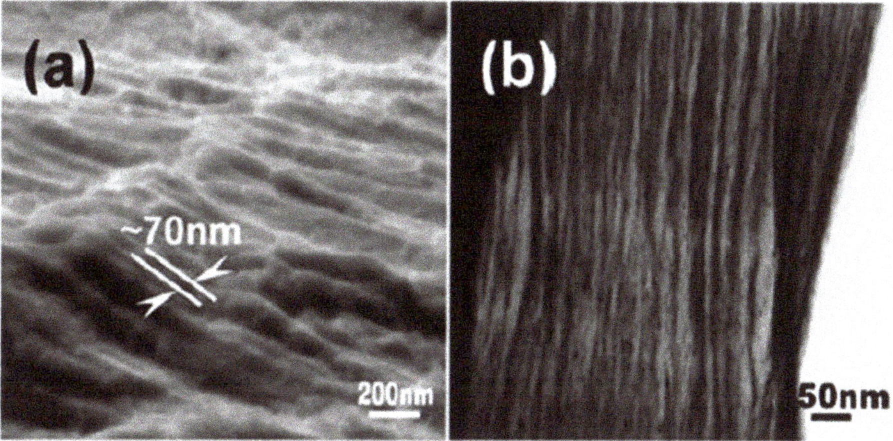

Figure 1. (a) SEM and (b) TEM images of a cross-section through the $Au_{49}Ag_{5.5}Pd_{2.3}Cu_{26.9}Si_{16.3}$ bulk metallic glass (BMG) after high pressure torsion (HPT), revealing a high density of shear bands (reprinted from [50], with permission from Elsevier).

It was also shown that after the high-energy ball milling of the $Zr_{70}Cu_{20}Ni_{10}$ metallic glass, the spacing between the shear bands ranged from 30 to 70 nm [61].

The introduction of shear bands using SPD has a significant effect on both the microstructure and atomic structure of metal glasses [41,50–55]. According to some works, excess free volume is accumulated in shear bands in BMGs formed at ambient temperature [62,63]. Hence, the high density of shear bands increases the free volume of an amorphous alloy.

Some other studies have shown that the structure of shear bands is heterogeneous [64,65] and, for example, in the bulk $Pd_{40.5}Ni_{40.5}P_{19}$ metallic glass, is probably a multilayer heterogeneous structure. It was concluded [65] that the observed inhomogeneous structure of shear bands can be associated with both short and medium ordering in a glassy system. In the study [64], a quantitative analysis showed variations in the local specific volume and local average order for different shear bands and even for different parts of the same shear band. Thus, the shear band differs in excess volume and in the local atomic order from the structure of the surrounding amorphous matrix, and the structure of an amorphous material after SPD is a composite of shear bands and the amorphous material between them [66]. Furthermore, under SPD, in the amorphous phase surrounding the shear bands, complex processes associated with the flow and transformation of point defects also occur [67]. As a result, the entire volume of the material transforms in a complicated manner under SPD.

The transformation path of the amorphous structure depends on the composition of the amorphous alloy and the SPD processing regimes. Among the interesting phenomena observed in amorphous alloys of some compositions is nanocrystallization during SPD processing at room temperature [41,51–54]. In a series of earlier publications, it was shown that nanocrystallization in the amorphous phase occurs under deformation [66–69]. In these works, the formation of separate nanocrystals in shear bands during deformation by upsetting, stretching, or microindentation was revealed. For the first time, nanocrystallization of the amorphous alloys during HPT was observed in the work [49,52]. We should note that whereas during conventional deformation, nanocrystallization was observed in only a small amount of the amorphous material in the shear bands, nanocrystallization during HPT was observed throughout the entire volume of the amorphous samples.

In a series of studies [41,42,49,51–54,70], nanocrystallization during HPT of initially amorphous melt-spun (MS) $Nd_{12}Fe_{82}B_6$, $Nd_9Fe_{84}B_7$ ribbons was investigated. In [52], the amorphous MS $Nd_{12}Fe_{82}B_6$ alloy ribbons were subjected to SPD by HPT at room temperature under a pressure of $P = 5$ GPa. As a result of this processing, monolithic 100%-density disc samples having a diameter of 10 mm and a thickness of 0.2 mm were produced from the initial ribbons. TEM investigations showed that the amorphous structure (Figure 2a) was typical for the initial MS Nd-Fe-B alloy. No crystallites were observed in the bright- or dark-field images, and the selected area electron diffraction (SAED) pattern showed diffusive rings typical for amorphous alloys [49]. Despite this, the structure of the initial alloy was not completely homogeneous, because magnetic measurements revealed a certain amount of crystalline phases of $Nd_2Fe_{14}B$ and α-Fe.

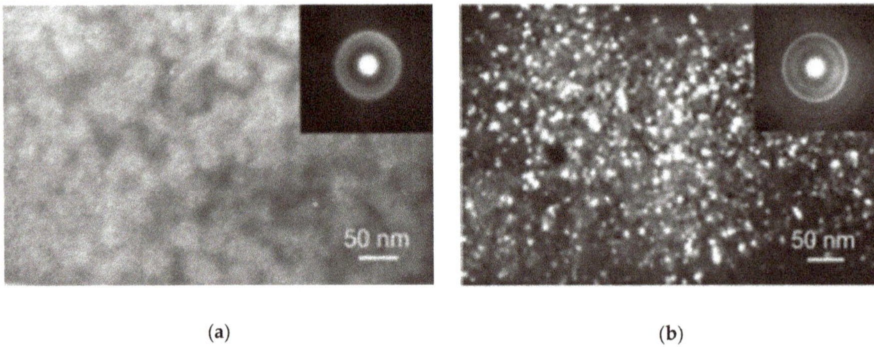

Figure 2. TEM images and SAED patterns for the $Nd_{12}Fe_{82}B_6$ alloy: (**a**) the initial melt-spun (MS) state, dark-field image; (**b**) dark-field image in α-Fe reflex after HPT $\varphi = 16\pi$ [49].

After HPT, the microstructure of the alloy changed considerably. In the SAED pattern, one may observe the appearance of the diffraction rings of the nanocrystalline α-Fe phase (Figure 2b). Particles of this phase of about 10 nm in size are clearly seen in the dark-field TEM images (Figure 2b) [49].

A change in the phase composition after HPT is also reflected by a change in the temperature dependences of the specific magnetization $\sigma(T)$. Figure 3a shows the $\sigma(T)$ dependence for the $Nd_{12}Fe_{82}B_6$ alloy in the initial MS state and after HPT. The MS state demonstrates a sharp decrease in magnetization during heating from 20 °C to 180 °C. Since the Curie temperature (T_c) of the amorphous Nd-Fe-B phase is about 200 °C, this behavior of the $\sigma(T)$ dependence points to a large amount of the amorphous phase in the initial MS alloy [52]. Another small kink in this curve at a temperature of 300 °C indicates the presence of the $Nd_2Fe_{14}B$ phase with T_c equal to 310 °C [71]. A drop in magnetization down to zero takes place at a temperature of about 770 °C and is related to the transition of α-Fe into the paramagnetic state. Thus, in the initial alloy, there are some amounts of the $Nd_2Fe_{14}B$ and α-Fe phases, except for the basic amorphous phase.

All the phases contained in the alloy are ferromagnetic at room temperature, and their volume fractions can be evaluated approximately, assuming the additive contribution of the phases into the total magnetization:

$$\sigma = \sigma_T V_T + \sigma_{Fe} V_{Fe} + \sigma_A V_A; \; V_T + V_{Fe} + V_A = 1, \tag{1}$$

where σ_T, σ_{Fe}, σ_A are the values of magnetization at 20 °C and V_T, V_{Fe}, V_A are the volume fractions of the $Nd_2Fe_{14}B$, α-Fe and amorphous phases, respectively. The σ_{Fe} and σ_T values are equal to 220 and 100 Am^2/kg for the measurement field $H = 9$ kOe (716.2 kA/m) [68]. The calculations of the alloy phase composition are presented in Table 1 [49].

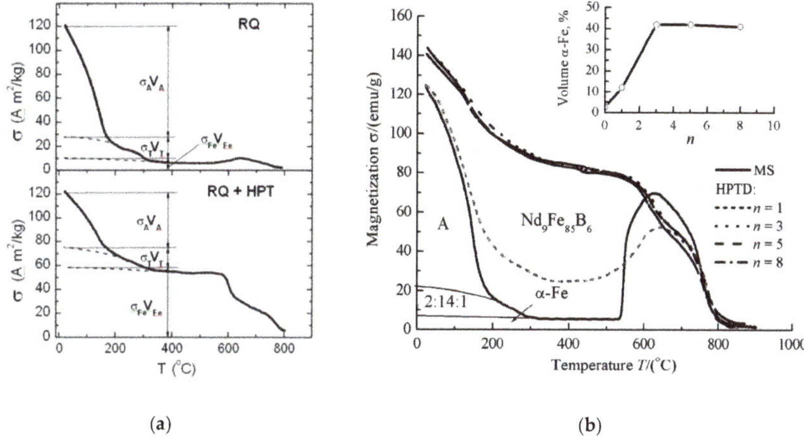

Figure 3. Temperature dependence of magnetization σ for: (**a**) the $Nd_{12}Fe_{82}B_6$ alloy in the rapidly quenched (RQ) state and after HPT $n = 5$ [49]; (**b**) the MS $Nd_9Fe_{85}B_6$ ribbons and samples after high pressure torsion deformation (HPTD) $n = 1$, $n = 3$, $n = 5$, $n = 8$; inset: the volume fraction of Fe for different numbers of revolutions (reprinted from [52], with permission from Elsevier).

The dependence of σ(*T*) for the MS alloy after HPT changed significantly (Figure 3a). The contributions of $Nd_2Fe_{14}B$ and amorphous phases to the total magnetization decreased, and the contribution of α-Fe increased greatly. This testifies to a decrease in the content of the amorphous and $Nd_2Fe_{14}B$ phases and an increase in the content of the α-Fe phase up to 25%, or almost 10-fold (see Table 1). Therefore, TEM, XRD, and magnetic studies testify that the HPT of the amorphous MS NdFeB alloy leads to the nanocrystallization of α-Fe to the enrichment of the amorphous phase with Nd [49].

It is interesting to note that the $Nd_2Fe_{14}B$ phase is decomposed into amorphous and α-Fe phases in the same way as occurs during the HPT of the cast Nd-Fe-B alloys [72–76].

In [52], the high-pressure torsion deformation of the MS $Nd_9Fe_{85}B_6$ ribbons was performed, varying the number of revolutions, *n*, from $n = 1$ to $n = 8$. Figure 3b shows the results of thermomagnetic analysis of the MS $Nd_9Fe_{85}B_6$ samples in a different state. The initial ribbons contained about 85% of the amorphous phase, 12% of the $Nd_2Fe_{14}B$ phase, and 3% of the bcc Fe-based phase. With increasing deformation rate to $n = 3$, the weight fraction of α-Fe increases to 40%, while that of the amorphous phase decreases (see Table 1). As the inset shows (Figure 3b), an increase in the HPT rate above $n = 3$ does not lead to a further increase in the volume fraction of α-Fe.

Table 1. Phase composition of the $Nd_{12}Fe_{82}B_6$ and $Nd_9Fe_{85}B_6$ alloys in the initial MS state and after HPT [49,52].

State	Amorphous Phase, vol. %	$Nd_2Fe_{14}B$, vol. %	α-Fe, vol. %
$Nd_{12}Fe_{82}B_6$ alloy			
MS	83	14	3
MS + HPT $n = 5$	65	10	25
$Nd_9Fe_{85}B_6$ alloy			
MS	83	15	2
MS + HPT $n = 1$	74	14	12
MS + HPT $n = 3$	45	13	42
MS + HPT $n = 5$	45	13	42
MS + HPT $n = 8$	47	12	41

Thus, the HPT processing of the amorphous MS $Nd_9Fe_{85}B_6$ and $Nd_{12}Fe_{82}B_6$ alloys leads to the same structural transformation. According to the XRD and thermomagnetic analyses, HPT processing leads to the decomposition of the amorphous phase and the formation of α-Fe nanocrystals. HPT leads to the decomposition of the initial crystalline phase of $Nd_2Fe_{14}B$ into nanocrystalline α-Fe and the amorphous phase enriched with Nd. With a further increase in strain (more than three revolutions), the amount of nanocrystalline α-Fe remains nearly unchanged: 40% for $Nd_9Fe_{85}B_6$ and 25% for $Nd_{12}Fe_{82}B_6$. Thus, there is certain strain, which corresponds to the stabilization of the structural state of the MS $Nd_9Fe_{85}B_6$ alloy. It should be noted that a similar stabilization of grain size and phase composition is also observed in crystalline materials [16].

The amorphous phase, which is formed by the HPT processing, is enriched in Nd in comparison with the overall amount of Nd in the alloy. An increase in the amount of Nd in the alloys results in an increase in the amount of the amorphous phase, which is formed by the HPT processing, in comparison with the amount of the α-Fe phase. It was shown by computation that the chemical composition of the amorphous phase is close to the equiatomic one: Nd 50% and Fe 50%. Thus, for Nd-Fe-B alloys, the chemical composition of the amorphous phase, produced by the HPT processing, weakly depends on the chemical composition of the initial alloy.

In [54], the structural evolution of amorphous $Nd_9Fe_{85}B_6$ subjected to HPT at room temperature was studied by measuring positron lifetimes. The study also included the determination of Doppler broadening of positron-electron annihilation photons.

Analysis of positron lifetime measurements showed the following. The ratio curve of pure defect-free Fe gives a horizontal line (Figure 4b). The curve of the freshly prepared MS $Nd_9Fe_{85}B_6$ alloy (in the region $p_L \geq 20 \times 10^{-3}$ m_0c) lies between the curves of defect-free pure Nd and Fe and differs greatly from the curve of pure Bor. The analysis shows that the chemical environment surrounding the structural free volumes (V) in the freshly prepared alloy MS $Nd_9Fe_{85}B_6$ is enriched in Nd and Fe atoms [54]. After HPT, the MS alloy curve moves to the line of defect-free pure Fe (horizontal line). This indicates that vacancy-type defects generated by HPT are mainly surrounded by Fe atoms. This clearly indicates the enrichment of the residual amorphous matrix with Fe atoms.

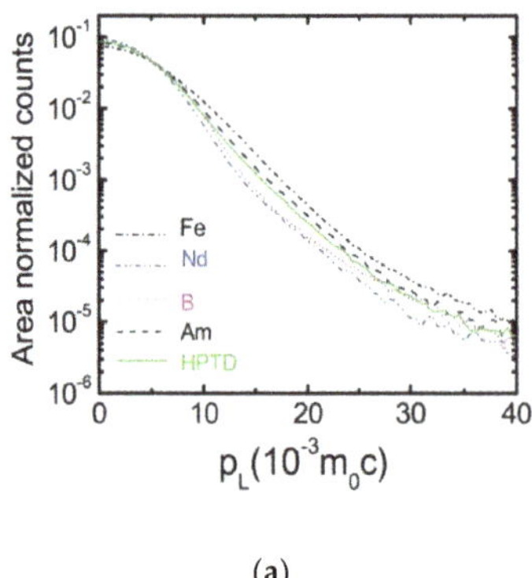
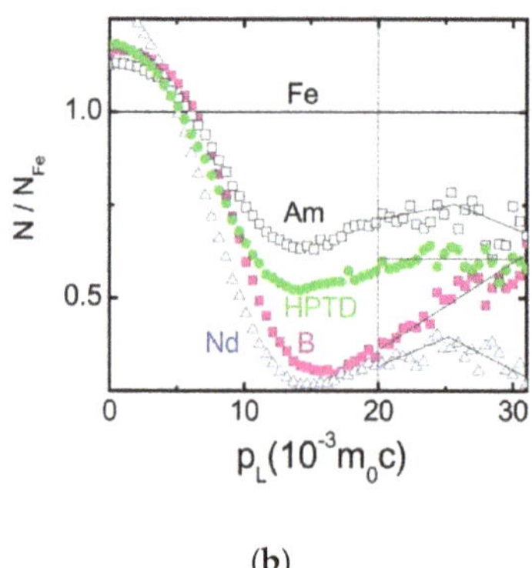

(a)   (b)

Figure 4. (**a**) Coincident Doppler broadening spectra: as-prepared (Am) MS $Nd_9Fe_{85}B_6$; HPTD-processed MS $Nd_9Fe_{85}B_6$; on defect-free pure Nd, Fe, and B at RT. These spectra are plotted after area normalizing processing. (**b**) Ratio curves of the coincidently measured Doppler broadening spectra (Am) MS $Nd_9Fe_{85}B_6$; HPT MS $Nd_9Fe_{85}B_6$; pure Nd, Fe, and B (at RT). All spectra are normalized to the spectrum of pure Fe (horizontal line). The shape of the ratio curve indicating that the HPT-induced vacancy-type defects are dominantly surrounded with Fe atoms (reprinted from [54], with the permission of AIP Publishing).

The following scheme was proposed in [54]: during HPT, the vacancy-type defects are induced in the amorphous phase, and deformation leads to the predominant association of Fe atoms with defects of the vacancy type and increases the mobility of Fe atoms. This contributes to the enrichment of some forms of amorphous matrices with Fe atoms and significantly reduces the crystallization activation energy. As a result, large-scale α-Fe nanocrystallization occurs during HPT treatment at room temperature.

The influence of HPT processing on the amorphous structure of MS $Ni_{44}Fe_{29}Co_{15}Si_2B_{10}$ was investigated in [77]. The HPT processing was performed under a pressure of 5 GPa through different numbers of revolutions (from 0.5 to 8) at temperatures of 293 and 77 K. Nanocrystallization under HPT at temperatures of 293 and 77 K is observed for this material. Nanocrystallization leads to the formation of the γ-phase (face-centered cubic) in the initial amorphous phase of Ni-Fe-Co alloys [77]. Both heterogeneous deformation through the shear bands and nanocrystallization in shear bands are observed in amorphous $Ni_{44}Fe_{29}Co_{15}Si_2B_{10}$ at the initial stage of HPT (for the 1st revolution). According to this work, a further increase in strain results in a change in the deformation mechanism. Shear bands are no longer observed. Instead of shear bands, a homogeneous distribution of nanoparticles of the crystalline phase with an average size up to 10 nm is observed. It was suggested that an increase in the number of revolutions (more than one) results in delocalization of the plastic flow overall and in shear bands in particular. It seems that the process of plastic deformation was transformed into a quasi-homogeneous one. This type of plastic flow corresponds to the deformation behavior of amorphous alloys at high temperatures, close to the glass transition temperature. In this case, the effects of a local temperature increase, which correspond to the shear deformation, are observed throughout the entire volume of the deformed material. The formation of the crystalline phase occurs not in localized shear bands, but in the entire volume.

As mentioned before, during the traditional deformation of amorphous alloys, nanocrystallization was observed in the areas of localized deformation in shear bands [66–69]. During SPD processing, nanocrystallization takes place in the entire volume of the deformed sample. This is due to a very high degree and high homogeneity of deformation during SPD. The mechanism of nanocrystallization is still unclear. Many studies show that during traditional deformation, this process is athermal in nature [66,69]. On the other hand, in [66], an assumption was made that the formation of nanocrystals in shear bands occurred due to the diffusion enhancement, which appeared as a result of defects produced by deformation of the amorphous alloy rather than due to adiabatic heating in local areas. This assumption seems to be well grounded, but the following should be noted. As we have found, α-Fe crystals are formed in an amorphous Nd-Fe-B alloy during HPT. As was revealed in [78], α-Al nanocrystals were formed during deformation of amorphous Al-based alloys. It appears that for these alloys, nanocrystallization during HPT is accompanied by decomposition of the amorphous alloy with the formation of nanocrystals of the main metal, whereas the neighboring areas become enriched with alloying elements. The decomposition of the amorphous phase with further consolidation of the main element in nanofields is observed prior to nanocrystallization. The phenomenon of a similar decomposition of alloys during ball milling was recently discussed in [79]. As was shown in [79], HPT processing can induce local redistribution of elements in alloys, as well as changes in the concentration of the main element in nanofields due to the formation of vacancy flows.

The following explanation of the observed effect was proposed [58]. The temperature of nanocrystallization of pure metals decreases significantly with a decrease in grain size; for the nanostructured state, nanocrystallization occurs at temperatures of 0.25 T_{melt} [16,31,80]. Moreover, due to the high values of the driven force of nanocrystallization, the crystallization temperature of amorphous films of pure metals (Fe, Ni) produced by deposition on a cold substrate is about 4 K [80]. Due to the relative simplicity of the crystalline lattice of a pure metal, the lattice can be easily restored, since to move adjacent atoms to the positions of the crystalline nodes, it is sufficient to move them to distances shorter than atomic radii. This process does not require diffusion activity. Thus, the crystallization of clusters with an increased concentration of the main metal formed during SPD occurs at room or even lower temperatures [58].

It was assumed [77] that with a local temperature growth, high values of local stresses cause crystallization, otherwise it would be difficult to explain the events of nanocrystallization that were observed at cryogenic temperatures (77K) [77]. It has been suggested that stress induces temperature-dependent processes: high values of stresses lead to the lowest temperature value of the temperature-activated process of crystallization.

The following hypothesis was proposed to explain the processes of stabilization of the structural state under HPT-induced nanocrystallization: the formation of crystals occurs during the processing, not after its completion [58,77]. During the HPT processing, nanocrystals are formed in the amorphous phase, and they affect the processes of further deformation. The crystallites interact with the newly formed shear bands in the amorphous matrix. As a result, the generation and accumulation of dislocations inside the crystallites occur. The highest dislocation density will obviously be near the border of the crystallite/amorphous matrix, where the interaction between the shear bands and crystalline boundaries will be the most active [77]. At some point (the moment of the highest dislocation density in the frontier zone), the frontier zone or even the entire crystalline particle spontaneously goes into the amorphous state due to the minimum of the free energy of the amorphous state in comparison with the high-defective crystal, that is the stress-induced process of crystals "dissolution" in the amorphous matrix will occur. Very small crystalline particles (less than 10 nm) are not able to accumulate dislocation-type defects [81]. Therefore, the nanocrystalline particles of the amorphous matrix smaller than 10 nm will not be dissolved under HPT processing because they will effectively push dislocations at the interface and remain defect-free. In other words, nanocrystals with a size less than 10 nm, arising in the amorphous matrix either directly upon crystallization in shear bands or under partial deformation "dissolving" of larger crystallites, will have structural stability during further stages of HPT processing. Thus, a new amorphous-nanocrystalline state, stable in the conditions of deformation, is formed. Based on the foregoing, it is easy to explain the fact that in all the studies, in which the process of crystal formation during the SPD of amorphous alloys was studied, the crystals were always nanoscale, i.e., only nanocrystallization was observed, and relatively large crystals were never observed [77].

Complex structural transformations take place during HPT in the amorphous MS Ti$_{50}$Ni$_{25}$Cu$_{25}$ alloy. According to [41,51], during the HPT processing of this alloy, very small (about 3 nm in size) nanocrystals of the B2 (Ti$_2$NiCu) phase similar to the composition of the amorphous phase are formed, and the volume fraction of nanocrystals is probably not large. However, in addition to some nanocrystallization in the Ti$_{50}$Ni$_{25}$Cu$_{25}$ alloy during the HPT processing, other transformations of the structure and, consequently, the properties of this alloy took place. In the studies [55,56], the microstructure of the amorphous MS Ti$_{50}$Ni$_{25}$Cu$_{25}$ ribbons subjected to HPT was investigated. HPT processing was performed for up to 10 revolutions under a pressure of 6 GPa at room temperature (RT) and at a temperature of 150 °C. As a result, solid samples with a thickness of 0.2–0.3 mm and a diameter of 10 mm were produced from the initial MS ribbons. According to the XRD and SAED patterns, the structure of the HPT-processed samples, as well as that of the initial MS ribbons was amorphous [55]. The bright-field image of the initial Ti$_{50}$Ni$_{25}$Cu$_{25}$ alloy exhibits in this case the characteristic "salt-pepper" contrast, which is typically observed in amorphous materials. Figure 5b shows a TEM image of the MS alloy subjected to HPT at RT. The SAED pattern contains an amorphous halo. In the bright-field image, brighter and darker regions with a size of about 20 nm can be distinguished [55].

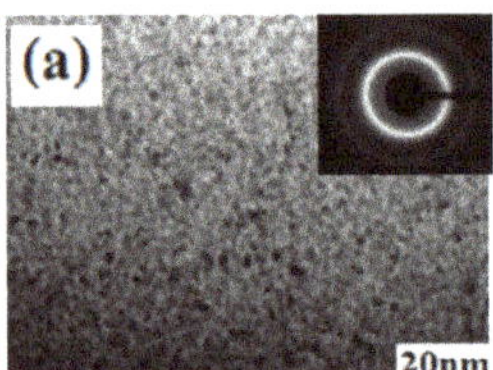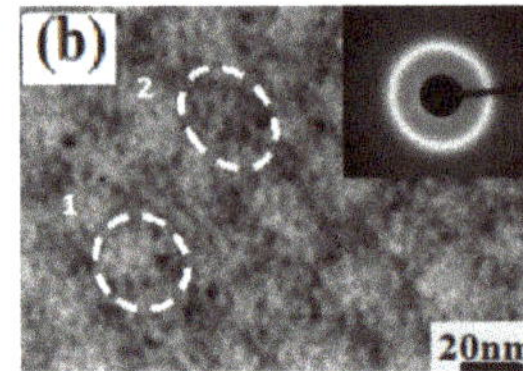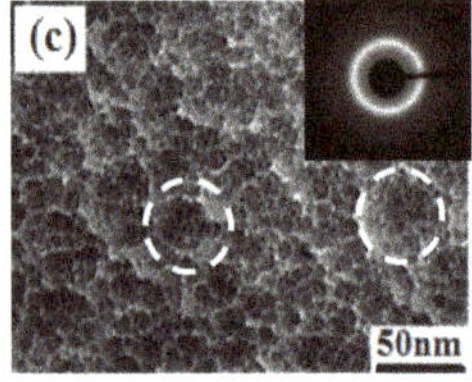

Figure 5. TEM images of the MS Ti$_{50}$Ni$_{25}$Cu$_{25}$ alloy: (**a**) the initial MS state, bright-field image, SAED pattern; (**b**) HPT at T = 20 °C, bright-field image, SAED pattern, (**c**) HPT at T = 150 °C, bright-field image, SAED pattern; the dashed lines encircle the dark and bright regions ("clusters") in the microstructure [55].

Similarly, the SAED pattern from the sample subjected to HPT at 150 °C contains an amorphous halo. The microstructure exhibits the following features observed using TEM (Figure 5c). Dark regions ("clusters" with a size of about 40 nm) become visible in the bright-field image (Figure 5c). These "clusters" are separated from each other by thin brighter interfaces, and they also have internal contrast. Small nanocrystals of the B2 phase about 3 nm in size in HPT-processed MS $Ti_{50}Ni_{25}Cu_{25}$ are observed in the dark-field TEM image. However, the size of nanocrystals is much smaller than the size of clusters visible in the bright-field image. Thus, it follows that nanocrystals and clusters are different structural elements.

Hence, the observed contrast in the TEM bright-field micrographs differs for the samples in different conditions: for the initial amorphous ribbons, for the MS alloy after HPT at 20 °C and after HPT at 150 °C, suggesting that the structure of the material depends on the temperature of deformation. Moreover, these "clusters" represent some kind of amorphous structure. This contrast could be the result of the existence in the amorphous phase of regions with a reduced free volume (bright regions, bright boundaries) and with an enhanced free volume (dark regions).

During the HPT processing of a number of other amorphous alloys, nanocrystallization was not observed [24,50,57,58]. The influence of HPT on the $Au_{49}Ag_{5.5}Pd_{2.3}Cu_{26.9}Si_{16.3}$ BMG samples was investigated in [50]. The atomic structure of the as-cast and SPD-treated specimens was analyzed by X-ray diffraction using a synchrotron radiation source at the BW5 station of HASYLAB. From X-ray diffraction analyses, the structure factor, $S(q)$, and the pair distribution function (PDF), $g(r)$, were deduced [82].

Investigation by X-ray diffraction using the synchrotron radiation detected the difference in the pair distribution function between the as-cast and SPD-treated Au-BMG samples recorded at ambient temperature and in a supercooled liquid state at a temperature of approximately 10 K below the onset of crystallization, T_x. As was shown in [50], the SPD treatment leads to the rearrangement of atoms in the BMG, e.g., atoms that were originally located in the first, second, and following nearest neighbor shells are displaced into the space between these shells. Consequently, the degree of short-range order (SRO) of atomic arrangements in the SPD-treated sample was reduced. This corresponds to an increase in free volume of approximately 0.6% and, consequently, to an increase in the lattice strain of about 0.2%. This implies that shear strain is mostly accumulated in high-density shear bands rather than between atoms. The differences in the atomic structures between the SPD-treated and as-cast BMGs is preserved even in the supercooled liquid region [50].

Comparing the PDFs for the as-cast samples with the SPD-treated below and/or above T_g, one can conclude that the temperature-dependent structural changes detected in the SPD-treated sample are weaker than those in the as-cast sample. These results indicate that the SPD-treated sample is relatively unstable at room temperature and during annealing, but it experienced less structural changes when it was heated at 2.5 K·min^{-1} up to the supercooled liquid (before crystallization). Thus, a new amorphous structure in the Au-based BMG formed as a result of HPT, consisting of an amorphous phase with high-density shear bands and a lower degree of short-range order [50]. The HPT processing of Au-based BMGs leads to the formation of a nanoglass-type structure with a reduced density and modified physical properties [83,84].

Thus, HPT can lead to complex structural modifications of the amorphous state even without nanocrystallization of the amorphous phase. Herewith, it can lead to the formation of a high density of shear bands, where the spacing between the bands is about 50–100 nm [50]. In other works [24,41,52–57,85], shear bands were not observed in the amorphous structure after HPT processing.

As mentioned earlier, in [77], it was shown that under HPT for the number of revolutions >1, the process of plastic flow became quasi-homogeneous. For some cases, TEM showed the formation of a cluster structure. For MS TiNiCu- and Zr-based BMGs, deformation proceeds through the formation of a high density of shear bands, which disappear in the final stages.

Therefore, it is possible to propose the following model of the transformation of an amorphous structure in MS TiNiCu during SPD. In the first stage, in initial amorphous structure of primary SBs appears, and the spacing between the SBs is several thousand nanometers (Figure 6). In the second stage, the subsequent increase in strain leads to the formation of secondary bands, with the spacing up to one hundred nanometers. Perhaps, in the third stage, a partial redistribution of the free volume and relaxed atoms in the amorphous state after SPD takes place, and a nanoglass-type structure is formed.

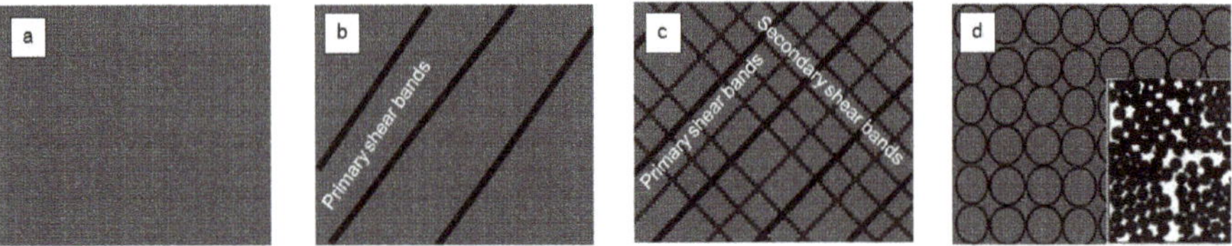

Figure 6. The proposed model of the transformation of the amorphous structure during SPD: (**a**) initial amorphous structure; (**b**) appearance of primary shear bands (SBs) under a low strain; the spacing between the SBs is several thousand nanometers; (**c**) subsequent increase in strain leads to the formation of secondary bands, with the spacing up to one hundred nanometers; (**d**) relaxed amorphous state after SPD (inset: enlarged view).

It is necessary to say a few words about the structure and properties of nanoglasses produced by the bottom-up technology.

Recent studies have revealed methods to introduce microstructural defects in the form of glass-glass interfaces into the homogenous structure of a glass so that amorphous solids result with microstructures that are comparable to the microstructures of polycrystalline materials. About 20 years ago, Herbert Gleiter and his group developed the idea of nanoglasses, in which nanometer-sized clusters with an amorphous structure comparable to the one of the corresponding melt-cooled glasses wre connected by interfaces with reduced atomic densities [83,84,86,87]. This idea may be understood by comparing the microstructure of nanoglasses with the microstructure of nanocrystalline solids. It is well known that one way of introducing a high density of interfaces into crystals is by consolidating nanometer-sized crystals with identical (Figure 7c) or different (Figure 7d) chemical compositions.

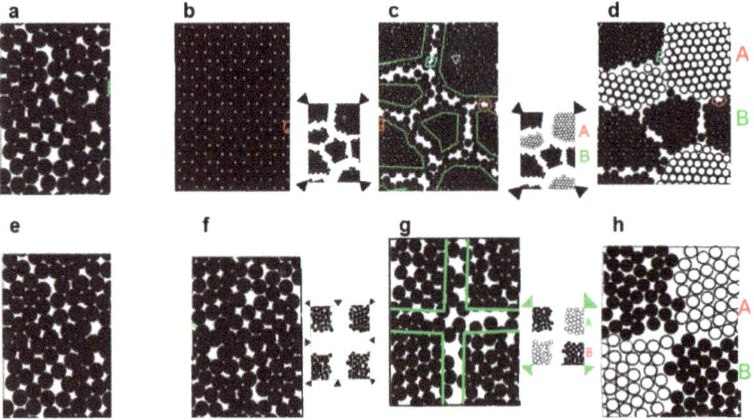

Figure 7. The analogy between the defect and the chemical microstructures of nanocrystalline materials and nanoglasses: (**a**) Melt of identical atoms and (**b**) a single crystal. The defect microstructure (**c**) and chemical microstructure (**d**) of nanocrystalline materials are compared with the corresponding defect microstructure (**g**) and the chemical microstructure (**h**) of nanoglasses. (**f**) displays the glassy structure obtained by quenching the melt shown in (**e**) (reprinted from [86], with permission from Elsevier).

Nanoglasses are based on the idea [83,84,86–91] of applying analogous approaches for creating glasses with controllable defect microstructures in the form of interfaces between adjacent glassy regions with identical or different chemical compositions. Again, if we start from a melt consisting of one kind of atom only (Figure 7e), a glass may be obtained by quenching the melt with a sufficiently high cooling rate (Figure 7f). By analogy to Figure 7c, it is proposed to introduce a high density of glass–glass interfaces by consolidating nanometer-sized glassy clusters (Figure 7g) with identical or different chemical compositions (Figure 7h).

Initially, nanoglasses were synthesized using the method of inert gas condensation (IGC), i.e., the consolidation of amorphous powder-clusters (IGC-samples). This production process involves the following two steps. During the first step, nanometer-sized glassy spheres are generated by evaporating (or sputtering) the material in an inert gas (e.g., He) atmosphere. The resulting glassy nanospheres are accumulated on the surface of a cold finger. After scraping from the cold finger, the material in the form of flakes is compacted into a pellet shaped nanoglass specimen. The compaction is performed at high pressures up to 5 GPa. This production procedure is basically identical to the one used to generate nanocrystalline materials. Until now, using this method, nanoglasses have been synthesized from various alloys: Au-Si, Au-La, Fe-Si, Fe-Sc, La-Si, Pd-Si, Ni-Ti, Ni-Zr, and Ti-P [86]. Using magnetron sputtering under certain conditions (gas pressure, target–substrate orientation), almost all amorphous alloys can be produced in the form of nanoglasses [86]. Both described bottom up methods have certain disadvantages, such as the possible oxidation and contamination of nanopowders during the preparation processes and the presence of pores in samples after consolidation; nanoglass materials processed by magnetron sputtering usually contain columnar "grains".

It is possible that the structures produced by SPD in several amorphous alloys are comparable in some aspects with the structures of the nanoglass-type produced by IGC. However, at present, there is no direct evidence for this. Special studies are required to answer this question.

3. Transformation of the Properties of Amorphous Alloys as a Result of SPD Processing

One of the widely used characteristics of amorphous materials is the free volume [5,8]. XRD is one of the possible methods for estimating the free volume ΔV in amorphous alloys, including those subjected to HPT [61,78,92,93]. X-ray diffraction makes it possible to determine the radius of the first coordination sphere (R_1) of the amorphous phase and the change in ΔV as a result of structural transformations [61]. Studies have shown that HPT leads to an increase in the value of R_1, and an increase in free volume [61].

Samples of HPT-processed BMG contain many SBs [50,94,95]. It is known that the atomic structure of the SBs and the regions nearest to them also differ from the amorphous matrix ΔV [13]. Therefore, the structure of BMG subjected to HPT can be represented as two phases: the structure in SB and the amorphous structure in the matrix. With large strains, ΔV of the matrix can also change. It is interesting to use direct density measurement to estimate ΔV and compare these values with ΔV obtained using XRD. An estimate of the change in ΔV in the HPT of amorphous alloys from a change in R_1 in accordance with XRD may show a significant error. However, this approach can also be used to analyze changes in the amorphous structure during HPT.

Usually, the hydrostatic weighing method is used to measure the density, and the density, in turn, depends on the free volume [96]. However, the application of this method to determine the density of samples subjected to HPT is difficult, due to the complex shape of the samples themselves. In [97], a new unique method of direct density measurement was proposed. In [98], XRD and this new method for density measurement of small samples [97] were used in order to measure changes in density and free volume of the amorphous alloy $Zr_{62}Cu_{22}Al_{10}Fe_5Dy_1$ subjected to HPT.

According to XRD, the structure of the as-cast and HPT-processed $Zr_{62}Cu_{22}Al_{10}Fe_5Dy_1$ BMG is amorphous (Figure 8a). The position of the first amorphous halo shifts towards the lower angles after HPT (Table 2). The values of the R_1 could be estimated using the Ehrenfest equation [78,93]:

$$2R_1 sin\theta = 1.23\lambda, \tag{2}$$

where θ is the scattering angle and λ is the radiation wavelength. The initial BMG and BMG after HPT $n = 5$ at RT and 150 °C have the R_1 2.999, 3.003, and 3.006 Å, respectively (Table 2) [98]. The variation of R_1 is correlated with the relative variation of the free volume ΔV by the equation [21]:

$$\frac{R_0^3}{R_{HPT}^3} = \frac{V_{HPT}}{V_0} \quad \Delta V = \frac{R_{HPT}^3 - R_0^3}{R_0^3} \times 100\%, \tag{3}$$

where R_0 and R_{HPT} are the R_1 sphere of the initial and HPT-processed BMG, respectively; V_0 and V_{HPT} are the mean atomic volumes of the initial and HPT-processed BMG, respectively. HPT at RT and 150 °C lead to an increase in the ΔV by 0.44 and 0.74%, respectively. HPT leads to an increase in the values of full width at half maximum (FWHM). The increase in FWHM was also explained by the structure transformation [92].

The formation of an asymmetrical amorphous halo reflects the separation of the amorphous phase into amorphous phases with different compositions [98]. The effects of separation of the amorphous phase after HPT in Nd-Fe-B and Al-Ni-RE amorphous MS alloys were observed in [41,93]. The amorphous halo of the initial $Zr_{62}Cu_{22}Al_{10}Fe_5Dy_1$ BMG retains its symmetrical shape after HPT processing [98], which means the absence of HPT-driven chemical separation in this BMG.

The density of the as-cast and HPT-processed $Zr_{62}Cu_{22}Al_{10}Fe_5Dy_1$ BMG was measured using the new technique [97]. The density measurements demonstrate that the initial $Zr_{62}Cu_{22}Al_{10}Fe_5Dy_1$ BMG has a density ρ equal to 6.98 kg/m^3 (Table 2) [98], which correlates well with data for bulk samples of this BMG produced by hydrostatic weighing [96]. HPT at temperatures of 20 °C and 150 °C leads to a decrease in the density values ($\Delta\rho$) by 2.1 and 1%, respectively, in comparison with the initial state (Table 2) [98].

A decrease in density values for samples after HPT treatment may be due to an increase in free volume. It was shown on the $Zr_{62}Cu_{22}Al_{10}Fe_5Dy_1$ BMG alloy, subjected to HPT at a temperature of 150 °C, that the $\Delta\rho$ value obtained from direct density measurements was close to the ΔV value obtained by the XRD method. The corresponding $\Delta\rho = 2.1\%$ for BMG treated with HPT at 20 °C is much larger than the $\Delta V = 0.44\%$ [98], which is probably due to the fact that HPT at a lower temperature can lead to the formation of pores or cracks in the HPT-treated sample, although SEM studies do not reveal the presence of pores and cracks in the samples. However, it should be borne in mind that pores and cracks may not be detected during microscopic studies, if they are nanometer in size. This feature introduces an additional error into the ΔV values determined by the direct method. It should be noted that the increase in ΔV values by 0.44 and 0.74% obtained in work [98] is close to the ΔV values (up to 1%) observed in other HPT-processed BMGs [61].

Table 2. Parameters of the amorphous structure of the $Zr_{62}Cu_{22}Al_{10}Fe_5Dy_1$ BMG in the initial state and after HPT ($n = 5$) processing at temperatures of 20 and 150 °C from XRD and direct density measurements [98].

State	2θ, deg	R_1, Å	FWHM, deg	ΔVXRD, %	ρ, g/cm^3	$\Delta\rho$, %
Initial BMG	36.84(3)	2.999	5.50	-	6.98	-
HPT at 20 °C	36.780(16)	3.003	6.106	0.44	6.83	2.14
HPT at 150 °C	36.743(15)	3.006	6.309	0.74	6.90	1.07

The value of $\Delta V = 0.44\%$ obtained by XRD for BMG after HPT at room temperature is less than the value of $\Delta V = 0.74\%$ for samples treated with HPT at 150 °C, i.e., processing at a higher temperature leads to a more expressed formation of free volume. It is possible that an increase in the temperature of HPT to 150 °C should contribute to the deformation of BMG and the generation of SB in the sample.

In [57], the behavior of $Zr_{62}Cu_{22}Al_{10}Fe_5Dy_1$ BMG after HPT during nanoindentation and its sensitivity to the strain rate were investigated. Samples in the form of disks were subjected to HPT $n = 5$ at room temperature and at 150 °C. According to XRD, differential scanning calorimetry (DSC), and TEM, the structure of the initial BMG was amorphous. Annealing at 500 °C caused partial crystallization of BMG, as shown in the X-ray diffraction patterns of Figure 8b.

The mechanical behavior of the as-cast BMG and samples after HPT is markedly different. Serration was visible on the nanoindentation curves for the as-cast BMG and annealed at 370 °C. In contrast, BMG after HPT at room temperature did not exhibit a serrated flow (Figure 9a). It can be assumed that the HPT treatment led to more homogeneous deformation compared to the as-cast BMG during nanoindentation due to an increase in the free volume and concentration of shear bands [99].

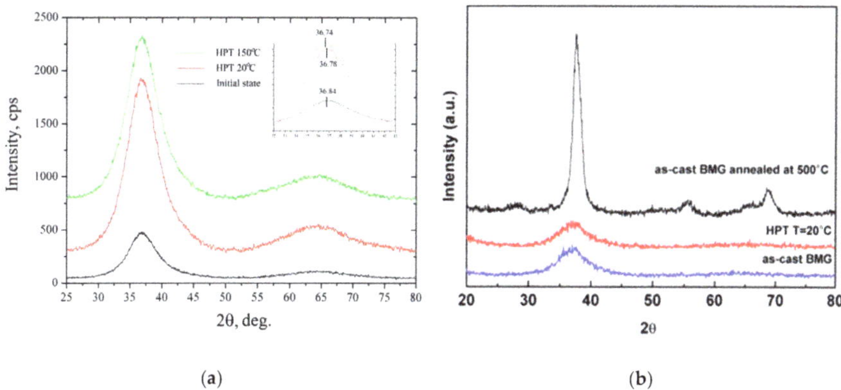

(**a**)   (**b**)

Figure 8. X-ray diffraction patterns of: (**a**) as-cast $Zr_{62}Cu_{22}Al_{10}Fe_5Dy_1$ and the specimens subjected to HPT at temperatures of 20 and 150 °C [98]; (**b**) as-cast $Zr_{62}Cu_{22}Al_{10}Fe_5Dy_1$ after HPT at RT and after annealing at temperatures 500 °C for 10 min (reprinted from [95], with the permission of AIP Publishing).

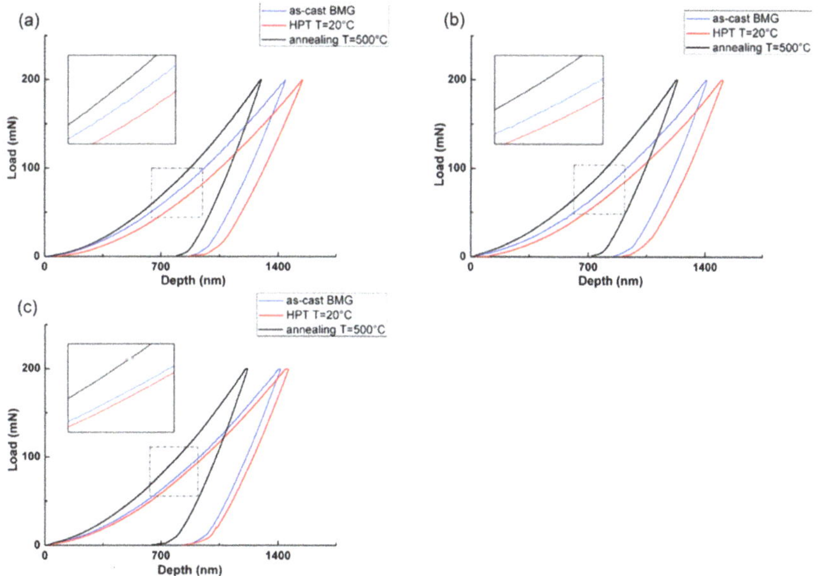

Figure 9. Typical load–penetration curves for the as-cast BMG, as-cast BMG subjected to HPT at 20 °C, and BMG annealed at 500 °C, tested at equivalent strain rates (**a**) 0.0025 s^{-1}, (**b**) 0.01 s^{-1} and (**c**) 0.05 s^{-1} (reprinted from [95], with the permission of AIP Publishing).

The strain rate sensitivity m was calculated from the nanoindentation data in [95]. HPT at room temperature leads to a significant 2.5-fold increase in the parameter m, from $m = 0.014$ to $m = 0.036$, while the elastic modulus decreases slightly, from 90 to 85 GPa; thus, the correlation between Young's modulus and m was found. Moreover, HPT leads to a decrease in hardness. Annealing at 500 °C leads to a decrease in m by 0.01. This is typical for RT straining of crystalline alloys. Interestingly, HPT at 150 °C leads to an increase in Young's modulus and hardness, but m increases as well (as compared to the initial BMG). The changes in m can be explained by changes in the ΔV and its distribution induced by HPT.

The following pattern of structural changes induced by HPT may be proposed. HPT introduces a significant amount of SB into the amorphous matrix and forms boundaries with an increased ΔV. Therefore, for HPT samples, the deformation area causes the propagation of many shear bands previously formed under HPT, which leads to a more uniform flow, which is expressed as the disappearance of the serrated flow on the indentation curve for the HPT sample and, consequently, an increase in m. The clusters in nanoglasses play the same role [86].

In [100], the effect of HPT on the tendency to plastic flow in the $Zr_{60}Cu_{18.5}Nb_2Ni_{7.5}Al_{10}Ti_2$ BMG was studied using microindentation and measurements of the relative height of the "pileup" at the base of the indent. The BMG are characterized by the formation of a "pileup" at the base of the indent during microindentation because of a tendency to local plastic flow. The plasticity of the deformed volume is determined from the "pileup" [101]. To assess plasticity, one should correlate the "pileup" height in the recovered indent with depth H [101]:

$$\delta_h = h/H. \tag{4}$$

The h had been adjusted by reducing it to constant HV in the following form:

$$\Delta_h = \delta_h\left(1 - \frac{\Delta HV}{HV_0}\right), \tag{5}$$

where $\Delta HV = HV_n - HV_0$; HV_n is the microhardness after HPT (n is number of anvil revolutions) and HV_0 is the microhardness in the initial state.

At HPT $n = 2$, the Δ_h of the $Zr_{60}Cu_{18.5}Nb_2Ni_{7.5}Al_{10}Ti_2$ BMG increases by 35–50%. At $n = 5$, the Δ_h abruptly decreases and becomes lower than that in the initial BMG [100].

Nanoglasses may exhibit unique properties as compared to their homogeneous amorphous alloy [83,84,86,87]. The properties of nanoglasses may be modified by controlling the size of the glassy regions and by varying their chemical composition. A $Fe_{90}Sc_{10}$ nanoglass is (at 300 K) ferromagnetic, whereas the corresponding MS glass is paramagnetic. Moreover, nanoglasses were more ductile, more biocompatible, and catalytically more active than the corresponding MS glasses. Thus, varying the structural state of amorphous alloys provides an opportunity to control the properties [86].

As shown in the above section, HPT leads to the transformation of the structure of amorphous alloys, the formation of SB, nanocrystallization, and the formation of inhomogeneities similar to the structure of nanoglasses. Correspondingly, HPT leads to essential transformations of the properties of amorphous alloys.

The results of studies of the effect of HPT on the structure of MS $Ti_{50}Ni_{25}Cu_{25}$ have already been presented above [54]. Tensile tests of the initial $Ti_{50}Ni_{25}Cu_{25}$ ribbon and the HPT-processed $Ti_{50}Ni_{25}Cu_{25}$ samples were conducted [102]. The vein-like pattern, typical for amorphous materials [103–105], is observable on the fracture of initial MS $Ti_{50}Ni_{25}Cu_{25}$. The distance between the expressed veins is about 10 μm (Figure 10a). The veins become more branched, and the distance between them is smaller: about 3 μm after HPT RT (Figure 10b). HPT at 150 °C leads to the appearance of areas with dimples with a size of about 3 μm [102]. Such fracture corresponds to a more ductile fracture behavior. The changes in the fracture surfaces can be interpreted as a successive increase of the material's local ductility due to HPT, which is connected with changes in the structure after HPT.

Figure 10. Fracture surface of specimens after tensile tests: (**a**) the initial amorphous TiNiCu ribbon; (**b**) HPT at T = 20 °C; (**c**) HPT at T = 150 °C [102].

Enhanced tensile ductility of the $Zr_{65}Al_{7.5}Ni_{10}Cu_{12.5}Pd_5$ BMG after HPT was reported in [85]. The BMG was subjected to a compression stage and HPT processing. The deformation behavior was analyzed using the digital image correlation (DIC) method and SEM. The changed ΔV and possible nanocrystallization were studied by DSC, positron annihilation spectroscopy, TEM, and synchrotron XRD. Studies showed that numerous SB nuclei were formed during tensile tests through shear transformation zones (STZs), which occurred in softer vacancy cluster regions, and that SB propagation was impeded by the typical free volume regions and SB interactions. The finite element method (FEM) was performed to investigate theoretically during HPT and residual stresses after HPT. The deformation behavior of the initial, compressed, and HPT specimens was investigated. The yield strengths of the initial and compressed BMG were similar (Figure 11a) [85]. The tensile deformation behaviors of the HPT specimens were different as compared with that of initial BMG. The values of yield strength of the HPT specimens were lower. The stress-strain curves demonstrated clear work-hardening behavior. The maximum strengths exceeded the yield stress of the initial BMG. HPT processing led to an increase of the plastic deformation region (DIC analyses), and the strain distribution became more homogeneous (Figure 11b) [85].

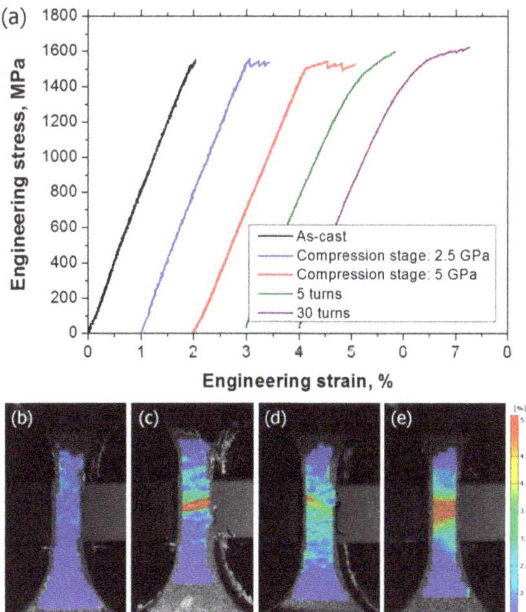

Figure 11. (**a**) Stress-strain curves of the as-cast and HPT-processed BMGs and the strain distribution of (**b**) the as-cast specimen, (**c**) the compression-stage HPT specimen, (**d**) the 5 turn HPT specimen, and (**e**) the 30 turn HPT specimen before fracture [85].

Investigations revealed that the HPT-processed specimens had a heterogeneous microstructure, consisting of a mixture of SBs and undeformed matrix. The heterogeneity prevented strain localization. In [85], it was also assumed that the work-hardening behavior with a tensile ductility of the $Zr_{65}Al_{7.5}Ni_{10}Cu_{12.5}Pd_5$ BMG resulted from the multiple SB caused by uniformly distributed heterogeneous microstructures without cracks or pores after HPT.

It was shown that the HPT of the Zr-based BMG led to a decrease in the elastic modulus E. The simultaneous effect of a decrease in the E and an increase in tensile ductility was crucial for applications of amorphous alloys.

4. Influence of HPT on the Behaviors of Amorphous Alloys during Annealing

In [24], the relation between hardness (*HV*), elastic modulus (*E*), and relaxation enthalpy (Δ*H*) of HPT-processed BMG was investigated. The $Zr_{50}Cu_{40}Al_{10}$ BMG was subjected to HPT under a pressure of 5 GPa at room temperature for up to 50 revolutions. Both elastic modulus E and hardness H, which were measured using nanoindentation, decreased with an increase in the number of revolutions (Figure 12a) [24].

The specimen for the DSC measurement was heated two times: the first time above the finishing of glass transition (743 K), and the second time above the crystallization temperature up to 873 K. The structural relaxation enthalpy ΔH was measured as the difference between the first and second heating DSC in the range of 400–730 K. The DSC curves for the HPT-deformed samples showed a pronounced exothermic peak due to structural relaxation in the temperature range below the glass transition temperature (T_g). An example of the DSC curve for the sample deformed by HPT for 50 revolutions is shown as an inset in Figure 12b. HPT did not alter T_g significantly, but led to a pronounced exothermic heat flow due to the relaxation of the deformed structure. Figure 12b shows an increase in the structural relaxation enthalpy with an increase in number of revolutions [24]. The relaxation enthalpy exhibited an excellent linear correlation with the observed decrease in the values of hardness and elastic modulus [24].

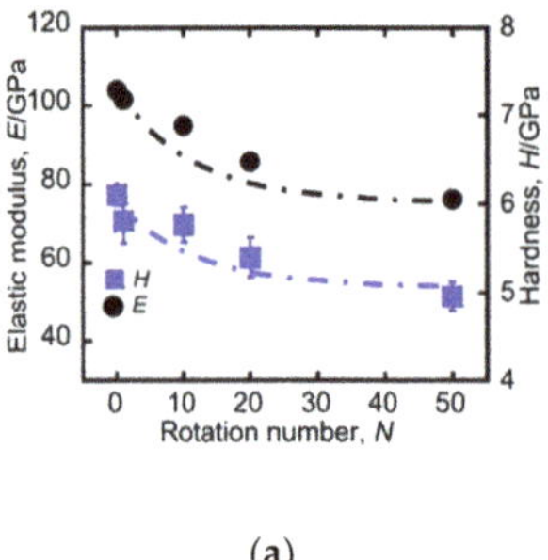
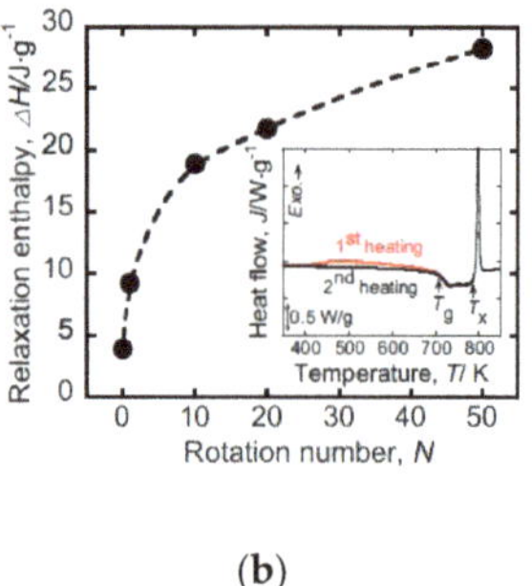

(a) (b)

Figure 12. (**a**) Elastic modulus and hardness as a function of the number of revolutions. The values for the initial BMG are plotted for $n = 0$. (**b**) Structural relaxation enthalpy as a function of the number of HPT revolutions. The value for the initial $Zr_{50}Cu_{40}Al_{10}$ BMG is plotted for $n = 0$. The inset shows the DSC curves of the sample deformed by HPT for 50 revolutions, where T_g and T_x denote the glass transition and crystallization temperature, respectively. The red and black lines denote the first and second heating curves, respectively (reprinted from [24], with the permission of AIP Publishing).

The observed decrease in hardness and elastic modulus can be attributed to the rejuvenated structure by HPT deformation. Local atomic environments in the rejuvenated structure BMG have a low local shear modulus. Thus, they are potential sites to become an STZ under the stress. The introduction of a rejuvenated volume decreases the overall mechanical strength. This is supported by the good linear correlation between *E* and *H* and the relaxation enthalpy [24]. It was shown that for the initial BMG state, the value of Δ*H* =5 J/g, and after HPT, it grew up to 25 J/g [24].

In [50], the DSC investigation of the as-cast and HPT-treated $Au_{49}Ag_{5.5}Pd_{2.3}Cu_{26.9}Si_{16.3}$ BMG was conducted. The increase of the free volume and ΔH was suggested by the broad exothermic peak prior to the T_g, (Figure 13). The T_g remained unaltered (404 K) for all samples (as-cast, HPT, and HPT-treated specimen stored for three months under RT), whereas the crystallization behavior was significantly altered after HPT. A considerable exothermic event in the below T_g was detected in the HPT sample as compared to the initial state (inset of Figure 13). The ΔH was about 0.33 kJ mol^{-1} for the HPT sample, and only $\Delta H = 0.06$ kJ mol^{-1} for the initial state. The area of this peak decreased for the HPT sample aged at RT for three months by 55% (H ~ 0.15 kJ mol^{-1}). These observations suggested that the HPT enhanced the free volume of the BMG. During annealing at RT for three months, the free volume was recovered. The crystallization temperature T_c for the HPT sample was about 12 K lower than that measured for the initial state. Furthermore, the crystallization signal in the DSC curve HPT sample split into two peaks, indicating that the mode of crystallization was modified by HPT.

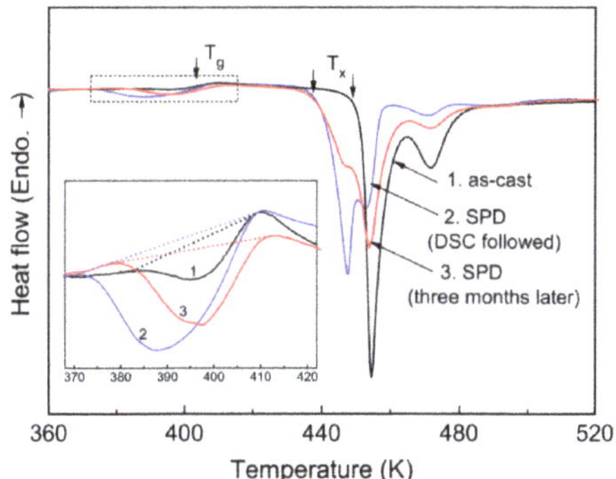

Figure 13. DSC traces of the as-cast and SPD-processed $Au_{49}Ag_{5.5}Pd_{2.3}Cu_{26.9}Si_{16.3}$ BMG. The third DSC curve was obtained for the SPD-treated sample after aging for three months at ambient temperature. The inset is the local magnification of the selected regime below T_g. A heating rate of 20 K·min^{-1} was used in all DSC measurements, endotermic effects up (Endo.) (reprinted from [50] with permission from Elsevier).

In [54], the relaxation enthalpy was measured during heating in the range of 20–400 °C for the initial MS TiNiCu and for the sample after HPT processing $n = 5$ at RT and 150 °C. T_c and ΔH are shown in Table 3. As a result of HPT at RT, the T_c temperature dropped down by about 40 °C in comparison with the initial state, and T_c dropped down by 20 °C as a result of HPT at 150 °C (Table 3). HPT led to a decrease in ΔH in comparison with the initial ribbons (Table 3). The value of ΔH remained virtually the same for the samples processed by HPT at RT, 50, and 100 °C (Table 3). For the sample, subjected to HPT at 150 °C, ΔH dropped 1.5 times in comparison with the HPT-treated samples at RT and approximately two times in comparison with the initial ribbons. The reduction of the ΔH may be associated with the reduced free volume ΔV in the amorphous phase [24].

Table 3. Crystallization temperatures and energy of structural relaxation occurring during heating for the as-spun and HPT-processed MS $Ti_{50}Ni_{25}Cu_{25}$ samples [55].

Sample	Crystallization Peak, °C	Crystallization Start, °C	Crystallization Finish, °C	Relaxation Energy, J/g
as-spun	460	444	463	41
HPT20	432	401	450	37
HPT50	437	408	453	37
HPT100	437	414	455	35
HPT150	441	421	455	24

Thus, the HPT processing results in an increase in the relaxation energy for the $Zr_{50}Cu_{40}Al_{10}$ and $Au_{49}Ag_{5.5}Pd_{2.3}Cu_{26.9}Si_{16.3}$ BMGs [24,50] and in a decrease for MS $Ti_{50}Ni_{25}Cu_{25}$ [54]. According to our data, the value of relaxation enthalpy for MS $Ti_{50}Ni_{25}Cu_{25}$ is significantly higher than the value of relaxation enthalpy of the $Zr_{50}Cu_{40}Al_{10}$ BMG. We can speculate that the HPT processing of the state with increased (probably corresponding to the maximum possible stored energy due to high quenching rates of producing) values of relaxation enthalpy (melt-spun ribbon) leads to a decrease in these values due to processes of relaxation. However, for BMGs, apparently, the opposite tendency for changes in energies is observed. From the initial low values, relaxation enthalpy increases due to the accumulation of energy via HPT processing. On the other hand, SPD leads to partial nanocrystallization in the $Ti_{50}Ni_{25}Cu_{25}$ alloy [41,51], which should lead to a decrease in relaxation energy. At the same time, in Zr- and Au-based BMGs, HPT-nanocrystallization does not occur. The question of whether SPD-induced nanocrystallization is the only way to reduce the relaxation energy under SPD is still open.

The process of crystallization taking place during annealing of amorphous alloys after HPT significantly differs from this of the non-deformed analogues [41,51]. For instance, in the MS $Ti_{50}Ni_{25}Cu_{25}$ alloy, HPT and annealing result in markedly finer nanograins than the annealing of the initial amorphous alloy [15,50]. The grain size in the MS $Ti_{50}Ni_{25}Cu_{25}$ alloy after annealing at 450 °C 10 min is 1.5 μm (Figure 14a), whereas in the sample subjected to HPT and subsequent annealing, $d_g = 100$ nm (Figure 14b).

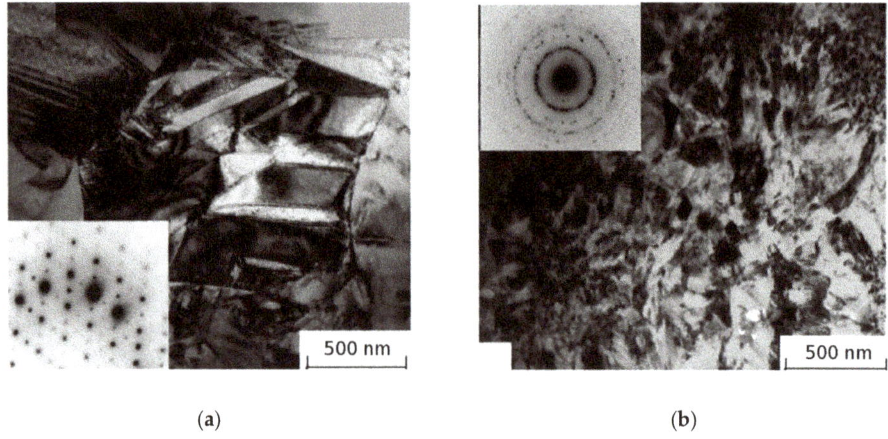

(a)   (b)

Figure 14. Bright-field TEM image and the corresponding SAED pattern for the MS $Ti_{50}Ni_{25}Cu_{25}$ alloy: (a) after annealing at 450 °C for 10 min, note the martensite plates inside the grains; (b) subjected to HPT ($T = 200$ °C) and subsequent annealing at 450 °C for 10 min [55].

It has been also revealed that the crystallization during annealing of the MS $Nd_{12}Fe_{82}B_6$, $Nd_9Fe_{85}B_6$, and $Nd_9Fe_{84}B_7$ alloys subjected to HPT differs from crystallization of the non-deformed MS alloy [41,52–54]. As a result of HPT-induced nanocrystallization, a high density of α-Fe nanocrystals with a size 10 nm has been induced in amorphous Nd-Fe-B after HPT at RT. These α-Fe nanocrystals are the nuclei for crystallization of the $Nd_2Fe_{14}B$ phase. As a result, a homogeneous structure of the nanocrystalline α-Fe and nanocrystalline $Nd_2Fe_{14}B$ phases is formed after annealing in the HPT sample, as compared with the non-deformed MS alloy [53]. In the results, the values of H_c and σ_r are ~30% higher for the annealed MS+HPT alloy in comparison to the annealed non-deformed MS $Nd_{12}Fe_{82}B_6$ alloy (Table 4) [52].

Table 4. Hysteresis properties of the MS $Nd_{12}Fe_{82}B_6$ alloy after HPT and annealing at 600 °C, 10 min [76].

State	H_c (kA/m)	σ_r (Am^2/kg)
MS + annealing	336	72
MS + HPT + annealing	528	86.5

The $Nd_9Fe_{85}B_6$ magnets prepared by a combination of HPT and subsequent thermal annealing show enhanced magnetic properties with an increase by 13% in B_r, 19% in H_c, and 30% in BH_{max} as compared with the magnets prepared by annealing of amorphous $Nd_9Fe_{85}B_6$ [53].

4.1. The Strain Achieved by HPT and Special Monitoring Schemes for the Formation of Shear Bands

Some recent results on the HPT processing of BMGs are shown in the following papers. In [106], the shear band evolution with the HPT deformation was examined in the Vit105 ($Zr_{52.5}Cu_{17.9}Ni_{14.6}Al_{10}Ti_5$) bulk metallic glass. For this purpose, a new HPT deformation scheme was developed. Under this scheme, two halves of disks of the bulk metallic glass were joined together and processed by high-pressure torsion for various strains (Figure 15).

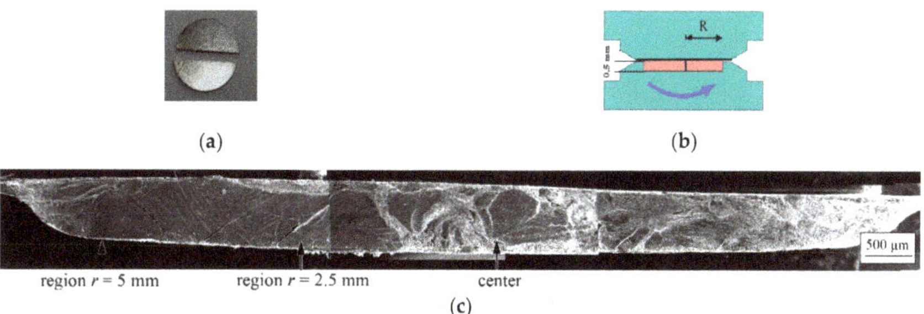

(a)   (b)

(c)

Figure 15. (**a**) Image of the two halves of the as-cast HPT disk. (**b**) Scheme of constrained HPT processing: the lower anvil has a groove with a depth of 0.5 mm, and the upper anvil is flat. (**c**) SEM image of the internal surface of the sample subjected to HPT for one revolution (reprinted from [106] with permission from Elsevier).

The images of the Vit105 and Cu samples after the HPT processing of two halves of a disk are shown in Figure 16.

The SEM examination samples of internal surfaces were subsequently used to study the formation of SPD under HPT. An increase in deformation led to a significant increase in SBs' density. The highest density was observed for the HPT state $n = 5$. The distance between the shear bands was down to 500 nm (Figure 17). The maximum density of shear bands was observed at the edges of HPT-treated samples and in the areas adjacent to the upper anvils (Figures 15 and 17) [107].

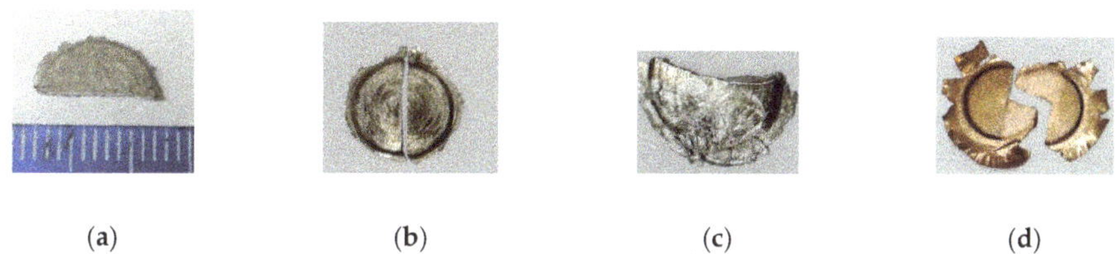

(a)   (b)   (c)   (d)

Figure 16. Optical images of samples: (**a**) BMG after compression only; (**b**) BMG after HPT for $n = 1/4$ revolutions; (**c**) BMG after HPT for $n = 5$ revolutions; (**d**) two halves of pure Cu after HPT for $n = 1/4$ revolutions (reprinted from [106] with permission from Elsevier).

From the obtained shear angle of the lower part relative to the upper part (Figure 16), one can see that the shear strain by HPT for $n = 5$ can be estimated, which equals $\gamma_{r5} \approx 1$, whereas the well-known formula $\gamma = 2\pi n R/h$ (6) predicts $\gamma_{r5} = 290$. Thus, HPT introduces into this BMG a much smaller strain in comparison with the prediction of Formula (6). A soft metal–pure Cu was subjected to HPT for 1/4 revolution under a similar scheme, and in this case, the shear angle was consistent with that predicted by Formula (6). It was also shown in [107–109] that the HPT of high-strength BMGs actually produced a smaller strain than predicted by Formula (6). In [107], it was shown that in Vitreloy LM-1B BMG, HPT with rotation $n = 0.1$ led to plastic strain of $\gamma_{R=2} = 0.12$ instead of the nominal $\gamma_{R=2} = 2.0$ value predicted by Formula (6). This discrepancy can be explained by the effect of "slippage" during HPT [17].

However, the density of the shear bands in the Vit105 BMG changed considerably as a result of HPT. According to XRD, HPT $n = 5$ led to an increase in the free volume (by about 1.3%) [106]. Thus, though the actual SB in the HPT samples was significantly lower in comparison with the expected value, the BMG under HPT experienced significant transformations in the structure. This fact requires further study.

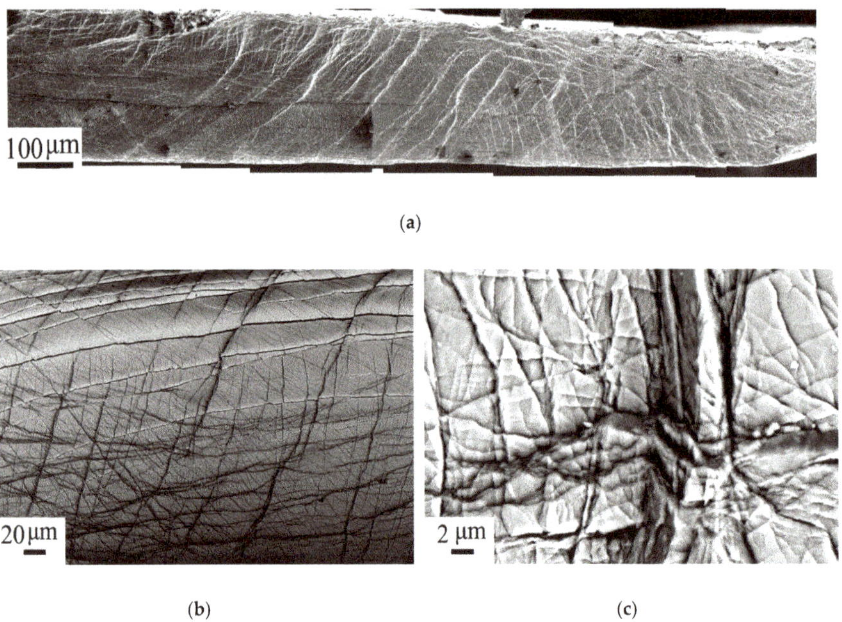

Figure 17. SEM images of the internal surface of the Vit105 BMG after HPT for $n = 5$: (**a**) general view of the sample; (**b**) region r-2.5 mm, magnification ×390; (**c**) magnification ×5000 (reprinted from [106] with permission from Elsevier).

The temperature increase that occurs during the HPT of BMG is important. The problem of the temperature increase during the HPT of various metallic materials is reflected in some studies [110–117]. The data available in the papers are rather contradictory, but in most papers, the heating of metal samples during HPT was estimated from 30 to 200 °C. The data on the temperature increase during the HPT of amorphous alloys are even more contradictory. According to [36,118], the temperature of the alloy increased to T_g during HPT for $n = 1$ (at an anvil rotation speed of 1 rpm) and reached 400 °C for CuZr(Al) [118]. However, these provisions are questionable. If the temperature during the HPT increases to T_g, then homogeneous deformation should take place in amorphous alloys, without the formation of shear bands, and the relaxation of the material should occur (the free volume decreases, and the relaxation energy decreases) [65,119]. However, in many works SBs were observed in amorphous alloys after HPT [61], and the free volume and relaxation energy increased. It can be assumed that during the HPT of BMGs, they are heated up to approximately the same temperatures as during the HPT of crystalline materials—to 200 °C [110–116].

The structural transformations of amorphous alloys during HPT are caused by several factors, including an increase in the temperature of the sample as a whole. However, it can be argued that the main physical mechanism that causes the transformation of the structure of amorphous alloys during HPT is the formation and motion of a very large density of shear bands in the samples. According to [106], the distance between the bands, calculated from SEM images, can be about 500 nm. Estimation of the gaps between the shear bands in amorphous materials after HPT based on TEM data showed that they reached 30 nm [94,120–122].

The mechanism of the structural transformation of an amorphous material during the propagation of a shear band can be described as follows: in the band itself (representing the shear plane), at the moment of motion, the material becomes heated to the melting temperatures [123]. However, this heating occurs in just microseconds, and the material heated in the narrow zone of the shear band (about 10 nm wide) also cools down in just microseconds due to the outflow of heat into the material surrounding the band [123]. Therefore, in the zone of the shear band, both relaxation/crystallization of the amorphous phase due to the heating and a high density of quasi-vacancies–defects of the atomic volume and the growth of non-equilibrium (growth of the free volume and internal energy) due to very rapid cooling of the material heated in the band take place.

According to [124], the structure of an amorphous material transforms even at a distance of 10 μm from the shear band that has passed. Thus, at the density of shear bands observed during the HPT processing, all of the amorphous material subjected to deformation should transforms.

Furthermore, the following notes can be made. During the HPT processing, the effect was observed when the height of the materials under processing, h, decreases with increasing number of anvil revolutions, n, due to the outflow of the material from under the anvils into the flash. The sample experiences a complex deformation by compression under high pressure, combined with torsion and the flow of the material into the flash. Correspondingly, at the initial stages of deformation, the structure is significantly influenced by the deformation. However, as the number of revolutions increases and n reaches a certain critical value (roughly estimated for BMGs as $n = 3$–5), the values of h become practically stabilized. "The deformation by compression combined with material flow into the flash" ceases or grows minimally with further increasing n. As shown above, the "torsional deformation", as such, of BMG samples due to the displacement of the lower part of the sample with respect to its upper part does not take place. The effect of sliding of the anvil surface on the sample surface under high pressure occurs. The frictional work is transformed into heat, and consequently into the heating of the sample and the anvils (as mentioned above, to the temperatures around 200 °C, or somewhat higher in the sliding mode). No significant deformation of the sample occurs here. Hence, in the HPT processing of BMGs and other solid materials, the initial stage can be distinguished, with a noticeable actual strain and strain-induced structural transformation. After n reaches a certain critical value, the strain as such decreases with increasing n, the heating makes a considerable contribution into the structural transformation, and the increasing number of revolutions leads rather to an increase in the time of exposure (annealing) at the temperature of self-heating during HPT processing (although this is undoubtedly subject to discussion). On the other hand, in some works, it is reported that when n during the HPT processing exceeds $n = 5$ (up to $n = 10$ and higher), structural changes also occur in BMG, as in crystalline alloys: the internal energy increases, which indicates a higher strain with increasing n [77]. These issues certainly require further study and analysis.

4.2. Accumulative HPT Procedure

As shown in [106], the actual shear strain in samples subjected to HPT is significantly lower in comparison with the expected value. In order to achieve high strain in BMGs, the authors proposed a new method: "accumulative HPT [125]. In the accumulative HPT procedure, the sample undergoes several cycles of HPT processing, then cutting, stacking, pressing, and subsequent HPT, as shown in Figure 18. Finally, the stacked segments are subjected to HPT with $n \geq 3$ as a result of which the fragments are consolidated into a monolithic disk [125]. The total number of revolutions during the "accumulative HPT" processing of Vit105 BMG was $n = 5$, and the total deformation of the BMG by pressing and shear was roughly estimated as $\gamma_\Sigma = 6$ [125].

Some of the BMG disks were processed by conventional HPT with $n = 5$. In this case, the total deformation of the BMG could be estimated as $\gamma_\Sigma = 2$. These estimates are very approximate, but they show that the total deformation at the "accumulative HPT" was 2.5 times greater than during conventional HPT.

The structure of the initial BMG is amorphous, as shown by the XRD method, and the position of the amorphous halo of the BMG after HPT shifts towards lower angles (Table 5). This means an increase in R of the first coordination sphere (R_1) (Table 5) [82,98]. The relative changes in ΔV after HPT could be estimated from R_1 according to [82]. The increase in free volume (ΔV) after conventional HPT $n = 5$ was $\Delta V \approx 1\%$ and ΔV after accumulative HPT was $\approx 2.5\%$. The $\Delta V = 2.5\%$ is too large compared to the ΔV usually observed during HPT of a BMG [82]. This is due to the large error in determining ΔV from the XRD [82]. However, these results indicate that the structure of BMG after accumulative HPT transforms much more significantly than after conventional HPT. The increase in the FWHM after HPT (Table 5) was also explained by changes in the structure of BMG [92]. The growth of FWHM after "accumulative HPT" is greater than after conventional HPT (Table 5), which also indicates the efficiency of accumulative HPT.

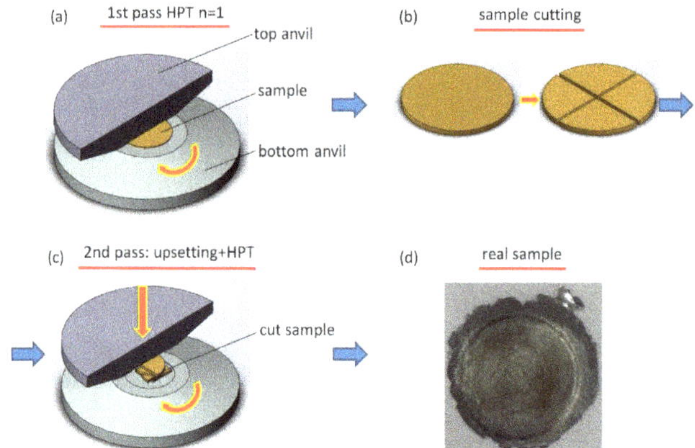

Figure 18. Principle of the "accumulative HPT" process: (**a**) disk-shaped sample subjected to HPT $n = 1$; (**b**) sample cut into four pieces; (**c**) pieces are stacked on top of the HPT anvil and HPT $n = 1$ applied again; (**d**) BMG sample obtained according to this procedure (reprinted from [125] with permission from Elsevier).

Table 5. Parameters of the amorphous structure of the BMG in different states from XRD (reprinted from [125] with permission from Elsevier).

BMG state	2θ (deg)	R_1 (ang.)	FWHM (deg)	ΔV
Initial	37.57(4)	2.942	6.34(4)	-
conventional HPT $n = 5$	37.42(4)	2.953	6.69(4)	1.1
accumulative HPT	37.22(4)	2.969	7.25(4)	2.5

5. Conclusions

Thus, recent studies have shown that SPD significantly affects the atomic structure and properties of amorphous alloys. The variation of microstructure resulting from SPD processing is closely related to processing parameters (amount of shear strain, temperature of processing, imposed pressure).

In a number of publications, it was shown that nanocrystallization in the amorphous phase occurs under SPD. We should note that under conventional schemes of deformation, nanocrystallization was observed in a small fraction of amorphous material in shear bands, whereas during HPT, it takes place throughout the entire volume of amorphous samples. It is interesting to note that during HPT of an amorphous alloy (for instance Nd–Fe-B), the amorphous phase is decomposed into the amorphous and crystalline phases of basic pure metals. In some amorphous alloys (for instance, MS $Ti_{50}Ni_{25}Cu_{25}$), besides nanocrystallization during HPT processing other complex transformations of the structure were observed. Amorphous "clusters" become visible, and these "clusters" represent some kind of amorphous structure. This contrast could be the result of the existence in the amorphous phase of regions with reduced free volume and with enhanced free volume.

During the HPT processing of a number of other amorphous alloys, nanocrystallization was not observed, but a new amorphous structure could be formed as a result of HPT, depending on changes in the short-range order, the total amount, and the redistribution of free volume. Perhaps, structures produced by the SPD methods are in some aspects comparable to the nanoglass-type structures produced by IGC.

Correspondingly, as a result of the HPT processing of amorphous alloys, essential transformations occur in their properties, in particular mechanical properties. For instance, as a result of preliminary HPT processing, the fracture fractography changes. Nanoindentation studies have shown that HPT processing leads to a significant increase in the values of the strain rate sensitivity in comparison with the initial state. At the same time, the course of change of the elastic modulus in a Zr-based BMG depends on the temperature of the HPT processing (20 or 150 °C). In some cases, HPT leads to a decrease in the values of Young's modulus. The first work, indicating the emergence of tensile ductility in some BMGs ($Zr_{65}Al_{7.5}Ni_{10}Cu_{12.5}Pd_5$) after HPT processing, has been published. The emergence of tensile ductility can be explained by the formation of a high density of nanoscale inhomogeneities in the amorphous state. High tensile strength, high hardness, and low elastic modulus provide the great potential of BMGs for various commercial applications; however, these applications are limited by the brittleness of amorphous materials. Thus, a decrease in the elastic modulus and an increase in tensile ductility via HPT processing can provide wide applications for amorphous alloys.

It has been also revealed that the process of crystallization that occurs during the annealing of the MS amorphous alloys subjected to HPT differs significantly from the crystallization of the non-deformed analogues. In the case of MS Nd-Fe-B alloys, this enabled producing higher magnetic properties via a combination of HPT processing and annealing than via annealing of non-deformed analogues. Therefore, the combination of HPT processing and annealing in some cases can lead to the formation of specific nanostructured states with improved functional properties. However, currently, many aspects of the nature of the transformation of the structure and properties of amorphous alloys subjected to HPT are still unclear and require further research.

A discrepancy between the experimentally observed and predicted shear strains has been detected. The actual strain is significantly smaller than the predicted one.

The authors proposed a new method, "accumulative HPT", to achieve high strain in hard materials, including BMGs. The study showed that the structure of the Zr-based BMG during accumulative HPT transforms much more significantly than in the case of conventional HPT with the same number of revolutions.

Author Contributions: Conceptualization, D.G.; writing, original draft preparation, D.G.; writing, review and editing, V.A. All authors read and agreed to the published version of the manuscript.

Funding: The work is funded by RFBR IND-a Research Project 19-58-45014.

Acknowledgments: The authors are grateful for the productive collaboration of Ruslan Valiev, Aleksandr Glezer, Jing Tao Wang, Horst Hahn, Dmitry Louzguine-Luzgin, Aleksandr Aronin, Galina Abrosimova, Yulia Ivanisenko, Evgeniy Ubyivovk, Evgeniy Boltynjuk, Andrey Bazlov, Roman Sundeev, Anna Churakova, Askar Kilmametov, Almir Mullayanov, Alfred Sharafutdinov, Ilshat Sabirov, Julia Bazhenova, and many other people.

Conflicts of Interest: The authors declare no conflict of interest.

References

1. Greer, A.L.; Ma, E. Bulk Metallic Glasses: At the Cutting Edge of Metals Research. *MRS Bull.* **2007**, *32*, 611–619. [CrossRef]
2. Inoue, A. Stabilization of metallic supercooled liquid and bulk amorphous alloys. *Acta Mater.* **2000**, *48*, 279–306. [CrossRef]
3. Abrosimova, G.E. Evolution of the structure of amorphous alloys. *Physics-Uspekhi* **2011**, *54*, 1227–1242. [CrossRef]
4. Goncharova, E.V.; Konchakov, R.A.; Makarov, A.S.; Kobelev, N.P.; Khonik, V.A. On the nature of density changes upon structural relaxation and crystallization of metallic glasses. *J. Non. Cryst. Solids* **2017**, *471*, 396–399. [CrossRef]
5. Greer, A.L. Metallic Glasses. *Science* **1995**, *267*, 1947–1953. [CrossRef]
6. Inoue, A.; Nishiyama, N. New Bulk Metallic Glasses for Applications as Magnetic-Sensing, Chemical, and Structural Materials. *MRS Bull.* **2007**, *32*, 651–658. [CrossRef]
7. Axinte, E. Metallic glasses from "alchemy" to pure science: Present and future of design, processing and applications of glassy metals. *Mater. Des.* **2012**, *35*, 518–556. [CrossRef]
8. Louzguine-Luzgin, D.V.; Inoue, A. Bulk Metallic Glasses. In *Handbook of Magnetic Materials*; Elsevier: Sendai, Japan, 2013; pp. 131–171. [CrossRef]
9. Wang, Y.B.; Xie, X.H.; Li, H.F.; Wang, X.L.; Zhao, M.Z.; Zhang, E.W.; Bai, Y.J.; Zheng, Y.F.; Qin, L. Biodegradable CaMgZn bulk metallic glass for potential skeletal application. *Acta Biomater.* **2011**, *7*, 3196–3208. [CrossRef]
10. Cao, Q.P.; Liu, J.W.; Yang, K.J.; Xu, F.; Yao, Z.Q.; Minkow, A.; Fecht, H.J.; Ivanisenko, J.; Chen, L.Y.; Wang, X.D.; et al. Effect of pre-existing shear bands on the tensile mechanical properties of a bulk metallic glass. *Acta Mater.* **2010**, *58*, 1276–1292. [CrossRef]
11. Park, K.-W.; Lee, C.-M.; Kim, H.-J.; Lee, J.-H.; Lee, J.-C. A methodology of enhancing the plasticity of amorphous alloys: Elastostatic compression at room temperature. *Mater. Sci. Eng. A* **2009**, *499*, 529–533. [CrossRef]
12. Zhang, Q.S.; Zhang, W.; Xie, G.Q.; Louzguine-Luzgin, D.V.; Inoue, A. Stable flowing of localized shear bands in soft bulk metallic glasses. *Acta Mater.* **2010**, *58*, 904–909. [CrossRef]
13. Ma, E.; Ding, J. Tailoring structural inhomogeneities in metallic glasses to enable tensile ductility at room temperature. *Mater. Today* **2016**, *19*, 568–579. [CrossRef]
14. Ketov, S.V.; Sun, Y.H.; Nachum, S.; Lu, Z.; Checchi, A.; Beraldin, A.R.; Bai, H.Y.; Wang, W.H.; Louzguine-Luzgin, D.V.; Carpenter, M.A.; et al. Rejuvenation of metallic glasses by non-affine thermal strain. *Nature* **2015**, *524*, 200–203. [CrossRef] [PubMed]
15. Valiev, R.Z.; Estrin, Y.; Horita, Z.; Langdon, T.G.; Zechetbauer, M.J.; Zhu, Y.T. Producing bulk ultrafine-grained materials by severe plastic deformation. *JOM* **2006**, *58*, 33–39. [CrossRef]
16. Valiev, R.Z.; Zhilyaev, A.P.; Langdon, T.G. *Bulk Nanostructured Materials*; John Wiley & Sons, Inc: Hoboken, NJ, USA, 2013; ISBN 9781118742679.
17. Zhilyaev, A.; Langdon, T. Using high-pressure torsion for metal processing: Fundamentals and applications. *Prog. Mater. Sci.* **2008**, *53*, 893–979. [CrossRef]
18. Kovács, Z.; Henits, P.; Zhilyaev, A.P.; Révész, Á. Deformation induced primary crystallization in a thermally non-primary crystallizing amorphous Al85Ce8Ni5Co2 alloy. *Scr. Mater.* **2006**, *54*, 1733–1737. [CrossRef]
19. Kovács, Z.; Henits, P.; Zhilyaev, A.P.; Chinh, N.Q.; Révész, Á. Microstructural characterization of the crystallization sequence of a severe plastically deformed Al-Ce-Ni-Co amorphous alloy. *Mater. Sci. Forum* **2006**, *519–521*, 1329–1334. [CrossRef]
20. Henits, P.; Révész, Á.; Schafler, E.; Szabó, P.J.; Lábár, J.L.; Varga, L.K.; Kovács, Z. Correlation between microstructural evolution during high-pressure torsion and isothermal heat treatment of amorphous Al 85 Gd 8 Ni 5 Co 2 alloy. *J. Mater. Res.* **2010**, *25*, 1388–1397. [CrossRef]

21. Boucharat, N.; Hebert, R.J.; Rösner, H.; Wilde, G. Deformation-Induced Nanocrystallization in Al-Rich Metallic Glasses. *Solid State Phenom.* **2006**, *114*, 123–132. [CrossRef]
22. Sarac, B.; Spieckermann, F.; Rezvan, A.; Gammer, C.; Krämer, L.; Kim, J.T.; Keckes, J.; Pippan, R.; Eckert, J. Annealing-assisted high-pressure torsion in Zr55Cu30Al10Ni5 metallic glass. *J. Alloys Compd.* **2019**, *784*, 1323–1333. [CrossRef]
23. Huang, J.Y.; Zhu, Y.T.; Liao, X.Z.; Valiev, R.Z. Amorphization of TiNi induced by high-pressure torsion. *Philos. Mag. Lett.* **2004**, *84*, 183–190. [CrossRef]
24. Meng, F.; Tsuchiya, K.; Seiichiro; Yokoyama, Y. Reversible transition of deformation mode by structural rejuvenation and relaxation in bulk metallic glass. *Appl. Phys. Lett.* **2012**, *101*, 121914. [CrossRef]
25. Glezer, A.M.; Sundeev, R.V.; Shalimova, A.V. The cyclic character of phase transformations of the crystal ⇔ amorphous state type during severe plastic deformation of the Ti50Ni25Cu25 alloy. *Dokl. Phys.* **2011**, *56*, 476–478. [CrossRef]
26. Sundeev, R.V.; Glezer, A.M.; Shalimova, A.V. Are the abilities of crystalline alloys to amorphization upon melt quenching and severe plastic deformation identical or different? *Mater. Lett.* **2016**, *175*, 72–74. [CrossRef]
27. Sundeev, R.V.; Shalimova, A.V.; Glezer, A.M.; Pechina, E.A.; Gorshenkov, M.V.; Nosova, G.I. In situ observation of the "crystalline⇒amorphous state" phase transformation in Ti 2 NiCu upon high-pressure torsion. *Mater. Sci. Eng. A* **2017**, *679*, 1–6. [CrossRef]
28. Edalati, K.; Yokoyama, Y.; Horita, Z. High-pressure torsion of machining chips and bulk discs of amorphous Zr50Cu30Al10Ni10. *Mater. Trans.* **2010**, *51*, 23–26. [CrossRef]
29. Edalati, K.; Horita, Z. A review on high-pressure torsion (HPT) from 1935 to 1988. *Mater. Sci. Eng. A* **2016**, *652*, 325–352. [CrossRef]
30. Abrosimova, G.; Aronin, A.; Matveev, D.; Pershina, E. Nanocrystal formation, structure and magnetic properties of Fe–Si–B amorphous alloy afterdeformation. *Mater. Lett.* **2013**, *97*, 15–17. [CrossRef]
31. Abrosimova, G.E.; Aronin, A.S.; Dobatkin, S.V.; Kaloshkin, S.D.; Matveev, D.V.; Rybchenko, O.G.; Tatyanin, E.V.; Zverkova, I.I. The Formation of Nanocrystalline Structure in Amorphous Fe-Si-B Alloy by Severe Plastic Deformation. *J. Metastable Nanocrystalline Mater.* **2005**, *24–25*, 69–72. [CrossRef]
32. Abrosimova, G.; Aronin, A. Nanocrystal formation in Al- and Ti-based amorphous alloys at deformation. *J. Alloys Compd.* **2018**, *747*, 26–30. [CrossRef]
33. Czeppe, T.; Korznikova, G.; Morgiel, J.; Korznikov, A.; Chinh, N.Q.; Ochin, P.; Sypień, A. Microstructure and properties of cold consolidated amorphous ribbons from (NiCu)ZrTiAlSi alloys. *J. Alloys Compd.* **2009**, *483*, 74–77. [CrossRef]
34. Korznikova, G.F.; Korznikova, E.A. Production of bulk samples from Ni based melt-spun ribbons by consolidation on Bridgman anvils. *Lett. Mater.* **2012**, *2*, 25–28. [CrossRef]
35. Korznikova, G.F.; Czeppe, T.H.; Korznikov, A.V. On plastic deformation of bulk metallic glasses in Bridgman anvils. *Lett. Mater.* **2014**, *4*, 117–120. [CrossRef]
36. Révész, Á.; Kovács, Z. Severe Plastic Deformation of Amorphous Alloys. *Mater. Trans.* **2019**, *60*, 1283–1293. [CrossRef]
37. Valiev, R.Z.; Krasilnikov, N.A.; Tsenev, N.K. Plastic deformation of alloys with submicron-grained structure. *Mater. Sci. Eng. A* **1991**, *137*, 35–40. [CrossRef]
38. Valiev, R. Materials science: Nanomaterial advantage. *Nature* **2002**, *419*, 887, 889. [CrossRef]
39. Gunderov, D.V.; Raab, G.I.; Sharafutdinov, A.V.; Stolyarov, V.V.; Sellers, C. Cold consolidation of nanocrystalline NdFeB powders via a severe plastic deformation method. In Proceedings of the Fifteenth International Workshop on Rare-Earth Magnets and Their Applications, Dresden, Germany, 30 August–3 September 1998; Schultz, L., Mueller, K.H., Eds.; pp. 359–362.
40. Popov, A.G.; Ermolenko, A.S.; Gaviko, V.S.; Schegoleva, N.N.; Stolyarov, V.V.; Gunderov, D.V. Magnetic Hysteresis Properties and Structural Features of Nanocrystalline Nd9Fe84B7 Alloy Prepared by Melt-spinning and Severe Plastic Deformation. In *Proceedings of the Sixteenth International Workshop on Rare-Earth Magnets and Their Applications, Sendai, Japan, 2000*; Kaneko, H., Homma, M., Okada, M., Eds.; The Japanese Institute of Metals: Sendai, Japan, 2000; pp. 621–630.
41. Valiev, R.; Gunderov, D.; Zhilyaev, A.P.; Popov, A.G.; Pushin, V. Nanocrystallization Induced by Severe Plastic Deformation of Amorphous Alloys. *J. Metastable Nanocrystalline Mater.* **2004**, *22*, 21–26. [CrossRef]

42. Li, H.; Li, W.; Zhang, Y.; Gunderov, D.V.; Zhang, X. Phase evolution, microstructure and magnetic properties of bulk α-Fe/Nd2Fe14B nanocomposite magnets prepared by severe plastic deformation and thermal annealing. *J. Alloys Compd.* **2015**, *651*, 434–439. [CrossRef]

43. Straumal, B.B.; Mazilkin, A.A.; Protasova, S.G.; Gunderov, D.V.; López, G.A.; Baretzky, B. Amorphization of crystalline phases in the Nd–Fe–B alloy driven by the high-pressure torsion. *Mater. Lett.* **2015**, *161*, 735–739. [CrossRef]

44. Teitel', E.I.; Metlov, L.S.; Gunderov, D.V.; Korznikov, A.V. On the structural and phase transformations in solids induced by severe plastic deformation. *Phys. Met. Metallogr.* **2012**, *113*, 1162–1168. [CrossRef]

45. Gunderov, D.V.; Stolyarov, V.V. Bulk α-Fe/Nd2Fe14B nanocomposite magnets produced by severe plastic deformation combined with thermal annealing. *J. Appl. Phys.* **2010**, *108*, 053901. [CrossRef]

46. Korolev, A.V.; Kourov, N.I.; Pushin, V.G.; Gunderov, D.V.; Boltynjuk, E.V.; Ubyivovk, E.V.; Valiev, R.Z. Paramagnetic susceptibility of the Zr62Cu22Al10Fe5Dy1 metallic glass subjected to high-pressure torsion deformation. *J. Magn. Magn. Mater.* **2017**, *437*, 67–71. [CrossRef]

47. Gunderov, D.; Boltynjuk, E.; Ubyivovk, E.; Lukyanov, A.; Churakova, A.; Zamula, Y.; Batyrshin, E.; Kilmametov, A.; Valiev, R.Z. Atomic Force Microscopy Studies of Severely Deformed Amorphous TiNiCu Alloy. *Defect Diffus. Forum* **2018**, *385*, 200–205. [CrossRef]

48. Zhang, N.; Gunderov, D.; Yang, T.T.; Cai, X.C.; Jia, P.; Shen, T.D. Influence of alloying elements on the thermal stability of ultra-fine-grained Ni alloys. *J. Mater. Sci.* **2019**, *54*, 10506–10515. [CrossRef]

49. Gunderov, D.V.; Popov, A.G.; Schegoleva, N.N.; Stolyarov, V.V.; Yavary, A.R. Phase Transformation in Crystalline and Amorphous Rapidly Quenched Nd-Fe-B Alloys under SPD. In *Nanomaterials by Severe Plastic Deformation*; Wiley-VCH Verlag GmbH & Co. KGaA: Weinheim, Gemany, 2005; pp. 165–169.

50. Wang, X.D.; Cao, Q.P.; Jiang, J.Z.; Franz, H.; Schroers, J.; Valiev, R.Z.; Ivanisenko, Y.; Gleiter, H.; Fecht, H.-J. Atomic-level structural modifications induced by severe plastic shear deformation in bulk metallic glasses. *Scr. Mater.* **2011**, *64*, 81–84. [CrossRef]

51. Valiev, R.Z.; Pushin, V.G.; Gunderov, D.V.; Popov, A.G. The use of severe deformations for preparing bulk nanocrystalline materials from amorphous alloys. *Dokl. Phys.* **2004**, *49*, 519–521. [CrossRef]

52. Popov, A.G.; Gaviko, V.S.; Shchegoleva, N.N.; Shreder, L.A.; Gunderov, D.V.; Stolyarov, V.V.; Li, W.; Li, L.L.; Zhang, X.Y. Effect of High-Pressure Torsion Deformation and Subsequent Annealing on Structure and Magnetic Properties of Overquenched Melt-Spun Nd9Fe85B6 Alloy. *J. Iron Steel Res. Int.* **2006**, *13*, 160–165. [CrossRef]

53. Li, W.; Li, L.; Nan, Y.; Xu, Z.; Zhang, X.; Popov, A.G.; Gunderov, D.V.; Stolyarov, V.V. Nanocrystallization and magnetic properties of amorphous Nd9Fe85B6 subjected to high-pressure torsion deformation upon annealing. *J. Appl. Phys.* **2008**, *104*, 023912. [CrossRef]

54. Li, W.; Li, X.; Guo, D.; Sato, K.; Gunderov, D.V.; Stolyarov, V.V.; Zhang, X. Atomic-scale structural evolution in amorphous Nd9Fe85B6 subjected to severe plastic deformation at room temperature. *Appl. Phys. Lett.* **2009**, *94*, 231904. [CrossRef]

55. Gunderov, D.; Slesarenko, V.; Lukyanov, A.; Churakova, A.; Boltynjuk, E.; Pushin, V.; Ubyivovk, E.; Shelyakov, A.; Valiev, R. Stability of an Amorphous TiCuNi Alloy Subjected to High-Pressure Torsion at Different Temperatures. *Adv. Eng. Mater.* **2015**, *17*, 1728–1732. [CrossRef]

56. Gunderov, D.V.; Slesarenko, V.Y.; Churakova, A.A.; Lukyanov, A.V.; Soshnikova, E.P.; Pushin, V.G.; Valiev, R.Z. Evolution of the amorphous structure in melt-spun Ti50Ni25Cu25 alloy subjected to high pressure torsion deformation. *Intermetallics* **2015**, *66*, 77–81. [CrossRef]

57. Boltynjuk, E.V.; Gunderov, D.V.; Ubyivovk, E.V.; Lukianov, A.V.; Kshumanev, A.M.; Bednarz, A.; Valiev, R.Z. The structural properties of Zr-based bulk metallic glasses subjected to high pressure torsion at different temperatures. *AIP Conf. Proc.* **2016**, *1748*, 6. [CrossRef]

58. Gunderov, D.V. Some regularities in the amorphization and nanocrystallization under severe plastic deformation of crystalline and amorphous multicomponent alloys (In Russian). *Investig. Russ.* **2006**, *151*, 1404–1413.

59. Gunderov, D.; Boltynjuk, E.; Ubyivovk, E.; Churakova, A.; Kilmametov, A.; Valiev, R. Consolidation of the Amorphous Zr 50 Cu 50 Ribbons by High-Pressure Torsion. *Adv. Eng. Mater.* **2019**, 1900694. [CrossRef]

60. Gunderov, D.; Boltynjuk, E.; Churakova, A.; Batirshin, E.; Mullayanov, A.; Titov, V.; Ivanisenko, J. Effect of high-pressure torsion on the mechanical behavior of a Zr-based BMG. *IOP Conf. Ser. Mater. Sci. Eng.* **2019**, *672*, 012028. [CrossRef]

61. Shao, H.; Xu, Y.; Shi, B.; Yu, C.; Hahn, H.; Gleiter, H.; Li, J. High density of shear bands and enhanced free volume induced in Zr70Cu20Ni10 metallic glass by high-energy ball milling. *J. Alloys Compd.* **2013**, *548*, 77–81. [CrossRef]

62. Argon, A.S.; Kuo, H.Y. Plastic flow in a disordered bubble raft (an analog of a metallic glass). *Mater. Sci. Eng.* **1979**, *39*, 101–109. [CrossRef]

63. Yokoyama, Y. Ductility improvement of Zr–Cu–Ni–Al glassy alloy. *J. Non. Cryst. Solids* **2003**, *316*, 104–113. [CrossRef]

64. Rösner, H.; Peterlechner, M.; Kübel, C.; Schmidt, V.; Wilde, G. Density changes in shear bands of a metallic glass determined by correlative analytical transmission electron microscopy. *Ultramicroscopy* **2014**, *142*, 1–9. [CrossRef]

65. Mitrofanov, Y.P.; Peterlechner, M.; Binkowski, I.; Zadorozhnyy, M.Y.; Golovin, I.S.; Divinski, S.V.; Wilde, G. The impact of elastic and plastic strain on relaxation and crystallization of Pd–Ni–P-based bulk metallic glasses. *Acta Mater.* **2015**, *90*, 318–329. [CrossRef]

66. Jiang, W.H.; Pinkerton, F.E.; Atzmon, M. Deformation-induced nanocrystallization in an Al-based amorphous alloy at a subambient temperature. *Scr. Mater.* **2003**, *48*, 1195–1200. [CrossRef]

67. Chen, H.; He, Y.; Shiflet, G.J.; Poon, S.J. Deformation-induced nanocrystal formation in shear bands of amorphous alloys. *Nature* **1994**, *367*, 541–543. [CrossRef]

68. Lee, S.-W.; Huh, M.-Y.; Fleury, E.; Lee, J.-C. Crystallization-induced plasticity of Cu–Zr containing bulk amorphous alloys. *Acta Mater.* **2006**, *54*, 349–355. [CrossRef]

69. Kim, J.J.; Choi, Y.; Suresh, S.; Argon, A.S. Nanocrystallization during nanoindentation of a bulk amorphous metal alloy at room temperature. *Science* **2002**, *295*, 654–657. [CrossRef] [PubMed]

70. Popov, A.G.; Gaviko, V.S.; Shchegoleva, N.N.; Shreder, L.A.; Stolyarov, V.V.; Gunderov, D.V.; Zhang, X.Y.; Li, W.; Li, L.L. High-pressure-torsion deformation of melt-spun Nd9Fe85B6 alloy. *Phys. Met. Metallogr.* **2007**, *104*, 238–247. [CrossRef]

71. Buschow, K.H.J. New permanent magnet materials. *Mater. Sci. Reports* **1986**, *1*, 1–63. [CrossRef]

72. Stolyarov, V.V.; Gunderov, D.V.; Valiev, R.Z.; Popov, A.G.; Gaviko, V.S.; Ermolenko, A.S. Metastable states in R2Fe14B-based alloys processed by severe plastic deformation. *J. Magn. Magn. Mater.* **1999**, *196–197*, 166–168. [CrossRef]

73. Gaviko, V.S.; Popov, A.G.; Ermolenko, A.S.; Shchegoleva, N.N.; Stolyarov, V.V.; Gunderov, D.V. Decomposition of the Nd2Fe14B intermetallic compound upon severe plastic deformation by shear under pressure. *Phys. Met. Metallogr.* **2001**, *92*, 58–66.

74. Stolyarov, V.V.; Gunderov, D.V.; Popov, A.G.; Gaviko, V.S.; Ermolenko, A.S. Structure evolution and changes in magnetic properties of severe plastic deformed Nd(Pr)–Fe–B alloys during annealing. *J. Alloys Compd.* **1998**, *281*, 69–71. [CrossRef]

75. Popov, A.G.; Gaviko, V.S.; Shchegoleva, N.N.; Puzanova, T.Z.; Ermolenko, A.S.; Stolyarov, V.V.; Gunderov, D.V.; Raab, G.I.; Valiev, R.Z. Severe plastic deformation of R-Fe-B (R = Pr or Nd) hard magnetic alloys. *Phys. Met. Metallogr.* **2002**, *94*, S75–S81.

76. Popov, A.G.; Gaviko, V.S.; Shchegoleva, N.N.; Shreder, L.A.; Gunderov, D.V.; Stolyarov, V.V.; Li, W.; Li, L.L.; Zhang, X.Y. Effect of high-pressure torsion deformation and subsequent annealing on structure and magnetic properties of overquenched melt-spun Nd9Fe85B6 alloy. In Proceedings of the 19th International Workshop on Rare Earth Permanent Magnets and Their Applications, Beijing, China, 30 August–2 September 2006.

77. Glezer, A.M.; Plotnikova, M.P.; Shalimova, A.V.; Dobatkin, S.V. Severe plastic deformation of amorphous alloys: I. Structure and mechanical properties. *Bull. Russ. Acad. Sci. Phys.* **2009**, *73*, 1233–1239. [CrossRef]

78. Abrosimova, G.; Aronin, A. On decomposition of amorphous phase in metallic glasses. *Rev. Adv. Mater. Sci.* **2017**, *50*, 55–61.

79. Gapontsev, V.L.; Kesarev, A.G.; Kondrat'ev, V.V.; Ermakov, A.E. Phase separation in nanocrystalline alloys upon generation of nonequilibrium vacancies at grain boundaries. *Phys. Met. Metallogr.* **2000**, *89*, 430–434.

80. Handrich, K.; Kobe, S. *Amorphe Ferro- und Ferrimagnetika*; Physik-Verlag: Weinheim, Germany, 1980.

81. Gryaznov, V.G.; Kaprelov, A.M.; Romanov, A.E. Size effect of dislocation stability in small particles and microcrystallites. *Scr. Metall.* **1989**, *23*, 1443–1448. [CrossRef]

82. Yavari, A.R.; Le Moulec, A.; Inoue, A.; Nishiyama, N.; Lupu, N.; Matsubara, E.; Botta, W.J.; Vaughan, G.; Di Michiel, M.; Kvick, Å. Excess free volume in metallic glasses measured by X-ray diffraction. *Acta Mater.* **2005**, *53*, 1611–1619. [CrossRef]

83. Gleiter, H. Nanoglasses: A new kind of noncrystalline materials. *Beilstein J. Nanotechnol.* **2013**, *4*, 517–533. [CrossRef]

84. Gleiter, H.; Schimmel, T.; Hahn, H. Nanostructured solids–From nano-glasses to quantum transistors. *Nano Today* **2014**, *9*, 17–68. [CrossRef]

85. Joo, S.-H.; Pi, D.-H.; Setyawan, A.D.H.; Kato, H.; Janecek, M.; Kim, Y.C.; Lee, S.; Kim, H.S. Work-Hardening Induced Tensile Ductility of Bulk Metallic Glasses via High-Pressure Torsion. *Sci. Rep.* **2015**, *5*, 9660. [CrossRef]

86. Jing, J.; Krämer, A.; Birringer, R.; Gleiter, H.; Gonser, U. Modified atomic structure in a PdFeSi nanoglass. *J. Non. Cryst. Solids* **1989**, *113*, 167–170. [CrossRef]

87. Gleiter, H. Our thoughts are ours, their ends none of our own: Are there ways to synthesize materials beyond the limitations of today? *Acta Mater.* **2008**, *56*, 5875–5893. [CrossRef]

88. Gleiter, H. Nanocrystalline solids. *J. Appl. Crystallogr.* **1991**, *24*, 79–90. [CrossRef]

89. Weissmüller, J.; Schubert, P.; Franz, H.; Birringer, R.; Gleiter, H. The physics of non-crystalline solids. In Proceedings of the VII National Conference on the Physics of Non-Crystalline Solids, Cambridge, UK, 4–9 August 1991.

90. Fang, J.X.; Vainio, U.; Puff, W.; Würschum, R.; Wang, X.L.; Wang, D.; Ghafari, M.; Jiang, F.; Sun, J.; Hahn, H.; et al. Atomic Structure and Structural Stability of Sc 75 Fe 25 Nanoglasses. *Nano Lett.* **2012**, *12*, 458–463. [CrossRef] [PubMed]

91. Śniadecki, Z.; Wang, D.; Ivanisenko, Y.; Chakravadhanula, V.S.K.; Kübel, C.; Hahn, H.; Gleiter, H. Nanoscale morphology of Ni50Ti45Cu5 nanoglass. *Mater. Charact.* **2016**, *113*, 26–33. [CrossRef]

92. Cao, Q.P.; Li, J.F.; Zhou, Y.H.; Horsewell, A.; Jiang, J.Z. Effect of rolling deformation on the microstructure of bulk Cu60Zr20Ti20 metallic glass and its crystallization. *Acta Mater.* **2006**, *54*, 4373–4383. [CrossRef]

93. Aronin, A.; Matveev, D.; Pershina, E.; Tkatch, V.; Abrosimova, G. The effect of changes in Al-based amorphous phase structure on structure forming upon crystallization. *J. Alloys Compd.* **2017**, *715*, 176–183. [CrossRef]

94. Ubyivovk, E.V.; Boltynjuk, E.V.; Gunderov, D.V.; Churakova, A.A.; Kilmametov, A.R.; Valiev, R.Z. HPT-induced shear banding and nanoclustering in a TiNiCu amorphous alloy. *Mater. Lett.* **2017**, *209*, 327–329. [CrossRef]

95. Boltynjuk, E.V.; Gunderov, D.V.; Ubyivovk, E.V.; Monclús, M.A.; Yang, L.W.; Molina-Aldareguia, J.M.; Tyurin, A.I.; Kilmametov, A.R.; Churakova, A.A.; Churyumov, A.Y.; et al. Enhanced strain rate sensitivity of Zr-based bulk metallic glasses subjected to high pressure torsion. *J. Alloys Compd.* **2018**, *747*, 595–602. [CrossRef]

96. Churyumov, A.Y.; Bazlov, A.I.; Zadorozhnyy, V.Y.; Solonin, A.N.; Caron, A.; Louzguine-Luzgin, D.V. Phase transformations in Zr-based bulk metallic glass cyclically loaded before plastic yielding. *Mater. Sci. Eng. A* **2012**, *550*, 358–362. [CrossRef]

97. Kilmametov, A.; Gröger, R.; Hahn, H.; Schimmel, T.; Walheim, S. Bulk Density Measurements of Small Solid Objects Using Laser Confocal Microscopy. *Adv. Mater. Technol.* **2017**, *2*, 1600115. [CrossRef]

98. Gunderov, D.V.; Boltynjuk, E.V.; Sitdikov, V.D.; Abrosimova, G.E.; Churakova, A.A.; Kilmametov, A.R.; Valiev, R.Z. Free volume measurement of severely deformed Zr 62 Cu 22 Al 10 Fe 5 Dy 1 bulk metallic glass. *J. Phys. Conf. Ser.* **2018**, *1134*, 012010. [CrossRef]

99. Jiang, W.H.; Atzmon, M. Rate dependence of serrated flow in a metallic glass. *J. Mater. Res.* **2003**, *18*, 755–757. [CrossRef]

100. Glezer, A.M.; Louzguine-Luzgin, D.V.; Khriplivets, I.A.; Sundeev, R.V.; Gunderov, D.V.; Bazlov, A.I.; Pogozhev, Y.S. Effect of high-pressure torsion on the tendency to plastic flow in bulk amorphous alloys based on Zr. *Mater. Lett.* **2019**, *256*, 126631. [CrossRef]

101. Glezer, A.M.; Potekaev, A.I.; Cheretaeva, A.O. *Thermal and Time Stability of Amorphous Alloys*; CRC Press: Boca Raton, FL, USA, 2017; ISBN 9781315158112.

102. Gunderov, D.V.; Churakova, A.A.; Lukyanov, A.V.; Prokofiev, E.A.; Khasanova, D.A.; Zamanova, G.I. Thin microstructure of amorphous Ti-Ni-Cu alloy subjected to high pressure torsion. *Bull. Bashkir Univ.* **2015**, *20*, 406–407.

103. Bhowmick, R.; Raghavan, R.; Chattopadhyay, K.; Ramamurty, U. Plastic flow softening in a bulk metallic glass. *Acta Mater.* **2006**, *54*, 4221–4228. [CrossRef]

104. Jiang, F.; Zhang, D.H.; Zhang, L.C.; Zhang, Z.B.; He, L.; Sun, J.; Zhang, Z.F. Microstructure evolution and mechanical properties of Cu46Zr47Al7 bulk metallic glass composite containing CuZr crystallizing phases. *Mater. Sci. Eng. A* **2007**, *467*, 139–145. [CrossRef]

105. Huang, Y.; Shen, J.; Sun, J.; Zhang, Z. Enhanced strength and plasticity of a Ti-based metallic glass at cryogenic temperatures. *Mater. Sci. Eng. A* **2008**, *498*, 203–207. [CrossRef]

106. Gunderov, D.V.; Churakova, A.A.; Boltynjuk, E.V.; Ubyivovk, E.V.; Astanin, V.V.; Asfandiyarov, R.N.; Valiev, R.Z.; Xioang, W.; Wang, J.T. Observation of shear bands in the Vitreloy metallic glass subjected to HPT processing. *J. Alloys Compd.* **2019**, *800*, 58–63. [CrossRef]

107. Kovács, Z.; Schafler, E.; Szommer, P.; Révész, Á. Localization of plastic deformation along shear bands in Vitreloy bulk metallic glass during high pressure torsion. *J. Alloys Compd.* **2014**, *593*, 207–212. [CrossRef]

108. Dmowski, W.; Yokoyama, Y.; Chuang, A.; Ren, Y.; Umemoto, M.; Tsuchiya, K.; Inoue, A.; Egami, T. Structural rejuvenation in a bulk metallic glass induced by severe plastic deformation. *Acta Mater.* **2010**, *58*, 429–438. [CrossRef]

109. Adachi, N.; Todaka, Y.; Yokoyama, Y.; Umemoto, M. Cause of hardening and softening in the bulk glassy alloy Zr50Cu40Al10 after high-pressure torsion. *Mater. Sci. Eng. A* **2015**, *627*, 171–181. [CrossRef]

110. Figueiredo, R.B.; Pereira, P.H.R.; Aguilar, M.T.P.; Cetlin, P.R.; Langdon, T.G. Using finite element modeling to examine the temperature distribution in quasi-constrained high-pressure torsion. *Acta Mater.* **2012**, *60*, 3190–3198. [CrossRef]

111. Edalati, K.; Miresmaeili, R.; Horita, Z.; Kanayama, H.; Pippan, R. Significance of temperature increase in processing by high-pressure torsion. *Mater. Sci. Eng. A* **2011**, *528*, 7301–7305. [CrossRef]

112. Edalati, K.; Hashiguchi, Y.; Pereira, P.H.R.; Horita, Z.; Langdon, T.G. Effect of temperature rise on microstructural evolution during high-pressure torsion. *Mater. Sci. Eng. A* **2018**, *714*, 167–171. [CrossRef]

113. Yamaguchi, D.; Horita, Z.; Nemoto, M.; Langdon, T.G. Significance of adiabatic heating in equal-channel angular pressing. *Scr. Mater.* **1999**, *41*, 791–796. [CrossRef]

114. Zhilyaev, A.P.; García-Infanta, J.M.; Carreño, F.; Langdon, T.G.; Ruano, O.A. Particle and grain growth in an Al-Si alloy during high-pressure torsion. *Scr. Mater.* **2007**, *57*, 763–765. [CrossRef]

115. Todaka, Y.; Umemoto, M.; Yamazaki, A.; Sasaki, J.; Tsuchiya, K. Influence of High-Pressure Torsion Straining Conditions on Microstructure Evolution in Commercial Purity Aluminum. *Mater. Trans.* **2008**, *49*, 7–14. [CrossRef]

116. Zhilyaev, A. Energy Stored during High Pressure Torsion of Pure Metals. *Lett. Mater.* **2019**, *9*, 142–146. [CrossRef]

117. Lewandowski, J.J.; Greer, A.L. Temperature rise at shear bands in metallic glasses. *Nat. Mater.* **2006**, *5*, 15–18. [CrossRef]

118. Hóbor, S.; Kovács, Z.; Révész, Á. Macroscopic thermoplastic model applied to the high pressure torsion of metallic glasses. *J. Appl. Phys.* **2009**, *106*, 023531. [CrossRef]

119. Slipenyuk, A.; Eckert, J. Correlation between enthalpy change and free volume reduction during structural relaxation of Zr55Cu30Al10Ni5 metallic glass. *Scr. Mater.* **2004**, *50*, 39–44. [CrossRef]

120. Zheng, B.; Zhou, Y.; Mathaudhu, S.N.; Valiev, R.Z.; Tsao, C.Y.A.; Schoenung, J.M.; Lavernia, E.J. Multiple and extended shear band formation in MgCuGd metallic glass during high-pressure torsion. *Scr. Mater.* **2014**, *86*, 24–27. [CrossRef]

121. Boltynjuk, E.; Ubyivovk, E.; Gunderov, D.; Mikhalovskii, V.; Valiev, R.Z. Multiple Shear Bands in Zr-Based Bulk Metallic Glass Processed by Severe Plastic Deformation. *Defect Diffus. Forum* **2018**, *385*, 319–324. [CrossRef]

122. Gunderov, D.V.; Boltynjuk, E.V.; Ubyivovk, E.V.; Churakova, A.A.; Abrosimova, G.E.; Sitdikov, V.D.; Kilmametov, A.R.; Valiev, R.Z. High pressure torsion induced structural transformations in Ti- and Zr-based amorphous alloys. *IOP Conf. Ser. Mater. Sci. Eng.* **2018**, *447*, 012052. [CrossRef]

123. Greer, A.L.; Cheng, Y.Q.; Ma, E. Shear bands in metallic glasses. *Mater. Sci. Eng. R Rep.* **2013**, *74*, 71–132. [CrossRef]

124. Pan, J.; Chen, Q.; Liu, L.; Li, Y. Softening and dilatation in a single shear band. *Acta Mater.* **2011**, *59*, 5146–5158. [CrossRef]

125. Gunderov, D.V.; Churakova, A.A.; Astanin, V.V.; Asfandiyarov, R.N.; Hahn, H.; Valiev, R.Z. Accumulative HPT of Zr-based bulk metallic glasses. *Mater. Lett.* **2020**, *261*, 127000. [CrossRef]

Review

Specific Features of Structure Transformation and Properties of Amorphous-Nanocrystalline Alloys

Alexandr Aronin and Galina Abrosimova *

Structure Research Lab., Institute of Solid State Physics RAS, Chernogolovka 142432, Russia; aronin@issp.ac.ru
* Correspondence: gea@issp.ac.ru; Tel.: +74-9652-28462

Received: 6 February 2020; Accepted: 4 March 2020; Published: 9 March 2020

Abstract: This work is devoted to a brief overview of the structure and properties of amorphous-nanocrystalline metallic alloys. It presents the current state of studies of the structure evolution of amorphous alloys and the formation of nanoglasses and nanocrystals in metallic glasses. Structural changes occurring during heating and deformation are considered. The transformation of a homogeneous amorphous phase into a heterogeneous phase, the dependence of the scale of inhomogeneities on the component composition, and the conditions of external influences are considered. The crystallization processes of the amorphous phase, such as the homogeneous and heterogeneous nucleation of crystals, are considered. Particular attention is paid to a volume mismatch compensation on the crystallization processes. The effect of changes in the amorphous structure on the forming crystalline structure is shown. The mechanical properties in the structure in and around shear bands are discussed. The possibility of controlling the structure of fully or partially crystallized samples is analyzed for creating new materials with the required physical properties.

Keywords: metallic glasses; nanoglasses; structure evolution; properties

1. Introduction

Metallic alloys in an amorphous state (amorphous alloys or metallic glasses) were obtained for the first time by rapid melt quenching in 1960 [1]. The new material was of great interest, since it was determined at once that the properties of alloys in an amorphous state differ from those of crystalline materials of the same composition (for example, see [2]). Although many years have passed, interest in amorphous alloys persists. So far, there have not been any reliable models of an amorphous structure. There are a number of questions related to evolution of its structure and properties, as well as the possibility of obtaining nanocrystalline materials by thermal and deformation action on amorphous alloys. The properties of nanocrystalline materials and composite amorphous-nanocrystalline materials also differ from those of amorphous and traditional crystalline materials. Depending on chemical composition, nanocrystalline metallic materials are characterized by good plasticity and high viscosity, high strength and hardness, low moduli of elasticity, higher diffusion coefficients, as well as larger values of thermal expansion coefficient and better magnetic properties as compared with traditional crystalline materials [3–6]. Partially crystalline alloys containing nanocrystals have hysteresis magnetic properties at the level of the best crystalline and amorphous alloys; at the same time, they have high saturation induction comparable to that of the best high-silicon electrical steel. Light Al-based amorphous-nanocrystalline alloys have high strength, with the values of yield strength able to reach 1.6 GPa [7] at good plasticity.

This review analyzes changes in the structure of the amorphous phase from preparation to the initial stages of crystallization. The transformation of a homogeneous amorphous phase into a heterogeneous phase, the dependence of the scale of inhomogeneities on the component composition, and the conditions of external influences are considered. The effect of changes in the amorphous structure on the forming crystalline structure is shown. The crystallization processes of the amorphous phase, such as the homogeneous and heterogeneous nucleation of crystals, are considered. Particular attention is paid to an important feature of the processes of crystal formation in the amorphous phase, namely, compensation for volume mismatch. In the study of crystallization of metal glasses, this issue is usually not given sufficient attention. Meanwhile, the way of compensating for the volume mismatch has a decisive influence on the forming crystalline structure. When crystals form in the liquid phase (melt solidification), the difference in the specific volume of the liquid and solid phases is quickly compensated due to the high mobility of atoms at high temperatures. In the amorphous phase, such a process is hampered due to significantly lower diffusion coefficients at room temperature or crystallization temperature, which does not exceed 500 °C for most amorphous alloys, that is noticeably lower than the melting temperature. As will be shown below, in this case, volume mismatch can be compensated by changing the sequence of formation of crystalline phases, changing the morphology of the crystals formed, the formation of nanocrystals with subsequent recrystallization, or the formation of pores. Data on the influence of thermal or deformation effects on the parameters of the crystal structure are presented. The effect of plastic deformation on the formation of nanocrystals and the features of the mechanical properties of the amorphous phase near the zones of plastic deformation localization are shown. In conclusion, there is evidence of the possibility of restoring the amorphous structure using the cryogenic cycling method.

Studies of the structure and crystallization processes of metal glasses were carried out on a lot of systems. The main attention in this review is focused on the alloys of such compositions in which the above features are most clearly manifested.

2. Structural Changes Occurring during Heating and Deformation

An amorphous state is unstable, and the amorphous phase decays with the formation of the crystalline phases under heat treatment. In a number of cases, transformation of the structure is more complex, and the structure changes of the amorphous phase are observed before the crystallization onset: separation to regions differing in the chemical composition and short-range order. At the same time, a heterogeneous, amorphous structure is formed. The formed amorphous regions generally do not have a sharp interface, the transition may exhibit peculiarities of spinodal decomposition [8,9]. Structure changes in metallic glasses before the crystallization onset were discovered in a large number of alloys [10–18]. A lot of works [19–26] are devoted to the study of the separation of the amorphous phase and heterogeneous amorphous structure formation.

A heterogeneous amorphous structure can be observed both in an initial state immediately after melt quenching and after various external actions: heat treatment, deformation, irradiation, and others. For example, as-prepared amorphous Pd-Au-Si alloy is uniform, but regions differing in the chemical composition were formed in the sample after annealing at 400 °C [10]. In amorphous $(Mo_{0.6}Ru_{0.4})_{100-x}B_x$ such regions were revealed before the crystallization onset [15], and in amorphous $Fe_{67}Co_{18}B_{14}Si_1$ alloy having a uniform amorphous structure after the quenching, heating to 400–600 K resulted in the formation of regions enriched with boron and regions with Fe–Co composition [16]. Of course, regions with different chemical compositions are characterized by different radii of the first coordination sphere (different shortest distances between the atoms) and, correspondingly, different short-range order. Thus, short-range order in the amorphous phase, of course, depends on the chemical composition, and it changes with the composition change. An amorphous structure can change significantly with the change in the component concentration; for example, the structure of amorphous $(Zr_{0.667}Ni_{0.333})_{1-x}B_x$ alloy changes with boron concentration change from 0 to 25 at.% [12]. It was shown that short range order type changes when x ≈ 0.05. In Fe-Zr system, the phase diagram of which contains two eutectic regions (at 10 and 76 at.% of Zr), the formation of the amorphous phases is possible near each of the eutectic points. Naturally, the amorphous Fe-based phase and amorphous Zr-based phase will differ from each other in the radius of the first coordination sphere and short-range order.

The amorphous phase structure can change significantly not only under heating, but under deformation. Structure changes under deformation action were discovered at the beginning of the investigation of amorphous alloys. In [27], it was found out that under rolling of amorphous $Pd_{80}Si_{20}$ alloys, a shift of the first peak of the structure factor towards smaller angles occurs, which indicates a change in the radius of the first coordination sphere. The studies showed also that the plastic deformation of metallic glasses at low temperatures and moderate loads is strongly localized, occurring in narrow regions, with shear bands, almost not affecting the main part of the amorphous phase. In these bands the structure, of course, should change.

Notably, in another group of works [28–30], the authors of which showed by the example of amorphous Zr-based alloys that in the absence of plastic deformation, tension leads to the change of a distance between the atoms, with these changes depending on the applied stress orientation. Amorphous Zr-based alloys become brittle and hardly undergo plastic deformation, that is why the mentioned works considered the region of elastic deformation only. Later the investigations of structure changes under plastic deformation were carried out for $Pd_{40}Ni_{40}P20$ alloy. The authors of [31] found out that after rolling deformation the amorphous alloy structure becomes anisotropic: after deformation, the distance between the atoms along the rolling direction increases, while that in a perpendicular direction almost does not change. The observed effect decreases with time. The structure change is accompanied by a decrease in the transverse sound velocity in the deformed samples compared with the initial samples. All the obtained results [28–31] are an evidence of the elliptical nature of the first coordination "sphere" in the deformed samples. Based on the studies carried out using X-ray radiation with different wavelengths [31], it was also discovered that the structure changes are more pronounced in a subsurface region of the sample where deformation is maximum. Thus, the structure of the deformed samples may differ in cross-section. Structure changes during deformation were observed also under the use of small-angle X-ray scattering. Thus, for example, in $Pd_{80}Si_{20}$ alloy, deformation of 20% resulted in a significant increase in the intensity of small-angle scattering [32], and in $Fe_{40}Ni_{40}P_{14}B_6$ alloy [33], the existing inhomogeneities with a size of 32 Å increased up to 35 Å after 2-h annealing at 100 °C. Changes in the amorphous phase structure bring changes in the physical properties: the Curie temperature increase [34] or there is an emergence of two Curie temperature values (in the case of separation of the amorphous phase and formation of areas differing in the chemical composition and type of short-range order [35]), embrittlement [36,37], etc.

3. Formation of Nanoglasses

As was shown, both heat treatment and deformation can result in the formation of a heterogeneous amorphous structure. In recent years, the term "nanoglasses" has been used to describe a heterogeneous amorphous structure. Although, as stated above, a heterogeneous amorphous structure was observed in a number of systems by a lot of researchers, the term "nanoglass" was introduced in the work by Gleiter [38] quite recently. Today nanoglasses are the particular focus in world scientific literature. These materials, which differ from homogeneous amorphous, and nanocrystalline materials in the structure, can display unique functional properties: mechanical, magnetic, catalytic, and others.

For the first time nanoglasses were synthesized as a bulk sample by consolidation of amorphous powders under pressure. However, the samples obtained by this method had small geometric sizes, and under their obtaining it was extremely difficult to avoid oxidation of the surface of amorphous nanoparticles and formation of micropores embrittling the samples. Action on an amorphous alloy obtained by melt quenching in the form of a ribbon, can be used as an alternative method of obtaining nanoglasses. The method of obtaining metallic glasses by melt quenching onto a moving substrate is one of the main methods of obtaining. Thereby ribbons with a thickness of 10–50 μm, width of several millimeters to tens of centimeters, and length of several meters are formed. Such alloys have significantly large sizes for both investigation of the structure and properties and use in industry. As stated above, heterogeneity regions resulting in a nanoglass state can be formed in the amorphous phase under different actions: heat treatment of different types, deformation (rolling, hydrostatic, and quasi-hydrostatic compression, high-pressure torsion, and others) at different temperatures, and irradiation. The separation of the amorphous phase can occur at different scales. Thus, for example, the heating of metallic $Zr_{40}Ti_{10}Cu_{50}$ glass before the onset of crystallization results in the separation of an amorphous matrix at the scale of about several nanometers [39], which subsequently results in the formation of a composite amorphous–nanocrystalline structure with an extremely small grain size (2–5 nm). At the same time, the separation scale is significantly larger in a number of other systems. It was found that, in the $Fe_{90}Zr_{10}$ alloy [9], long-term low-temperature annealing (at a temperature of 100 °C) leads to the decomposition of the amorphous phase with the size of the heterogeneity regions of about 25 nm. In the annealed $Ni_{70}Mo_{10}P_{20}$ alloy [17], heterogeneities sized of about 30 nm are formed below the glass transition temperature; and in $Ni_{70}Mo_{10}B_{20}$ metallic glass [40] after annealing above the glass transition temperature, regions of an amorphous phase of different compositions up to 50 nm in size are formed.

The irradiation of amorphous alloys, for example, by fast neutrons, can also lead to the formation of an inhomogeneous amorphous structure consisting of regions and different short-range orders. When the amorphous $Pd_{80}Si_{20}$ alloy was irradiated with fast neutrons with a dose of 5×10^{20} neutrons/cm^2, the formation of structural inhomogeneities was observed, which are clusters with increased electron density, surrounded by boundary regions with a reduced electron density. The diameter of the clusters together with the shell was 10–20 Å [41]. Irradiation can lead not only to the formation of inhomogeneities, but also to their reduction. For example, amorphous alloys $(Mo_{60}Ru_{40})_{82}B_{18}$ and $Fe_{40}Ni_{40}P_{14}B_6$ became brittle after isothermal annealing, which is associated with the formation of inhomogeneities. After irradiation, the plasticity returned to its original value.

The peculiarities of structure and properties of nanoglasses are being investigated in alloys of different compositions [42–47].

4. Processes of Crystallization of Amorphous Alloys

Both homogeneous amorphous alloys and nanocrystals crystallize at an increase in the temperature, with the parameters of the formed structure (morphology, phase composition, sizes of structural components, etc.) depending on both heat treatment conditions and sample history.

Crystallization of amorphous $(Ni_{70}Mo_{30})_{90}B_{10}$ alloy [48] is an example of the typical transformation from an amorphous to a nanocrystalline state. Figure 1 illustrates bright-field (a) and dark-field (b) images of the microstructure of $(Ni_{70}Mo_{30})_{90}B_{10}$ sample which was annealed for 144 h at 873 K. One can see that the nanocrystals nucleate uniformly over the sample and are randomly distributed over an amorphous matrix. An average nanocrystal size is 16 nm. Figure 2 shows a high-resolution image of the structure of $(Ni_{65}Mo_{35})_{90}B_{10}$ alloy sample annealed at 873 K for 72 h. A characteristic feature of the nanocrystal arrangement in an amorphous matrix is that the nanocrystals do not have regions contacting with each other. The crystals are isolated from each other by the regions of an amorphous matrix. The minimum thickness of a layer of the amorphous phase between the nanocrystals is ~2 nm.

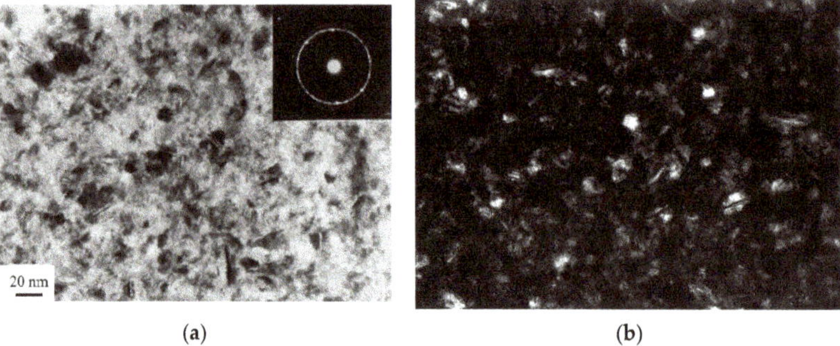

(a)   (b)

Figure 1. (a) Bright-field and (b) dark-field electron microscope images of the microstructure of (Ni70Mo30)90B10 sample after annealing at 873 K for 144 h.

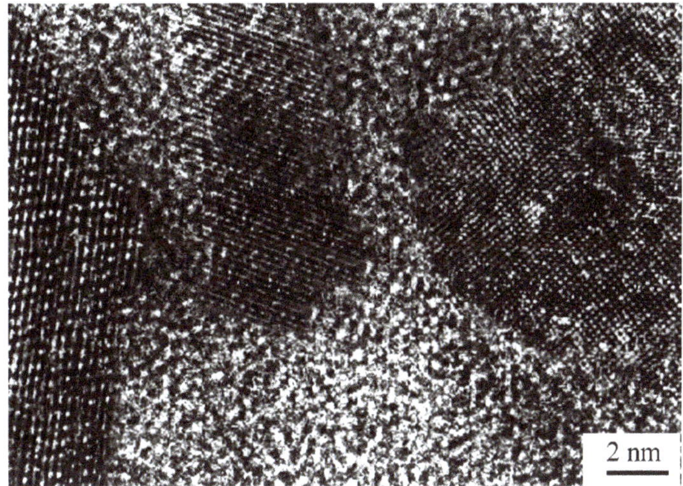

Figure 2. High-resolution electron microscope image of the microstructure of (Ni70Mo30)90B10 sample after annealing at 873 K for 72 h.

The fraction of the crystalline phase increases gradually during heat treatment. The formed nanocrystals are a solid solution of Mo in Ni. A nanocrystal size is several nanometers. As annealing duration increases, the grain size increases insignificantly, and then almost does not change. The composition of regions of the amorphous matrix changes at that; it gets enriched with components which are insoluble or have limited solubility in nanocrystals (Mo, B). A composition change of the amorphous phase leads to a change in its crystallization temperature, which results in the completion of nanocrystal growth. Such component redistribution resulting in a composition change in intercrystalline regions of the amorphous phase was observed in alloys of other compositions (Fe-Zr-B, Fe-Si-B-Cu-Nb) [49,50]. In some cases, the dependence of nanocrystal size on the distance from the surface was observed under the crystallization of the ribbons of amorphous alloys [51]. Thus, for example, in the subsurface regions of the above mentioned $(Ni_{70}Mo_{30})_{90}B_{10}$ alloy annealed for 72 h at 873 K, an average size of the nanocrystalline phase grains is 20 nm and decreases to 17 nm at a depth of about 8 μm. The revealed difference in the size of nanocrystals (and their lattice parameter) along the sample depth is related to different chemical compositions of the subsurface and deep regions of the alloy.

A nanocrystalline alloy is two-phase; the structure consists of an amorphous matrix and nanocrystals uniformly distributed over the matrix, which have no direct contact with each other. The detailed studies of crystallization processes of amorphous alloys of Ni-Mo-B system [52,53] showed that the lattice parameters of nanocrystals with the fcc lattice in this system change depending on the isothermal annealing duration. The analysis of changes occurring demonstrated that, under primary crystallization, the composition of formed crystals differs from matrix composition, the composition of the remaining amorphous phase changes as crystallization proceeds. In the case of the alloys under study, the amorphous phase gets enriched with a refractory component. On the other hand, the composition of the crystals precipitated during annealing also undergoes some changes. Changes in the nanocrystal composition lead to changes in the lattice parameters. The dissolution of Mo in Ni leads to an increase in the lattice parameter of a solid solution, and dissolution of B leads to its decrease. As is well known [54], at 873 K the equilibrium concentration of Mo in Ni is 17 at.%, and that of B is less than 1 at.%. At the moment of formation, nanocrystals are a supersaturated solid solution of Mo and B in Ni. Under isothermal exposure, codiffusion of Mo and B from nanocrystals occurs as grains grow. Due to different chemical compositions of initial alloys at the initial moment, the compositions of nanocrystals in the alloys also turn to be different. Different values of supersaturation of solid solution nanocrystals with Mo and B result in different values of the lattice parameter, starting with exposure for 5 h. The maximum value of the lattice parameter is observed in the case of the highest content of Mo in $(Ni_{65}Mo_{35})_{90}B_{10}$ alloy and the minimum content of B in $(Ni_{70}Mo_{30})_{95}B_5$ alloy. At an increase in the concentration of B in $(Ni_{70}Mo_{30})_{90}B_{10}$ alloy as compared with $(Ni_{70}Mo_{30})_{95}B_5$ alloy or a decrease in the concentration of Mo in $(Ni_{70}Mo_{30})_{90}B_{10}$ alloy as compared with $(Ni_{65}Mo_{35})_{90}B_{10}$ alloy, nanocrystals are formed with a low value of the lattice parameter. The codiffusion of Mo and B into the surrounding matrix occurs under annealing. Loss of the highest amount of Mo (together with B) from the nanocrystals in $(Ni_{65}Mo_{35})_{90}B_{10}$ alloy (as compared with other alloys) leads to an insignificant decrease in the lattice parameter. Since a lower amount of Mo leaves from the nanocrystals in $(Ni_{70}Mo_{30})_{90}B_{10}$ alloy, one should expect the lattice parameter to increase due to B diffusion into the amorphous matrix, occurring at the same time, which is observed in the experiment.

The formed nanocrystalline structure in the above alloys has high thermal stability. The analysis carried out demonstrated that the nanocrystalline structure exists until the nanocrystals are isolated from each other by the amorphous phase. As previously stated, the amorphous matrix is enriched with B and a refractory component (Mo) and has an increased stability as compared with that of the amorphous phase of an initial composition. As soon as the amorphous matrix between the crystals disappears, they begin to grow rapidly. Such a case is shown in Figures 3 and 4. Figure 3 displays bright-field (a) and dark-field (b) images of a large crystal. Figure 4 demonstrates a high-resolution electron microscope image of the crystals contacting with each other. One can see that in this case the boundary is a region with a thickness of several Å, which divides two adjacent crystals. An arrow marks the region of a direct contact of the nanocrystals with each other.

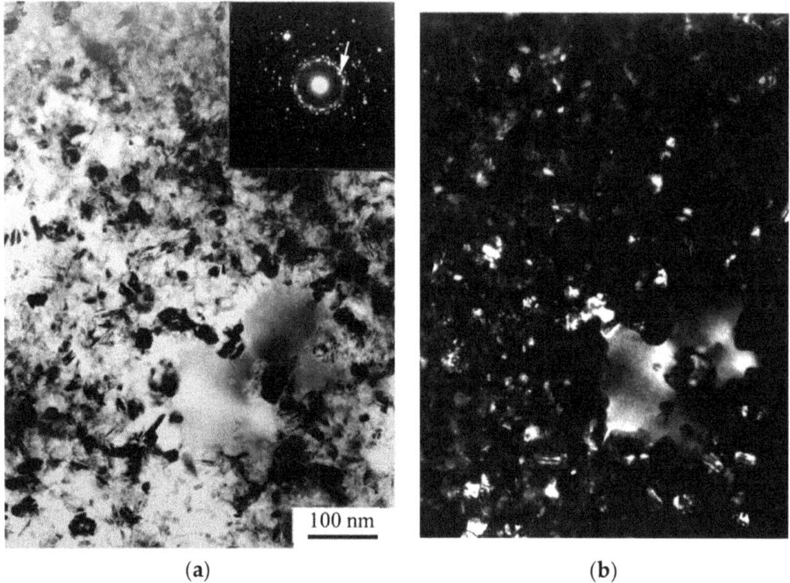

(a)   (b)

Figure 3. Decomposition of the nanocrystalline phase in (Ni65Mo35)90B10 alloy (873 K, 240 h; (**a**) bright-field and (**b**) dark-field images are in the reflection shown in the electron diffraction pattern).

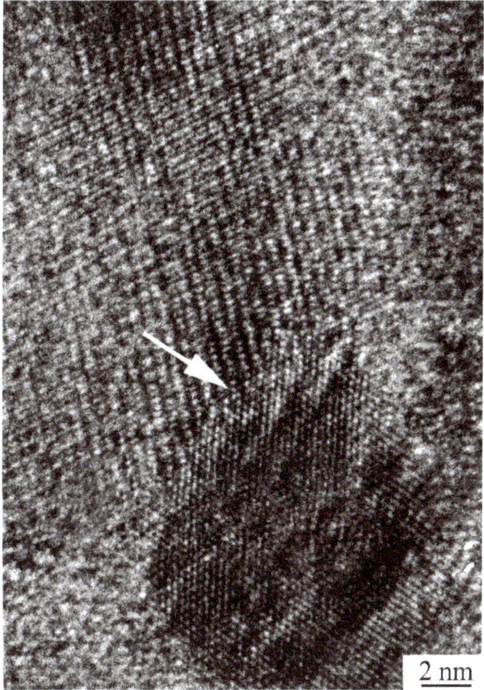

Figure 4. High-resolution image of the structure of annealed (Ni65Mo35)90B10 sample under the decomposition of the nanocrystalline structure. An arrow marks the region of a direct contact of the nanocrystals with each other.

5. Heterogeneous Nucleation

Another group of nanocrystalline alloys under the most extensive study is light Al-based alloys. These alloys are light high-strength materials, which opens up extensive possibilities of their practical application. These alloys also crystallize by the nucleation and growth mechanism, and nanocrystals are formed by the primary crystallization reaction. Nanocrystal growth under the primary crystallization was investigated in a number of works, and it was shown, for example, for alloys of Al-Ni-Ce system [55] that it is diffusion-controlled. In literature, there are controversial data on the mechanisms of nanocrystal nucleation. There are literature data [56] stating that the process of Al nanocrystal nucleation is homogeneous. According to [57], this process is heterogeneous.

It is known that the homogeneous crystallization related to the fluctuation nucleation of nuclei with a critical size can take place only above the glass transition temperature T_g. At $T < T_g$, the viscosity of an amorphous phase is too high for such fluctuations, and nucleation can occur by heterogeneous mechanism only [58]. In light metallic glasses, it is difficult to define in which temperature range relative to T_g the process of nucleation occurs since the value of T_g in these alloys is unknown. Consequently, it is impossible to conclude correctly on the nucleation mechanism, based on the temperature range of the transformation.

To conclude reliably on the mechanism of nucleation under nanocrystallization, the data on nanocrystal distribution over size under different durations of isothermal exposure and the corresponding analysis of the distribution are required. This issue was studied in most details for alloys of $Al_{86}Ni_{11}Yb_3$ system [59]. As a result of primary crystallization, a structure is formed in an alloy, which consists of an amorphous matrix containing Al fcc crystals randomly distributed over it.

Al fcc crystals have a size of several nanometers. An example of this structure is shown in Figure 5. Al nanocrystals are generally isolated by the amorphous matrix from each other.

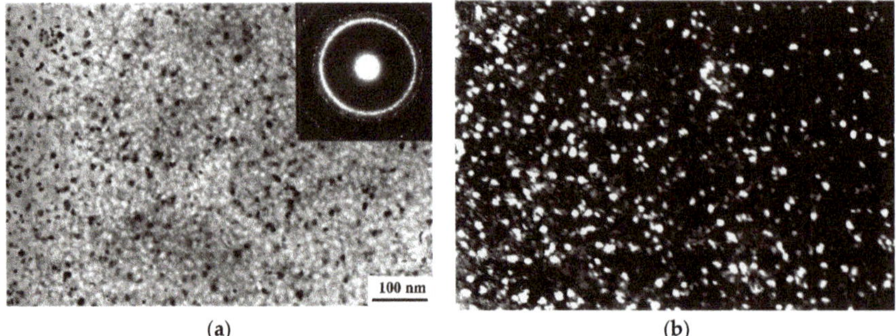

(a)   (b)

Figure 5. Microstructure of Al86Ni11Yb3 sample after annealing at 473 K for 30 min: (**a**) bright-field and (**b**) dark-field images.

A change in the average nanocrystal size measured by dark-field electron microscope images depending on the exposure duration is presented in Figure 6. The average size changes from 8 nm (under exposure for 5 min) to 12 nm (under exposure for 60 min). As one can see, the sharpest changes in the average nanocrystal size occur at initial transformation stages. The experimentally obtained nanocrystal distributions over size for exposures for 5 and 15 min are illustrated in Figure 7. Note that the fraction of the smallest crystals may be too low due to difficulty of the observation. This is particularly important for the obtained distribution over size under exposure for 5 min, when the distribution is shifted towards the region of small sizes.

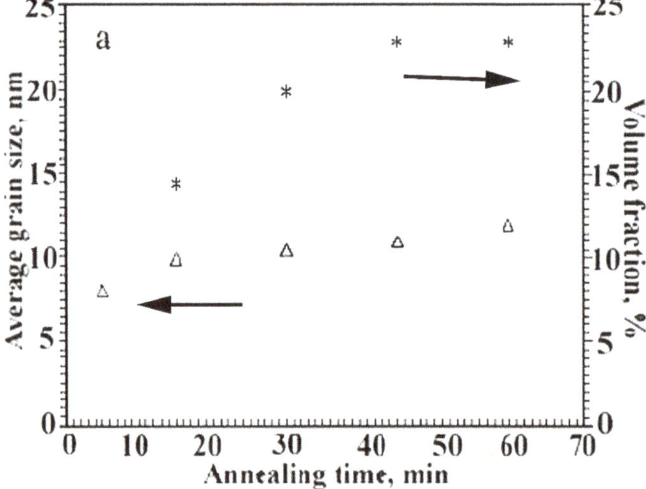

Figure 6. Dependence of an average nanocrystal size (triangles) and fraction of the nanocrystals (asterisks) on the duration of annealing at 473 K for Al86Ni11Yb3 sample [59] [reproduced from Physics of The Solid State 2001, 43, 2003 with permission from Pleiadis Publishing, 2020].

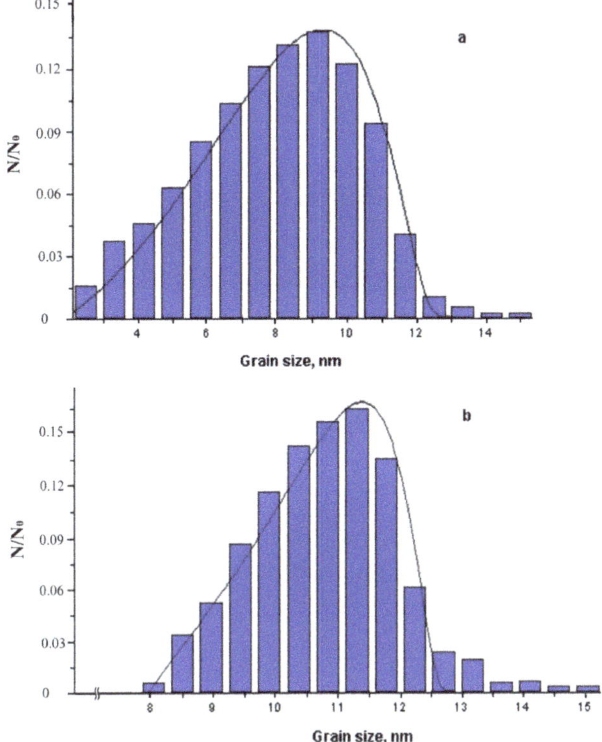

Figure 7. Nanocrystal distribution over size after annealling for (**a**) 5 and (**b**) 15 min. The theoretical data are marked with solid curves the experimental data are marked with columns [59] [reproduced from Physics of The Solid State 2001, 43, 2003 with permission from Pleiadis Publishing, 2020].

The obtained experimental distributions of nanocrystals over size have some specific features. To analyze the features, let us consider theoretically possible crystal distributions over size and compare them with those observed experimentally. Figure 8 shows all the possible size distributions of crystals, which crystallize by the nucleation and growth mechanisms, for homogeneous and heterogeneous nucleation [60].

If one compares these distributions with those observed experimentally, it is clear that heterogeneous nucleation with a latent period takes place in the alloys under study. This is indicated by the following:

- the absence of small crystals in the distribution under exposure for 15 min;
- a significant decrease in the region of small sizes under exposure for 5 min;
- a gradual decrease in the large particle fraction (the right branch of the distribution), which is typical for the nonstationary rate of nucleation of the nanocrystals (with the latent period).

Thus, at the annealing beginning, there is a time period (τ), during which the stationary distribution of subcritical nuclei over size is reached that corresponds to classical theory. According to [61], in this case, the time-dependent rate of nucleus formation I(t) is determined by the equation

$$I(t) = I_{st}\{1 + 2\sum(-1)^{n}\exp[-n^{2}(t/\tau)]\} \tag{1}$$

where the summation is performed over n in the range between 1 and ∞, τ is the latent period which increases sharply with decreasing temperature, and I_{st} is the nucleation rate under stationary conditions, which in turn is described by the equation

$$I_{st} = I_0 \cdot \exp(-L\Delta G_c /RT) \cdot \exp(-Q_N/RT) \tag{2}$$

where L is the Loschmidt number, Q_N is the activation energy of transfer of an atom through the crystallization front surface, ΔG_c is the free energy necessary for the formation of a critical nucleus. Equation (2) is an approximate solution of the Fokker–Planck equation which was obtained for the first time in [62].

At strong supercooling, the value of ΔG_c is very low, then

$$I_{st} = I_0 \exp(-Q_N/RT). \tag{3}$$

Let us examine the growth of the formed nanocrystals. Since, in the case under consideration, the concentration gradient of other alloy components arises in the amorphous matrix in front of growing Al nanocrystals, the matrix gets enriched with Ni and Yb, the atoms of which diffuse over larger distances. Then, the growth rate of the formed crystals decreases with annealing duration. At the same time, it is known [58] that under primary crystallization, the radius of the growing crystals depends parabolically on the duration of isothermal exposure. In our case, the growth of the crystals is determined by the bulk diffusion of Ni and Yb components in the amorphous matrix

$$R = \alpha (Dt)^{0.5} \tag{4}$$

where D is the bulk diffusion coefficient, t is the time of isothermal exposure, R is the radius of a growing crystal, and α is the dimensionless parameter of an order of unity. Meanwhile, we consider the parameter α to be independent of the fraction of the crystalline phase.

This time dependence of the crystal size results in that nanocrystal distribution over size in the case of heterogeneous nucleation becomes narrower with time. Thus, for the distribution shown in Figure 7a the dispersion was 15.76 nm^2, and for that presented in Figure 7b it was 4.96 nm^2. In principle, the Ostwald coalescence described by the Lifshitz–Slyozov theory [63] can also lead to narrowing of the histograms of nanocrystal distribution over size and shift it.

In order to specify the heterogeneous mechanism of nanocrystal nucleation and growth during isothermal exposure, it is necessary to perform computer calculations and plot the histograms of nanocrystal distribution over size for heterogeneous nucleation and diffusion-controlled growth with the purpose of comparing them with the experimental data.

To carry out these calculations using Equations (1)–(4), it is necessary to know the following parameters:

N_0, which is the number of nuclei, limited under heterogeneous crystallization;

τ, which is the latent period (duration of the nonstationary stage);

I_{st}, which is the rate of nucleation under stationary conditions;

I_0, which is the constant determining the stationary rate of nucleus formation, and

Q_N, which is the activation energy of transfer of an atom through the crystallization front surface.

The value of I_0 can be considered to be 3×10^{30} m s^{-1}; such a value is typical of the nucleation of Al nanocrystals in alloys of Al-Ni-REM systems [56]. The value of N_0 can be calculated from the experimental data as follows. Under exposures for more than 30 min, the fraction of the crystallize phase does not significantly change (it is about 0.23). An average size of the nanocrystals is about 12 nm. Then, the number of nanocrystals, N_0 (by assuming that all the regions of heterogeneous nucleation are implemented and by neglecting the possible nanocrystal coalescence), will be about 2×10^{23} m^{-3}. Such an estimate agrees well with the known literature data. Thus, in [56], it is reported that under nanocrystallization N_0 can reach 10^{25} m^{-3}. According to [64], in order to estimate nanocrystal distribution over size it is necessary to divide the time of isothermal exposure into short time intervals Δt and to calculate the number of nanocrystals crystallized during each interval Δt. Then, for heterogeneous crystallization at the limited number N_0 of active nuclei

$$N_i = I\,(t)(1 - x_{i-1})(1 - N_0^{-1}\sum N_j)\Delta t, \tag{5}$$

where the summation is performed over j in the range between 1 and i at $\sum N_j \le N_0$; $N_i = 0$ for all the other values of i. x_i is the volume fraction of a material crystallized during the time interval Δt by the mechanism of primary crystallization $(R = (Dt)^{1/2})$:

$$x_i \approx D^{3/2} \sum N_j \{\Delta t(i + 1 - j)\}^{3/2}, \tag{6}$$

where the summation is performed over j in the range between 1 and i.

The shape and location of the theoretical curve on the axes of size distribution depend on the values of the parameters substituted into the formulas. The correction of the parameters Q_N and τ and the diffusion coefficient (D) enables the best approximation of the theoretical curve to the distribution obtained experimentally.

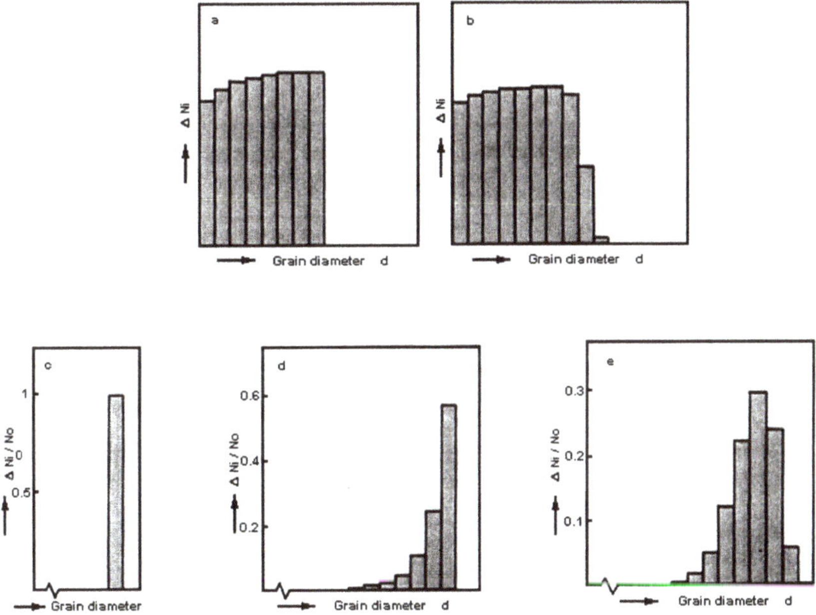

Figure 8. Histograms of the grain distribution over size for (**a,b**) homogeneous and (**c,d,e**) heterogeneous nucleation; (**b,e**) cases of nonstationary nucleation [59]. [reproduced from Physics of The Solid State 2001, 43, 2003 with permission from Pleiadis Publishing, 2020].

The two experimental curves of nanocrystal distribution over size for $Al_{86}Ni_{11}Yb_3$ alloy, presented above, should be compared with those calculated theoretically. When estimating both histograms, one should use the same crystallization parameters. A change in their shape and location should be related to the different process durations (5 and 15 min) only. This requirement is rather rigid. It was shown that both experimental curves can correspond quite well to the theoretical ones at the same values of Q_N and τ, but different diffusion coefficients D (Figure 9). For good correspondence of the curves, the diffusion coefficient D should diminish with a rise in the exposure time (and, correspondingly, in the fraction of the formed crystalline phase). A decrease in D under primary crystallization of amorphous alloys is rather usual at in increase in the crystallized material fraction [58]. The same phenomenon is apparently observed in the case under consideration. From the comparison of calculated and experimental data it follows that at 473 K the effective coefficient of diffusion of Ni and Yb in amorphous $Al_{86}Ni_{11}Yb_3$ alloy is 1.4×10^{-19} m^2s^{-1}, and the latent period is 150 s. Since usually the diffusion coefficient of Ni is significantly higher than that of Yb, one may suppose that it is the Yb removal rate from growing crystal that limits the nanocrystal growth. Then, the obtained diffusion coefficient value is related to Yb diffusion. The value of Yb diffusion coefficient which is 1.4×10^{-19} m^2s^{-1} at 473 K seems to be rather realistic. If one compares it with the available data on the diffusion coefficients of rare-earth metals in Al-based alloys, their similarity should be noted. Thus, the diffusion coefficient of Y in $Al_{88}Fe_7Y_5$ alloy is 9×10^{-20} m^2s^{-1} at 518 K [65]. This value was obtained using the Frank approach, where it is assumed that the parameter α in the equation $R = \alpha (Dt)^{0.5}$ is 1.5.

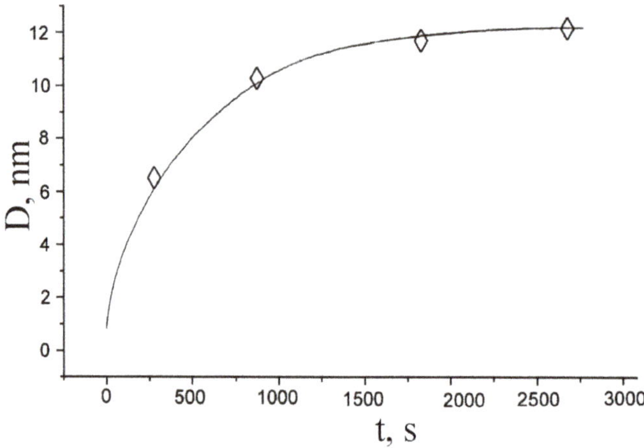

Figure 9. Dependence of an average size of the nanocrystals in Al86Ni11Yb3 alloy on the exposure duration at 473 K. The calculated results are marked with a solid line, the experimental values are shown with rhombi [59].

Note that Hono et al. [66] detected a rise in the Ce concentration in front of growing Al nanocrystals under the crystallization of alloys of Al-Ni-Ce systems, while Ni was distributed uniformly over the amorphous phase. Considering the similarity of the properties of Ce and Yb atoms, the above assumption on the ratio of the diffusion coefficients of Ni and Yb elements seems to be rather reasonable.

Another difference between the experimental and calculated distributions over size is, as one can clearly see in Figure 9, a "tail" of large particles presented in the experimental histograms. The existance of larger particles is probably related to the presence of a small amount of the so-called "frozen-in crystallization centers" in as-prepared alloy. The formation of crystals is facilitated in this case. The particles grow earlier (up to the completion of the latent period of attaining the stationary distribution of subcritical nuclei over size) and reach larger sizes. The particles grow earlier (by the end of the latent period of reaching the stationary distribution of subcritical nuclei) and grow to larger sizes.

When calculating above, it was assumed that the parameter α in the equation $R = \alpha (Dt)^{0.5}$ determines the time dependence of the radius of a growing crystal under primary crystallization, independently of the composition of the residual amorphous matrix. Therefore, when it turned out that the rate of crystal growth decreases with time more sharply than it follows from the dependence $R = \alpha (Dt)^{0.5}$, an agreement between the calculations and the experiment was reached by decreasing the diffusion coefficient with the process duration. However, it seems to be expedient to consider another approach which relates a more significant than it follows from the equation $R = \alpha (Dt)^{0.5}$ decrease in the growth rate with time to a decrease in the driving force of the process. If one takes into account a change in the matrix composition, according to the Ham's approach [67] the time dependence of the size will be determined as

$$R(t) = [2(C_m - C(t))/C_m - C_p]^{1/2} (D\,t)^{1/2} \tag{7}$$

where C_m and C_p are the concentration of a redistributed component in an amorphous matrix and precipitate at the interface; $C(t) = C_0/1 - x(t)$ is the concentration of a component in an amorphous matrix the time t after the process onset.

Ham's model considers a sequence of the identical particles; the initial sizes of a growing particle are negligibly small, the concentration of an admixture (alloying element) in a matrix near the particle is constant along the entire interface, at the interface the condition of diffusion equilibrium is satisfied, and an average size of the particle is about a half of the distance between the particles at the moment of reaction completion (i.e., at their maximum quantity). In [68], an expression was obtained which relates C_m, $C(t)$, C_p, t, and R_0 to each other, where R_0 is the distance between the particle centers.

$$(D\,t\,/(R_0)^2)\,[(C_0 - C_m)/\,(C_m - C_p)]^{1/3} = (1/6)\ln\,[(u^2 + u + 1)/(u^2 - 2u + 1)] - (1/3)^{1/2}\tan^{-1}\,[(2u + 1)/3^{1/2}], \tag{8}$$

where $u^3 = 1 - C(t)/C_0$.

When obtaining this expression, it was assumed that $C(t = 0) = C_0$, the initial radius of a particle is zero.

The essence of this consideration is that at an early crystallization stage the enrichment of a matrix is low due to the concentration redistribution, $(C(t) \approx C_0)$ and $R \sim (Dt)^{1/2}$. At times corresponding to the final stages of the reaction, an average matrix composition approaches C_m and $dR/dt \to 0$. The rate of precipitate growth converges to zero since the driving force of the precipitate process converges to zero. Taking into account the concentration dependence of the parameter α, the dependences of an average size of Al nanocrystals in $Al_{86}Ni_{11}Yb_3$ alloy were calculated. N_0 was 2×10^{23}, $D = 1 \times 10^{-19}$ m^2 s^{-1}, x(t), the fraction of a crystallized material, was determined. The obtained results are demonstrated in Figure 9, where the results of the calculations based on Ham's approach are marked with a solid line and the experimental points are shown as rhombi. It is seen that a good agreement between the calculated and experimental data is observed. Note also that, in this case, the correspondence is observed when using one value of the diffusion coefficient for all durations of isothermal exposure in calculations. This value almost coincides with that obtained earlier. A slight difference is caused only by the coefficients used in the estimates. In the first case we assumed $R = (Dt)^{1/2}$ and in the second case we calculated $R = \alpha\,(Dt)^{1/2}$, where

$$\alpha = [2(C_m - C(t))/C_m - C_p]^{1/2}. \tag{9}$$

In principle, Ham's analysis is related to binary systems. However, based on the data presented in [69] one can consider it to be rather applicable to the case of light three-component alloys.

One should also analyze a possible effect of the Ostwald coalescence on the observed histograms of nanocrystal distribution over size. In the Lifshitz–Slyozov theory, the evolution of crystal sizes is described by the following equation:

$$R^3 - \overline{R}_0^{\,3} = 8\,D\,\sigma\,V_m\,C(\infty)\,t\,/\,9\,N\,k\,T \tag{10}$$

where $\overline{R}$ is the average particle size, R_0 is the initial average size, V_m is the molar volume of precipitates, σ is the energy of the particle-matrix interface, $C(\infty)$ is the equilibrium solubility of a component at a great distance from a particle, k is the Boltzmann constant, and N is the Avogadro number.

The maximum growth rate caused by the coalescence is

$$(dR/dt)_{max} = 8\,D\,\sigma\,V_m\,C(\infty)\,/\,27\,N\,k\,T\,R^2 \tag{11}$$

The maximum growth rate estimated by Equation (11), which was caused by the coalescence, was about 0.2 nm/h (for 473 K, $R \approx 4$ nm after the exposure for 5 min, and $D = 1.4 \times 10^{-19}$ m^2/s). For $R \approx 5.5$ nm (after the exposure for 15 min), the maximum growth rate is <0.03 nm/h. One can see that these rates are insignificant in the considered time interval of the nucleation and evolution of the nanocrystals (exposure for up to 60 min). Ardell [70] made the corrections in Equation (8), allowing for the volume fraction of crystals. Ardell obtained an equation which differs from the Lifshitz–Slyozov equation in the parameter K, which is the function of the crystal volume fraction only:

$$\overline{R}^3 - \overline{R_0}^3 = 8\,K\,D\,\sigma\,V_m\,C(\infty)\,t\,/\,9\,N\,k\,T \tag{12}$$

$K = 1$ at zero volume fraction of precipitates, $K \approx 6$ at 15% fraction (exposure of the alloy under investigation for 5 min), and $K = 10$ at 25% fraction, which approximately corresponds the exposure for 15 min in our case. Then, the maximum growth rates caused by the coalescence are ~1 nm/h and 0.3 nm/h for nanocrystals in the alloy after the exposure for 5 and 15 min, respectively. Note that these estimates are too high since the diffusion coefficient decreases with time (in this calculation, the dependence of the parameter α in the equation $dR/dt = (\alpha/2)(D/t)^{1/2}$ on the crystalline phase fraction was considered, and it was taken to be constant and equal to 1). The obtained values can be regarded as the estimates "above" the values of an instantaneous growth rate (under the exposure for 5 and 15 min) decreasing with time. Therefore, in this case, even consideration of the parameter K related to the volume fraction of precipitates does not make these rates significant for the evolution of nanocrystal distribution over size.

Furthermore, in the Lifshitz–Slyozov theory it is considered that the system is in equilibrium, ib this case the formation and growth of particles of the second phase do not occur due to the matrix (the fraction of particles of the second phase is constant). Thus, a change in the nanocrystal size, caused by coalescence processes, which is described by the Lifshitz–Slyozov theory, may be significant after the completion of nanocrystal growth from the amorphous phase during the existence of metastable equilibrium of the nanocrystals – amorphous matrix. For coalescence processes to occur, this equilibrium should be persist for a long time, and no further crystallization of the amorphous phase should take place. It is important to note that the particle distributions over size were obtained for the stages at which a metastable equilibrium between the amorphous and nanocrystalline phases was not reached yet and the fraction of the nanocrystalline phase continued to increase.

Thus, nanocrystal nucleation under the crystallization of amorphous Al$_{86}$Ni$_{11}$Yb$_3$ alloy occurs by the heterogeneous mechanism from "frozen-in" crystallization centers. A good agreement between the experimental data on a change in the nanocrystal size with the time of isothermal annealing and the calculations by the method described above was observed also for alloys of Al-Ni-Y system [71].

6. Some Features of Nanocrystal Formation (Free Volume)

The interrelation of an amorphous state and the crystalline phases emerging under its decay would be more evident if under crystallization there were no effects having a significant influence on the formed structure. The bulk crystallization effect belongs to these effects. The point is that the density of amorphous metallic alloys is 1–5% lower than their density in a crystalline state. Larger differences in the density of the amorphous phase and the arisen crystalline phases are observed in the case when the amorphous phase has semiconductor properties, for example, in the Al–Ge system. Therefore, it seems to be important to study how the compensation of bulk phase mismatch occurs under crystallization and how this affects the formed structure.

Several methods of the compensation of bulk mismatch are possible:

- using elastic deformation;
- by viscous flow of an amorphous matrix;
- by plastic deformation of the crystalline phases;

- due to diffusion escape of excess volume with formation of pores in the reaction front and/or diffusion of the elementary carriers of free volume to the surface and their annihilation on it;
- by disruption.

Obviously, since the densities of the amorphous and emerged crystalline phases differ, elastic stresses and, consequently, deformation of both the crystalline and amorphous phases occur in the front. The values of the elastic stresses and deformation can determine the relative position of the phases, their morphology and structure, as well as the formation sequence. Obviously, the methods of compensation of the bulk effect act in an integrated way.

Let us consider some examples which demonstrate different methods of the bulk effect compensation.

6.1. Dependence of the Sequence of Phase Formation and Crystal Morphology on the Bulk Effect Demonstration in Alloys of Fe-B, Fe-Co-Si-B Systems

As stated in [72], the method of compensation of bulk mismatch under crystallization can determine the morphology of precipitates. In order to investigate the bulk effect demonstration, a comparative study of crystallization in the samples of "thick" (with an initial thickness) and "thin" cross-sections was carried out. Polished samples for electron microscopy are meant by thin cross-sections. Under the in-situ studies of crystallization in an electron microscope column, phase transformations are usually investigated in samples with a thickness of $\approx$100 nm. The thin cross-sections were studied to get rid of the influence of effects related to thick cross-sections on crystallization. The remoteness of the sinks of elementary free volume carriers from the reaction front, difficulty of the compensation of bulk mismatch, and others belong to these effects.

The investigations of samples of $Fe_{100-x}B_x$ (16 < x < 20) alloys showed [73] that at the first stage of crystallization of samples of all the studied compositions, α-Fe crystals (or rather crystals of a solid solution of B in α-Fe) are precipitated. In the thin cross-sections, precipitates of α-Fe have the form of commas, needles, plates, differing in the thickness, of an irregular shape with a length of several tens of nanometers and thickness of 5–10 nm (Figure 10). Precipitate chains are formed, and a correlation in the arrangement of first and subsequent precipitates is observed. Then, grains of Fe_3B arise, and a structure consisting of Fe and Fe_3B grains is formed.

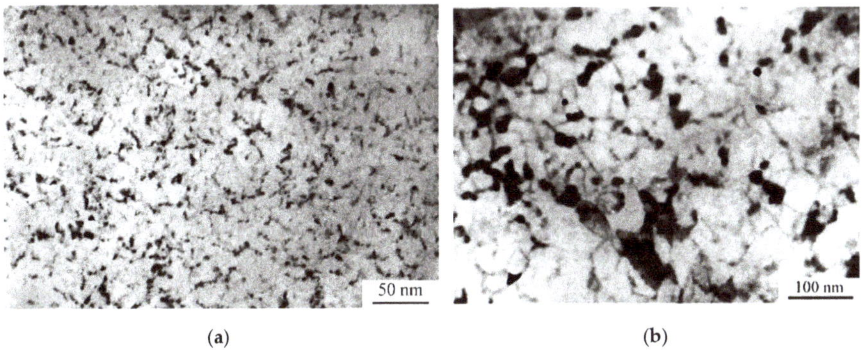

(**a**)   (**b**)

Figure 10. Microstructure of $Fe_{84}B_{16}$ alloy sample annealed in an electron microscope (in-situ) at 623 K for 90 min: (**a**) bright-field and (**b**) dark-field images [reproduced from Mat. Sci. [73]. Eng., 1978, 36, 193 with permission from Elsevier, 2020].

Under the crystallization of ribbons with an initial thickness, the sequence of phase precipitation and morphology of the precipitates differ. At the first stage, in alloys of a hypoeutectic composition (at.% of B < 17.5), as well as in the thin cross-sections, α-Fe crystals are the first to nucleate and grow. They have faceting and form crystallites with a distinctly dendritic shape (Figure 11). The crystallites are randomly distributed over the amorphous matrix, and no correlation in their arrangement is observed. At the second crystallization stage, colonies are formed, which consist of α-Fe and Fe$_3$B.

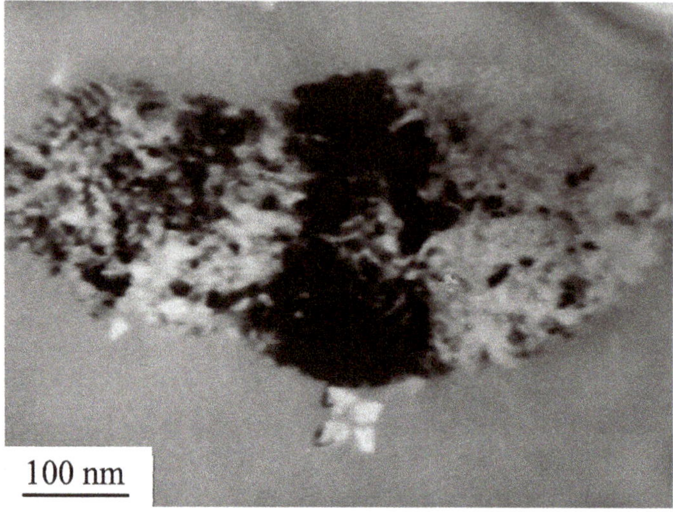

Figure 11. Microstructure of Fe$_{84}$B$_{16}$ alloy sample with an initial thickness, annealed for 60 min at 623 K.

Thus, in hypoeutectic alloys, primary α-Fe crystals were not formed in the thick cross-sections. In the thin cross-sections, α-Fe crystals were always first to be precipitated. In the thick cross-sections, colonies were formed, which consisted of α-Fe and Fe$_3$B, while in the thin cross-sections no colonies (i.e., simultaneous formation of α-Fe and Fe$_3$B) were observed, and Fe$_3$B grains emerged at the second crystallization stage. Fully crystallized samples of thick cross-sections contained dendritic α-Fe crystals and colonies consisting of α-Fe and Fe$_3$B, and in the thin cross-sections there was a structure consisting of Fe and Fe$_3$B grains.

It is reasonable to relate the observed difference in the morphology and arrangement of crystals in the thin and thick cross-sections to the proximity in thin surface cross-sections. In the thick cross-sections, the compensation of the bulk effect occurs, most probably, with the viscous flow of a matrix and deformation of precipitates and a matrix, and in some cases, probably, with the formation of micropores in the reaction front.

The formed α-Fe crystals in the thick cross-sections grow anisotropically in the field of tensile stresses, which explains their growth in the {111} planes in the {110} direction, but not in the close-packed {110}, as was observed in [74]. In the case of thin cross-sections, the compensation of the bulk effect has time to proceed by the diffusion of excess volume carriers to the surface due to proximity of the surface. Deformation fields around α-Fe have lower values, and the crystal shape is more equilibrium than that in the case of thick cross-sections. According to the literature data, the density of Fe crystals is 7.48×10^3 kg m^{-3} [74], and that of Fe$_3$B is 7.48×10^3 kg m^{-3} [75]. The density of amorphous Fe$_{83}$B$_{17}$ alloy is $7.31 10^3$ kg m^{-3} [75]. Consequently, under the formation of the same fraction of these phases, the arisen level of stresses, related to crystallization, will be higher in the case of Fe formation than that in the case of Fe$_3$B formation. However, the compensation of bulk mismatch and, consequently, a decrease in the internal stresses are carried out more easily in the thin cross-sections. Therefore, the formation of primary Fe crystals and Fe$_3$B occurs in all the alloys under study after the completion of Fe precipitation. The formation of colonies in the thick cross-sections, which have a complex structure and structural components nano-sized in two directions, may be associated, in this case, with a decrease in the stress level in the crystallization front and avoidance of the disturbance of material continuity.

In principle, there can be another explanation of the effects observed, i.e., chemical composition change in the thin cross-sections as compared with the thick ones. However, the study of distribution of element concentrations over sample depth, performed by Auger electron spectroscopy, did not reveal any significant differences.

6.2. Formation of a Nanocrystalline Structure as Demonstration of the Bulk Effect

A vivid demonstration of the influence of the bulk crystallization effect on the morphology and structure of the formed phases is the formation of a nanocrystalline structure in $Al_{32}Ge_{68}$ alloy. A feature of this material is that Al and Ge phases are formed under the decomposition of $Al_{32}Ge_{68}$ alloy, with Al being less dense and Ge being denser than the amorphous matrix. In this case, one could expect the formation of a specific structure in the region of the crystallization front. One can assume that the compensation of the bulk effects (with opposite signs) of phase formation should lead to the formation of a highly dispersed mixture of two formed phases in the reaction front in order to avoid the disturbance of sample continuity. In this case, bulk effects with opposite signs will be compensated, and the sample can remain continuous. The studies carried out [76] demonstrated that at an initial crystallization stage a nanocrystalline structure is formed, the nanocrystal size is about 10 nm. Crystallization begins in the depth and propagates to the sample edge. The formed nanocrystalline structure is very unstable. At that, equilibrium phases of Al and Ge are formed. Recrystallization under isothermal exposure occurs very quickly; the nanocrystalline region is followed immediately by the region of larger crystals. According to the data of differential scanning calorimetry, crystallization proceeds in one stage. Thus, at initial crystallization stages, a nanocrystalline structure is formed, which was detected near the crystallization front only. A significant grain growth occurs at a distance of 100 nm from the front.

The formation of a nanocrystalline structure, as a consequence of the compensation of bulk mismatch under the amorphous phase crystallization, is caused by the necessity of elastic stress compensation at a nanolevel in the crystallization front in order to avoid the disturbance of material continuity (the emergence of additional free surfaces under the decomposition will increase the free energy of a system and can make this process unprofitable). In this case, such demonstration of the bulk mismatch compensation is related to large differences in the density of an initial amorphous phase and formed crystalline phases. The specific molar volume of an amorphous matrix is, according to different data, within the limits of $(11.2–12.6)\ 10^{-6}\ m^3mol^{-1}$, and that of crystalline Al and Ge is $10.0 \times 10^{-6}\ m^3mol^{-1}$ and $13.6 \times 10^{-6}\ m^3mol^{-1}$, respectively. That is, the density of an amorphous matrix is in an intermediate position between Al and Ge densities, and the difference is approximately 20% for both elements. Thus, the formation of one phase under crystallization results in a very high level of stresses in the reaction front, which is compensated immediately under the formation of the second phase. Thereby, one succeeds in avoiding sample decomposition. The formation of phases in a nanocrystalline state results in the stress compensation not only at macrolevel, but at the microlevel (nanolevel). An important fact is that the amorphous phase exists in a narrow concentration rage only, and precipitation of one of the phases will result in a change in its composition and, consequently, to the decomposition. Therefore, the simultaneous formation of two phases is observed, and their size is determined by the necessity of compensation of the bulk effect of phase formation in order to preserve the sample continuity.

The formed nanocrystalline structure turns to be very unstable. In this structure, the area of interphase interfaces and the corresponding energy are high, that is why grain coarsening is necessary to diminish the free energy of the system, which happens in reality. In this case, nothing impedes the coalescence processes: the nanocrystals are separated by high-angle boundaries, but not by the special low-energy ones, and there is no layer of an amorphous matrix isolating the crystals from each other. Under the formation of a nanocrystals by the primary crystallization mechanism, the amorphous matrix changes its composition as the nanocrystals nucleate and grow. In the case of amorphous Al-Ge alloy, this process is impossible. The reason is that, as stated, the amorphous alloy of $Al_{32}Ge_{68}$ nominal content has a very narrow concentration region of existence [77], and the layers of an amorphous phase with a changed composition cannot exist between the grains. Consequently, there are no obstacles to coalescence, and the nanocrystalline structure will decompose easily. Al should be the leading phase here, for which the corresponding homologous temperature is lower. As shown experimentally, the prevailing growth of Al crystals is observed under the nanocrystalline phase separation.

6.3. Compensation of Structural Mismatch by Pore Fformation and Nanocrystal Formation in the Shear Bands

A number of works, for example, [78–80] were devoted to the study of shear band formation and structure in amorphous alloys. Plastic deformation in alloys is strongly localized and is realized by the formation and propagation of different shear bands. The rate of shear band propagation does not depend on the deformation rate in a range of 2×10^{-4}–10^{-2} s^{-1}. High shear stresses are localized in shear bands, as a result, a large amount of free volume is concentrated in them; the shear bands have more random structure compared to the surrounding amorphous phase [78,81]. The structure of the main part of an amorphous matrix also can change under deformation; it becomes anisotropic under certain conditions [29–31]. The number of shear bands depends on the alloy chemical composition and deformation conditions. Their size can be from several tens of nanometers to several micrometers in width and from tens to hundreds of micrometers in length.

As a result of plastic deformation, the amorphous phase structure becomes non-uniform: an amorphous structure in the region of a shear band differs from the structure in a surrounding amorphous matrix. The relaxation of a structure in the shear band can lead to different effects.

In literature there are several models which describe the processes responsible for the shear band formation. All of them include, to varying degrees, the concept of free volume. Thus, for example, in the work by Spaepen [82] it is assumed that the motion of a material in a shear band consists of the formation and disappearance of free volume regions. An interest in shear bands is caused by several factors. Firstly, a lot of mechanical properties of amorphous alloys depend on the presence and characteristics of shear bands. Secondly, as was demonstrated in [83–85], the nucleation of nanocrystals begins in shear bands or their vicinity. In [86], it was demonstrated that pores also grow in shear bands (Figure 12). The arrow indicates the reflex in which the dark-field image is obtained.

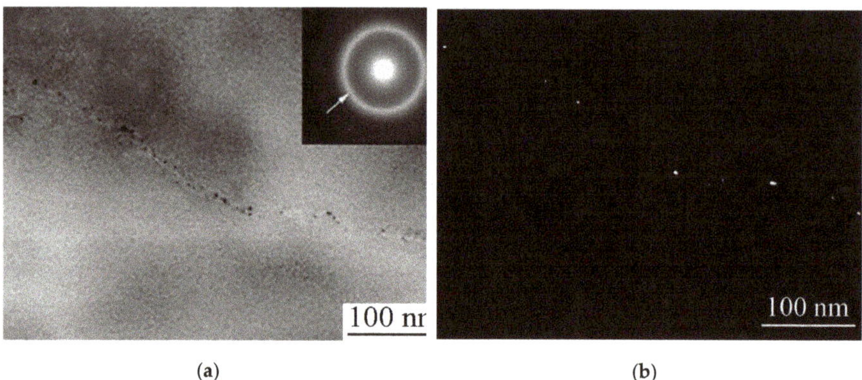

Figure 12. Microstructure of deformed $Al_{88}Ni_2Y_{10}$ sample (40%): (**a**) bright-field and (**b**) dark-field images.

The analysis of pore growth kinetics (Figure 13) allowed estimating the effective diffusion coefficient. When calculating, it was assumed that pores grow due to diffusion of free volume from a shear band to the bulk of the surrounding matrix. It was determined that the diffusion coefficient decreases with time at room temperature RT (from 3×10^{-24} m^2/s for ageing for 1.78×10^7 s, to 7×10^{-25} m^2/s for ageing for 2.95×10^7 s). A decrease in the diffusion coefficient with time is caused by free volume depletion in shear zones adjacent to pores. As mentioned above, shear bands can be the regions of facilitated nanocrystal formation. The formation of nanocrystals in a shear band can be observed in Figure 12.

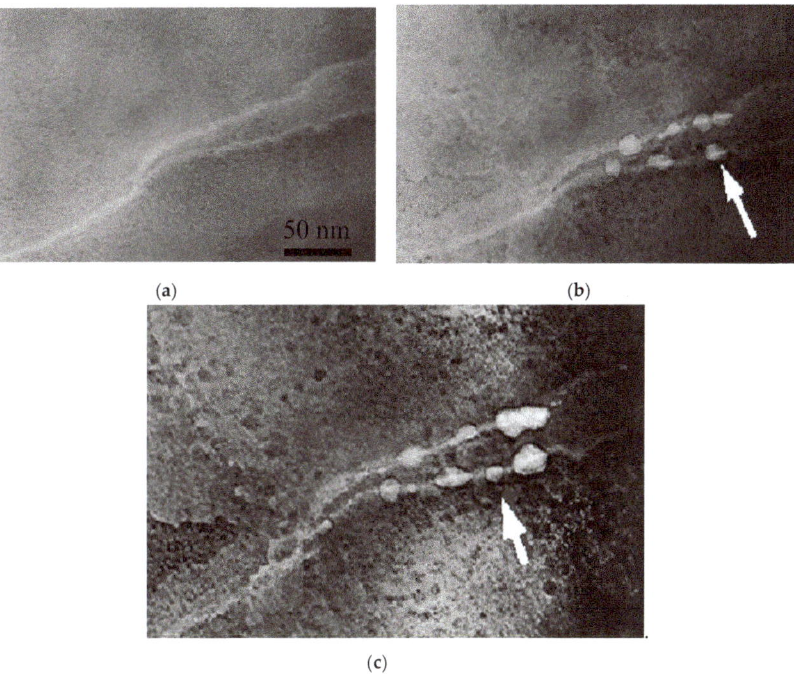

(a)

(b)

(c)

Figure 13. Shear bands in amorphous $Al_{88}Ni_2Y_{10}$ after deformation at room temperature (**a**), nanopores (marked with arrows) in shear bands after ageing at room temperature for 1.78×10^7 s (**b**), nanopores in shear bands after ageing at RT for 2.96×10^7 s (**c**) [reproduced from Mechanics of Materials, 2017, 113, 19 with permission from Elsevier, 2020] [86].

A number of experimental results also are an evidence of high diffusion coefficient values in these regions. In [30,87] the formation of nanocrystals was observed in deformation bands under the subsequent exposure of samples at room temperature. The preferred crystallization in shear bands was observed also in [88]. The authors found out the preferred nanocrystal formation in shear bands in $Al_{88}Y_7Fe_5$ alloy samples deformed by tension. The authors related the nanocrystal formation in shear bands to an increase in the local mass transfer rate in these regions. Similar results were obtained in [89], where the authors demonstrated the formation of Al nanocrystals in rolled amorphous $Al_{85.1}Ni_6Co_2Gd_6Si_{0.9}$ alloy at room temperature. Figure 14a shows the microstructure of an alloy deformed by multiple rolling. Figure 14b illustrates the microstructure of the same alloy after ageing at room temperature for ~6000 h. In Figure 14a, one can clearly see brighter regions with an extended shape, which are oriented along some direction. The size of the regions is up to 100 nm in length and 20–40 nm in width. These regions have rather sharp boundaries with the amorphous matrix and represent regions (shear bands) with a high level of deformation, i.e., shear bands. They contain a small number of nanocrystals formed during deformation. The surrounding matrix remains amorphous. Thus, the formation of Al nanocrystals under rolling occurred in the places of plastic deformation localization. Under the ageing of a sample at room temperature, its structure changed significantly. In shear bands, the number of nanocrystals increased strongly, and shear bands became completely filled with nanocrystals. Meanwhile, a small number of nanocrystals emerged in other parts of the amorphous matrix, too. The average size of nanocrystals in shear bands increased slightly.

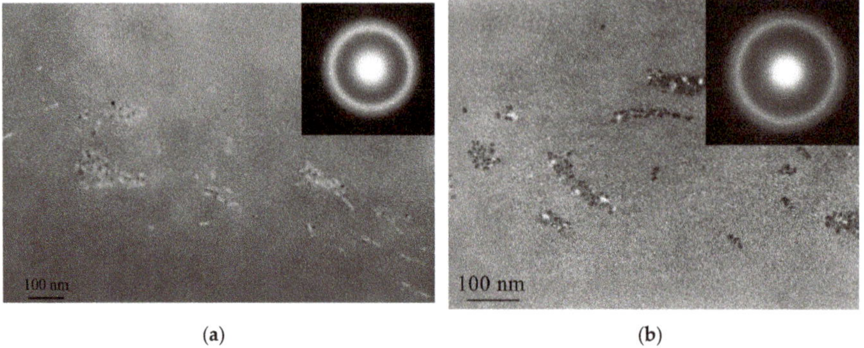

(a)   (b)

Figure 14. Microstructure of Al85.1Ni6Co2Gd6Si0.9 alloy sample (**a**) after rolling and (**b**) after ageing at room temperature) [reproduced from Acta Materialia, 2008, 56, 2834 with permission from Elsevier, 2020] [83].

For comparison, Figure 15 shows the microstructure of a sample after rolling and heating to 245 °C. One can see that a considerable number of nanocrystals were formed in the sample during heat treatment; however, it is seen that their major part is concentrated in the region of the shear bands. The rates of diffusion providing a slight growth of Al nanocrystals at about 60 °C were estimated to be 10^{-24} m^2/s. These results agree with the data of [90], where it was demonstrated that the formation of excess volume during plastic flow can lead to an increase in the diffusion coefficient in shear bands by 4–6 orders of magnitude. The formation of nanocrystals in local shear bands in deformed amorphous alloys was observed in both Fe–B alloys (Figure 16) and deformation bands of $Al_{88}Ni_2Y_{10}$ alloy (Figure 12). An important conclusion of the authors of the above works is that high atomic mobility in the zones of plastic deformation localization makes a crucial contribution in the nanocrystallization process stimulated by deformation.

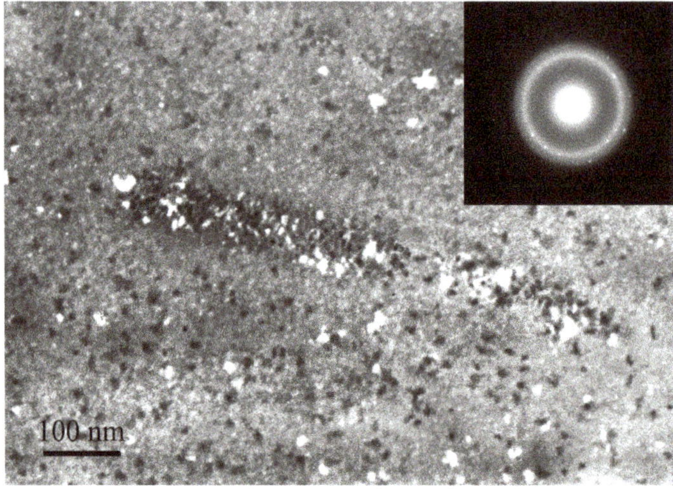

Figure 15. Microstructure of a sample after rolling and heating to 245 °C [reproduced from Physics of The Solid State 2011, 53, 229 with permission from Pleiadis Publishing, 2020] [89].

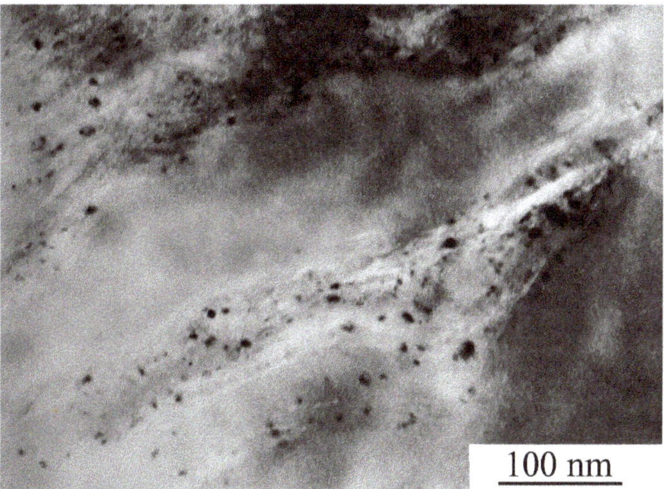

Figure 16. Nanocrystals in deformed amorphous Fe-B alloy.

7. Nanocrystal Formation in Amorphous Phase

The parameters of the crystallized structure depend on the processing conditions. As a result, many properties of materials change dramatically with a change in structure. So, for example, the properties of nanocrystalline alloys differ both from the properties of polycrystals and from the properties of amorphous alloys [3,4]. As a rule, the nanocrystalline structure forms by the primary crystallization reaction. The nanocrystalline structure in most cases is two-phase and consists of nanocrystals formed by the primary crystallization reaction, and interlayers of the remaining amorphous phase of a changed composition. The nanocrystalline structure was first obtained in an alloy of the Fe-Cu-Nb-Si-B system, called Finemet [91].

The principle of obtaining a nanocrystalline structure in this alloy was based on the fact that small amounts of copper and niobium were added to the base composition of the amorphous Fe–Si –B alloy. An amorphous Fe–Si–B alloy crystallizes upon heating to form a usual structure with a grain size noticeably exceeding nanosizes. Copper addition leads to the formation of microsegregations, which serve as sites of facilitated nucleation of crystals, and the addition of slowly diffusing niobium helps to slow down the growth of crystals. As a result, during crystallization of the amorphous Fe–Cu–Nb–Si–B alloy, a microstructure with a crystal size of about 10 nm was obtained. Nanocrystalline alloys of the Fe-CuNb–Si–B system have excellent magnetic properties and a huge amount of work has been devoted to their study [92–97]. Later, a number of alloys were obtained in the nanocrystalline state by the controlled crystallization of the amorphous phase. To date, the nanostructure has been obtained in a wide group of metal systems; there is a number of data on the parameters of the nanocrystalline structure obtained by different methods [98–101]. As was mentioned above, depending on chemical composition, nanocrystalline materials have good plasticity and high viscosity, high strength and hardness, low moduli of elasticity, higher diffusion coefficients, larger values of thermal expansion coefficient, and better magnetic properties as compared with traditional crystalline materials. Research on nanocrystalline materials is actively being carried out at the present time.

8. Nanocrystal Formation under Heating and Deformation

As already mentioned, the amorphous phase is crystallized at an increase in the temperature or duration of heat treatment. At that, metastable phases are generally formed at an initial crystallization stage, with some of these phases being formed only under the crystallization of the metallic glasses and not being formed under other conditions [102–104]. A transition from the amorphous to the equilibrium state is often carried out by successive structural transformations [35,105]. An important feature of the amorphous phase crystallization is the formation of the crystalline phases at an initial stage, the short-range order of which corresponds to that of these ordered regions. Therefore, a structural state of amorphous phase before beginning of the crystallization can have a decisive effect on the morphology, phase composition, crystallographic characteristics of a structure formed under crystallization. In [106–108], the effect of an initial amorphous state on the parameters of a crystalline structure formed under the crystallization of the metallic glasses was investigated. It was determined that deformation or annealing within an amorphous state leads to the formation of a non-uniform amorphous structure (nanoglass). This affects the parameters of a crystalline structure formed under the subsequent treatment. It was shown that the state of the amorphous phase before the crystallization onset can significantly affect the characteristics of the formed crystalline structure. The formation of an inhomogeneous amorphous structure in Al-based alloys accelerates the crystallization processes, affects the nanocrystal size and the fraction of a nanocrystalline component in amorphous-nanocrystalline alloys. The history of samples turned to be important, too, that is, under which conditions nanocrystals were nucleated: under deformation or heat treatment. The investigations of a large group of Al-based alloys shows that the highest fraction of the nanocrystals and the smallest nanocrystal size were observed in the case when the heterogeneous amorphous phase was formed during deformation. Table 1 lists some data on the parameters of the structure formed under the crystallization of the uniform and non-uniform amorphous phases [106–108].

Table 1. The nanocrystal size (D) and the nanocrystalline phase fraction (f) formed in the uniform and non-uniform amorphous phases.

A Structure before the Crystallization Onset	After the First Stage of	Crystallization
	D, nm	f, %
	$Al_{87}Ni_8Y_5$	-
The uniform amorphous phase	23	18
The non-uniform amorphous phase after heat treatment	15	20
The non-uniform amorphous phase after rolling deformation	13	25
	$Al_{87}Ni_8La_5$	-
The uniform amorphous phase	32	12
The non-uniform amorphous phase after heat treatment	30	13
The non-uniform amorphous phase after rolling deformation	26	14
	$Al_{87}Ni_8Gd_5$	-
The uniform amorphous phase	26	23
The non-uniform amorphous phase after heat treatment	24	-
The non-uniform amorphous phase after rolling deformation	21	-
HPT 1 rev	6	22
HPT 5 rev	6	25
	$Al_{88}Ni_{10}Y_2$	-
The uniform amorphous phase	19	28
The non-uniform amorphous phase after rolling deformation	18	35

As one can see from the table, the largest Al nanocrystals formed under the separation of the amorphous phase and their lowest volume fraction are observed under the crystallization of the homogeneous amorphous phase (without any treatment of the amorphous phase before the crystallization onset). Under nanocrystallization of the heterogeneous amorphous phase with preliminary heat treatment, the listed parameters were intermediate.

Thus, the formation of the nanocrystals depends essentially on the conditions of an action on an amorphous structure, and the use of combined treatments permits obtaining structure with different structural parameters.

Note that the use of heat treatment allows obtaining a nanocrystalline structure not in all systems. For a nanostructure to be formed from the amorphous phase, a high rate of crystal nucleation and a low rate of crystal growth are necessary. This depends, particularly, on the diffusion rate and, of course, cannot be provided in all materials. Severe plastic deformation turned to be the other method, which is effective in view of the nanocrystallization initiation [84,89,102,109–113]. The use of this method enabled obtaining an amorphous-nanocrystalline structure in alloys where it is not formed under the crystallization by heat treatment [112,113]. The formation of a nanostructure under plastic deformation generally occurs in the zones of plastic deformation localization (shear bands) or in the regions surrounding them. The formation of nanocrystals in these regions is caused by high values of the parameters of diffusion mass transfer. The reasons for an increase in the diffusion coefficient by several orders of magnitude are not fully understood yet. In general, an increase in the diffusion coefficient in deformation bands is related to one of the two processes (or their combination): a local strong, but short (~30 ps), increase in the temperature in this region [114–118] and a change in the structure of the amorphous phase in a shear band (an increase in the free volume fraction) [119–122]. Today, it is unclear which of the reasons constitutes a deciding factor [123]. Both of the factors obviously promote the diffusion acceleration, and one of them can prevail in different cases. The formation of shear bands occurs under the action of shear stresses. Therefore, shear stresses play an important role in nanocrystalline structure formation.

9. Mechanical Properties of Metallic Glasses near Shear Bands

The results above and the available literature data show that a material in shear bands is softened due to high concentration of free volume [78] and a disordered structure. This results in a significant diminution in the diffusion coefficient in an amorphous phase (by several orders of magnitude), and shear bands become the regions of facilitated nanocrystal formation under subsequent heating or even aging at room temperature. Since these factors turn to be important to find out the conditions of the nanocrystal formation [124,125], the issue of definition of the mechanical properties of the material near shear bands has become prominent. As is well known, a thickness of shear bands is from several tens to hundreds of nanometers; steps with the sizes depending on the deformation level and elastic parameters of alloys formed on the sample surface in the places where shear bands come to the surface [78,126]. The mechanical properties of the alloys with shear bands were studied in a number of works by nanoindentation [125–127]. At that, it was determined [127] that near a shear band, there is a region of low hardness. The size of this region is more than 100 μm, which is significantly greater than the size of the shear band. In [128] it was found out that both hardness and the Young's modulus diminish in the shear bands and in the zones around them. However, the size of the region with changed Young's modulus was smaller than that in [127].

The local distribution of zones with different Young's moduli in deformed amorphous $Al_{87}Ni_8La_5$ alloy was investigated in [128]. PeakForce QNM (DimensionFastScanTM Atomic Force Microscope (AFM), Bruker) was used for the investigation of local mechanical properties. The measurements were performed with standard probes, and the load was varied from 0.3 to 0.85 μN. After the deformation of an amorphous alloy by multiple rolling, it was discovered that non-uniformity of the distribution of local mechanical properties over the alloy surface (non-uniform distribution of the effective Young's modulus) appears with a rise of the load. At that, the non-uniformity in a band system was also observed. The band thickness is 50–250 nm. Figure 17 shows the maps of mechanical property distribution, observed under different loads (in Figure 17 the bands are marked with arrows). The heterogeneities were most noticeable under the load of 0.85 μN. The values of thickness of the bands, as well as their shape and distribution over the surface, were fully consistent with the shear bands observed on the surface of deformed samples using a scanning electron microscope (Figure 18).

The results obtained in [128] are an evidence that the subsurface region of a deformed amorphous alloy contains a lot of deformation bands, where the material is characterized by a lower Young's modulus. The size of the zones with lower Young's modulus is tenths of a micrometer. These zones are distributed uniformly over the sample. The cross-sectional dimensions of such zones (bright bands in Figure 17) can change from tens nanometers up to 0.5 μm. A sharp contrast between the amorphous matrix and deformation bands indicates a significant change in the properties. It should be noted that the properties change insignificantly in the region with a width of several tenths of a micrometer. The difference between these results and the data from [125–127] is apparently caused by different deformation level: single shear bands in the mentioned works and numerous shear bands in [128]. Note, however that that the size of the zones with lower Young's modulus in vicinity of shear bands in amorphous $Al_{87}Ni_8La_5$ alloy is noticeably less than that observed in Zr-based alloys. This difference can be caused by different reasons, for example, different deformation conditions (compression of Zr-based alloys and rolling of Al-based alloy) or a different composition of the alloys.

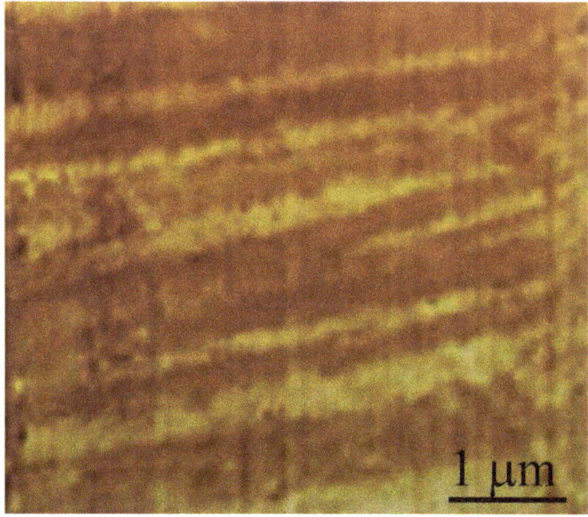

Figure 17. Maps of distribution of mechanical properties for Peak Force = 0.85 µN.

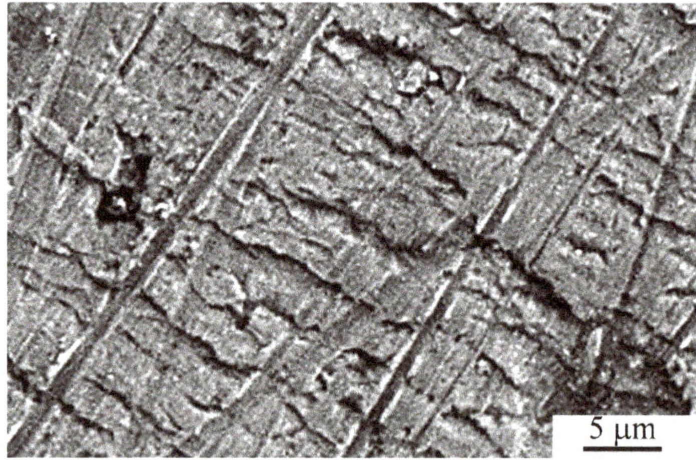

Figure 18. SEM image of the surface of deformed Al87Ni8La5 alloy [reproduced from Mater. Let. 2019, 252, 114 with permission from Elsevier, 2020] [128].

10. Amorphous Structure Rejuvenation

As already mentioned, amorphous alloys have a number of good properties [129–134]. Some of them (for example, magnetic properties) can be increased significantly by low-temperature annealing,

annealing in a magnetic field, etc. Such treatments, however, often result in the embrittlement of these materials, which limits dramatically the possibilities of their practical application. High-pressure torsion was used to recover plasticity [135–137]. It was shown in [137] that the pair distribution function of atoms changes after deformation. This indicates a noticeable redistribution of atoms, leading to structure change. The authors of [137] considered that this structure variation is caused by local heating because of the deformation. A rising degree of structure disordering after deformation was found also in [138].

Last years, a new method for recovery of plasticity was found; it is cycling in a temperature range between room or elevated temperature and the temperature of liquid nitrogen [139]. This method is called rejuvenation. The main idea of cryogenic thermal cycling is a variation of the structure under the action of stresses caused by a non-uniform change in the thermal expansion coefficient (TEC) in a sample. A heterogeneous amorphous phase contains regions with different chemical composition, density, short-range order type, etc. Since such non-uniform regions are characterized by different value of TEC, an abrupt change in the temperature under thermal cycling will induce stresses which cause irreversible local atomic restructurings [140,141]. The researchers succeeded in recovering plasticity in the $Zr_{55}Cu_{30}Al_{10}Ni_5$ alloy by this method [140]. The authors of this work, however, did not reveal any significant changes in the structure. The method of cryogenic cycling aroused a big interest [142–144]. The idea of induction of stresses in the alloy with the non-uniform distribution of the thermal expansion coefficients turned to be very promising. The use of this method allowed for the observation of a structural change in inhomogeneous amorphous $Zr_{46}Cu_{38}Al_8Ag_8$ alloy [145], as well as completely recovering the amorphous phase in partially crystalline $Al_{88}Ni_6Y_6$ and $Al_{87}Ni_8Gd_5$ alloys [146,147]. The authors of the latter work used the idea of increasing the difference in the thermal expansion coefficients due to the formation of a small number of nanocrystals in the amorphous phase and the subsequent cryogenic thermal cycling of a structure consisting of the amorphous phase and Al nanocrystals uniformly distributed over it. Figure 19 illustrates the initial parts of the X-ray diffraction patterns of $Al_{88}Ni_6Y_6$ alloy before (a) and after (b) the cryogenic thermal cycling (60 cycles). One can see that the intensity of the reflection corresponding to nanocrystals (curve 5) decreases significantly during cryogenic cycling. The authors of [146,147] did not observe any signs of the crystalline phases with an increase in the treatment duration. So, it was shown that the method of cryogenic rejuvenation actually enables recovering an amorphous structure in partially crystalline alloys, thus increasing the plasticity of a material.

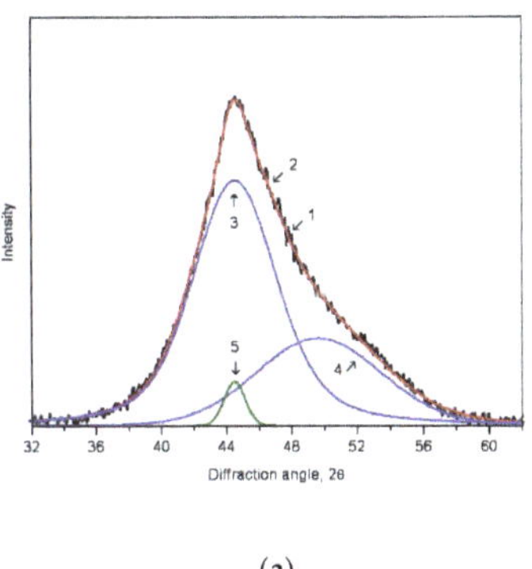
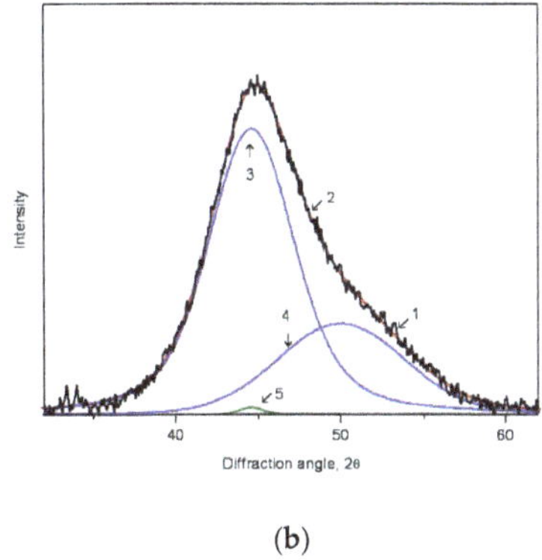

(a)(b)

Figure 19. An initial part of the X-ray diffraction pattern of Al88Ni6Y6 sample after deformation (a) and after deformation plus cryogenic cycling (b): (1) experimental spectrum, (3,4) diffuse reflections corresponding to 2 amorphous phases, (5) (111) Al nanocrystal reflection, (2) sum of 3–6 lines [reproduced from Mater. Let. 2019, 240, 159 with permission from Elsevier, 2020] [146].

11. Possibility of Controlling the Structure of Fully or Partially Crystallized Samples

An important consequence of the investigation results described above is the possibility of controlling a structure formed under crystallization. As shown above, both the phase composition [148] and parameters of the formed structure [106–108] depend on the history of a sample and the crystallization conditions. Also, it was found out that, for example, in amorphous alloys of Ni–Mo–P and Ni–Mo–B systems, quite different crystalline structures arise under the crystallization above or below the glass transition temperature [149–151]. This difference is caused by different structures of the amorphous phase in these temperature ranges. When carrying out crystallization under the conditions when it is preceded by separation of the amorphous phase, one can form a nanocrystalline structure which will not be formed under conventional annealing (for example, in Ni-Mo-B system [40]). Another example is ferromagnetic $Fe_{72}Al_5P_{10}Ga_2C_6B_4Si_1$ alloy with good magnetic properties, where a nanocrystalline structure is formed above the glass transition temperature only [152,153]. These structures also include the so-called SS-phase which was discovered under the crystallization of a number of Fe-, Co-, Ni-, and Pd-based amorphous alloys in 1976 [154], when there was no term "nanostructures". Another example is [155], where the possibility of obtaining an amorphous sample with the crystalline surface (Figure 20) or a crystalline sample with the amorphous surface (Figure 21) was demonstrated by an amorphous alloy of Fe-B-P system. The materials obtained by this approach will be characterized by different physical properties. The latter example shows how the knowledge of processes occurring within an amorphous state and under the crystallization of amorphous alloys allows obtaining different types of structures.

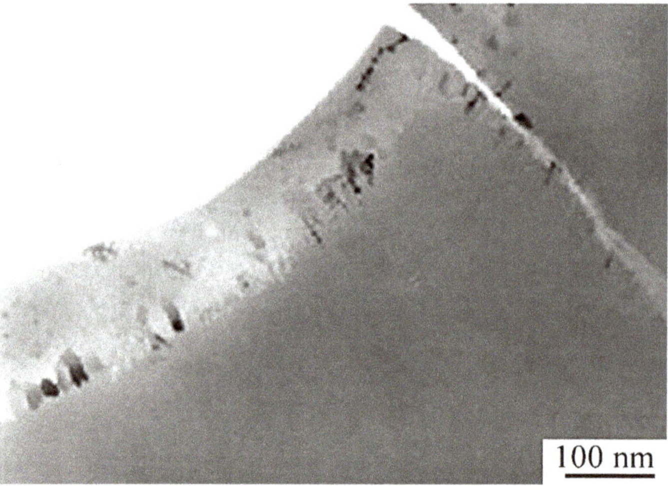

Figure 20. Microstructure of Fe83B10P7 alloy sample partially crystallized in-situ [reproduced from Physics of The Solid State 1991, 33, 2527 with permission from Pleiadis Publishing, 2020] [155].

Figure 21. Microstructure of pre-annealed Fe83B10P7 sample partially crystallized in situ [reproduced from Physics of The Solid State 1991, 33, 2527 with permission from Pleiadis Publishing, 2020] [155].

12. Conclusions

Summarizing the above, the following should be noted.

- The transformation of a homogeneous amorphous phase into a heterogeneous phase, including the scale of inhomogeneities and the effect of external influences, are considered.
- The influence of amorphous structure changes on the forming crystalline structure is shown.
- The crystallization processes of the amorphous phase, such as homogeneous and heterogeneous nucleation of crystals, are considered.
- Possible way of compensation for volume mismatch (changing the sequence of formation of crystalline phases, crystal morphology, formation of nanocrystals or pores) is analyzed.
- Nanocrystal formation is described.
- The effect of plastic deformation on the formation of nanocrystals and mechanical properties of the amorphous phase near the shear bands are shown.
- The possibility of restoring the amorphous structure is discussed.

The results presented above represent a serious step in developing the conception of the structure, its evolution, and some properties of amorphous and partially crystalline materials. The results of numerous studies show that the processes of crystallization of metallic glasses have a number of features, and different actions on a uniform amorphous structure can initiate processes resulting in the formation of materials with various structural parameters. These changes in the phase composition, microstructure, crystallization mechanism, as well as in the morphology and size of crystals, naturally affect the properties, which are important for the production of novel materials with the required physical properties.

Author Contributions: A.A. and G.A. conceived and designed the experiments; A.A. performed the experiments by all electron microscopy methods, studied mechanical properties etc., G.A. performed the experiments using X-ray diffraction methods, analyzed the data; both authors discussed the date and wrote the paper. All authors have read and agreed to the published version of the manuscript.

Funding: This research received no external funding.

Acknowledgments: The authors thank their colleagues from the Institute of Solid State Physics of the Russian Academy of Sciences for a useful discussion of the results.

Conflicts of Interest: The authors declare no conflict of interest.

References

1. Willens, R.H.; Klement, W.; Duwez, P. Continuous series of metastable solid solutions in silver-copper alloys. *J. Appl. Phys.* **1960**, *31*, 1136–1137. [CrossRef]
2. He, Y.; Poon, J.F.; Shiflet, G.Y. Synthesis and properties of metallic glasses that contain aluminum. *Science* **1988**, *241*, 1640. [CrossRef]
3. Gleiter, H. Nanocrystalline materials. *Prog. Mater. Sci.* **1989**, *33*, 223. [CrossRef]
4. Birringer, R. Nanocrystalline materials. *Mater. Sci. Eng. A* **1989**, *117*, 33–43. [CrossRef]
5. Abrosimova, G.E.; Aronin, A.S.; Kir'janov, Y.V.; Gloriant, T.F.; Greer, A.L. Nanostructure and microhardness of $Al_{86}Ni_{11}Yb_3$ nanocrystalline alloy. *NanoStructured Mater.* **1999**, *12*, 617–620. [CrossRef]
6. Inoue, A.; Ochiai, T.; Horio, Y.; Masumoto, T. Formation and mechanical properties of amorphous Al-Ni-Nd alloys. *Mater. Sci. Eng. A* **1994**, *179*, 649. [CrossRef]
7. Kim, Y.H.; Inoue, A.; Masumoto, T. Increase in mechanical strength of Al–Y–Ni amorphous alloys by dispersion of nanoscale fcc-Al particles. *Mater. Trans. JIM* **1991**, *32*, 331–338. [CrossRef]
8. Chou, C.-P.; Turnbull, D. Transformation behavior of Pd-Au-Si metallic glasses. *J. Non Cryst. Solids* **1975**, *17*, 169–188. [CrossRef]
9. Abrosimova, G.E.; Aronin, A.S. Phase segregation in the $Fe_{90}Zr_{10}$ amorphous alloy under heating. *Phys. Solid State* **1998**, *40*, 1603–1606. [CrossRef]
10. Chen, H.S.; Turnbull, D. Formation, stability and structure of palladium-silicon based alloy glasses. *Acta Metal.* **1969**, *17*, 1021. [CrossRef]
11. Osamura, K. SAXS study on the structure and crystallization of amorphous metallic alloys. *Colloid. Polym. Sci.* **1981**, *259*, 677. [CrossRef]
12. Terauchi, H. Heterogeneous structure of amorphous materials. *J. Phys. Soc. Jap.* **1983**, *52*, 3454–3459. [CrossRef]
13. Osamura, K. Structure and mechanical properties of a $Fe_{90}Zr_{10}$ amorphous alloy. *J. Mater. Sci.* **1984**, *19*, 1917. [CrossRef]
14. Mak, A.; Samwer, K.; Johnson, W.L. Evidence for two distinct amorphous phases in $(Zr_{0.667}Ni_{0.333})_{1-x}B_x$ alloys. *Phys. Lett. A* **1983**, *98*, 353. [CrossRef]
15. Mehra, M.; Schulz, R.; Johnson, W.L. Structural studies and relaxation behavior of (Mo0.6Ru0.4)100-xBxmetallic glasses. *J. Non Crystall. Solids* **1984**, *61*, 859–864. [CrossRef]
16. Nagarajan, T.; Asari, U.C.; Srinivasan, S.; Sridharan, V.; Narayanasamy, A. Amorphous phase separation in METGLAS 2605CO. *Hyperfine Interact.* **1987**, *34*, 491. [CrossRef]
17. Abrosimova, G.E.; Aronin, A.S.; Ignat'eva, E.Y.; Molokanov, V.V. Phase decomposition and nanocrystallization in amorphous $Ni_{70}Mo_{10}P_{20}$ alloy. *J. Magn. Magn. Mater.* **1999**, *203*, 169–171. [CrossRef]
18. Abrosimova, G.E.; Aronin, A.S. Evolution of the amorphous-phase structure in metal-metal type metallic glasses. *J. Surf. Investig.* **2015**, *9*, 887–893. [CrossRef]
19. Marcus, M.A. Phases separation and crystallization in amorphous Pd-Si-Sb. *J. Non Cryst. Solids* **1979**, *30*, 317–335. [CrossRef]
20. Yavari, A.R. New amorphous $Pb_{60}Pd_{40}$ with split first X-ray halo and possible unmixing. *Inter. J. Rapid Solidif.* **1986**, *2*, 47–54.
21. Inoue, A.; Bizen, Y.; Kimura, H.M.; Yamamoto, Y.; Tsai, A.P. Development of compositional short-range ordering in an Al50Ge40Mn10 amorphous alloy upon annealing. *J. Mater. Sci. Lett.* **1987**, *6*, 811–814. [CrossRef]
22. Inoue, A.; Yamamoto, M.; Kimura, H.M.; Masomoto, T. Ductile aluminum-based amorphous alloys with two separated phases. *J. Mater. Sci. Lett.* **1987**, *6*, 194–196. [CrossRef]
23. Yavari, A.R. On the structure of metallic glasses with double diffraction halos. *Acta Metall.* **1988**, *36*, 1863–1872. [CrossRef]
24. Kűndig, A.A.; Ohnuma, M.; Ping, D.H.; Ohkubo, T.; Hono, K. In situ formed two-phase metallic glass with surface fractal microstructure. *Acta Mater.* **2004**, *52*, 2441–2448. [CrossRef]
25. Mattern, N.; Kühn, U.; Gebert, A.; Gemming, T.; Zinkevich, M.; Wendrock, H.; Schultz, L. Microstructure and thermal behavior of two-phase amorphous Ni–Nb–Y alloy. *Scr. Mater.* **2005**, *53*, 271–274. [CrossRef]

26. Han, J.H.; Mattern, N.; Vainio, U.; Shariq, A.; Sohn, S.W.; Kim, D.H.; Eckert, J. Phase separation in $Zr_{56-x}Gd_xCo_{28}Al_{16}$ metallic glasses (0 <x <20). *Acta Mater.* **2014**, *66*, 262–272. [CrossRef]

27. Masumoto, T.; Maddin, R. Structural stability and mechanical properties of amorphous metals. *Mater. Sci. Eng.* **1975**, *19*, 1–24. [CrossRef]

28. Wang, X.D.; Bednarcik, J.; Saksi, K.; Franz, H.; Cao, Q.P.; Jiang, J.Z. Tensile behavior of bulk metallic glasses by in situ x-ray diffraction. *Appl. Phys. Lett.* **2007**, *91*, 081913. [CrossRef]

29. Stoica, M.; Das, J.; Bednarcik, J.; Franz, H.; Mattern, N.; Wang, W.H.; Eckert, J. Strain distribution in $Zr_{64.13}Cu_{15.75}Ni_{10.12}Al_{10}$ bulk metallic glass investigated by in situ tensile tests under synchrotron radiation. *J. Appl. Phys.* **2008**, *104*, 013522. [CrossRef]

30. Wang, X.D.; Bednarcik, J.; Franz, H.; Lou, H.B.; He, Z.H.; Cao, Q.P.; Jiang, J.A. Local strain behavior of bulk metallic glasses under tension studied by in situ x-ray diffraction. *Appl. Phys. Lett.* **2009**, *94*, 011911. [CrossRef]

31. Abrosimova, G.; Aronin, A.; Afonikova, N.; Kobelev, N. Influence of deformation on the structural transformation of the $Pd_{40}Ni_{40}P_{20}$ amorphous phase. *Phys. Solid State* **2010**, *52*, 1892–1898. [CrossRef]

32. Suzuki, C.K.; Doi, K.; Kohra, K. Small Angle and Vary Small Angle X-ray Scattreing from Amorphous Pd80Si20 before and after Cold Work. *Jpn. J. Appl. Phys.* **1981**, *20*, L271–L274. [CrossRef]

33. Walter, J.; Legrand, D.G.; Luborsky, F.E. Small Angle X-ray Scattering from the Amorphous Alloy $Fe_{40}Ni_{40}P_{14}B_6$. *Mater. Sci. Eng.* **1977**, *29*, 161–167. [CrossRef]

34. Liebermann, H.H.; Graham, C.D.; Flanders, P., Jr. Changes in curie temperature, physical dimensions, and magnetic anisotropy during annealing of amorphous magnetic alloys. *IEEE Trans. Magn.* **1977**, *13*, 1541–1543. [CrossRef]

35. Abrosimova, G.E.; Serebeyakov, A.V.; Sokolovskaya, Z.D. Change of structure and magnetic-properties of amorphous iron base-metal—metalloid alloys during heat-treatment. *Fiz. Met. I Metalloved.* **1988**, *66*, 727–730.

36. Chen, H.S. Ductile-brittle transition in metallic glasses. *Mater. Sci. Eng.* **1976**, *26*, 79–82. [CrossRef]

37. Naka, M.; Masumoto, T.; Chen, H.S. Effect of metalloidal elements on strength and thermal stability of Fe-base glasses. *J. De Phys. C* **1980**, *41*, C8-839–C8-842.

38. Gleiter, H. Nanoglasses: A new kind of noncrystalline materials. *Beilstein J. Nanotechnol.* **2013**, *4*, 517–533. [CrossRef]

39. Aronin, A.S.; Abrosimova, G.E.; Gurov, A.F.; Kir'janov, Y.V.; Molokanov, V.V. Nanocrystallization of bulk Zr–Cu–Ti metallic glass. *Mater. Sci. Eng. A* **2001**, *304*, 375–379. [CrossRef]

40. Abrosimova, G.; Aronin, A.; Ignatieva, E. Decomposition of amorphous phase in $Ni_{70}Mo_{10}B_{20}$ alloy above glass transition temperature. *Mater. Sci. Eng. A* **2007**, *449*, 485–488. [CrossRef]

41. Doi, K.; Kayano, H.; Masumoto, T. Small-angle scattering from neutron-irradiated amorphous $Pd_{80}Si_{20}$. *J. Appl. Cryst.* **1978**, *11*, 605. [CrossRef]

42. Fang, J.X.; Vainio, U.; Puff, W.; Würschum, R.; Wang, X.L.; Wang, D.; Ghafari, M.; Jiang, F.; Sun, J.; Hahn, H.; et al. Atomic structure and structural stability of $Sc_{75}Fe_{25}$ nanoglasses. *Nano Lett.* **2012**, *12*, 458–463. [CrossRef] [PubMed]

43. Hebert, R.J. Deformation-induced synthesis of Al-based heterogeneous nanoscale microstructures. *Rev. Adv. Mater. Sci.* **2004**, *6*, 120–130.

44. Gleiter, H. Are there ways to synthesize matcrials beyond the limits of today? *Metall. Mater. Trans. A* **2009**, *40*, 1499–1509. [CrossRef]

45. Ritter, Y.; Şopu, D.; Gleiter, H.; Albe, K. Structure, stability and mechanical properties of internal interfaces in $Cu_{64}Zr_{36}$ nanoglasses studied by MD simulations. *Acta Mater.* **2011**, *59*, 6588–6593. [CrossRef]

46. Adjaoud, O.; Albe, K. Microstructure formation of metallic nanoglasses: Insights from molecular dynamic simulations. *Acta Mater.* **2018**, *145*, 322–330. [CrossRef]

47. Cheng, B.; Trelewicz, J.R. Controlling interface structure in nanoglasses produced through hydrostatic compression of amorphous nanoparticles. *Phys. Rev. Mater.* **2019**, *3*, 035602. [CrossRef]

48. Aronin, A.S.; Abrosimova, G.E.; Zver'kova, I.I.; Kir'janov, Y.V.; Molokanov, V.V.; Petrzhik, M.I. The structure of nanocrystalline $Ni_{58.5}Mo_{31.5}B_{10}$ and structure evolution at heat treatment. *Mater. Sci. Eng. A* **1997**, *226*, 536–540. [CrossRef]

49. Zhang, Y.; Hono, K.; Inoue, A.; Sakurai, T. APFIM studies of nanocrystalline microstructural evolution in Fe-2%B(-Cu) amorphous alloys. *Mater. Sci. Eng. A* **1996**, *217*, 407–413. [CrossRef]

50. Hono, K.; Ping, D.H.; Ohnuma, M.; Onodera, H. Cu clustering and Si partitioning in the early crystallization stage of an $Fe_{73.5}Si_{13.5}B_9Nb_3Cu_1$ amorphous alloy. *Acta Mater.* **1999**, *47*, 997–1006. [CrossRef]

51. Abrosimova, G.E.; Aronin, A.S.; Zver'kova, I.I.; Kir'janov, Y.V.; Shekhtman, V.S. Nanocrystalline $Ni_{58.5}Mo_{31.5}B_{10}$ and $Ni_{63}Mo_{27}B_{10}$ alloys: Microstructure and its evolution at annealling. *J. Surf. Investig. X Ray Synchrotron Neutron Tech.* **1998**, *4*, 31–39. (In Russian)

52. Abrosimova, G.E.; Aronin, A.S.; Zver'kova, I.I.; Kir'janov, Y.V.; Gurov, A.F. Production, structure and microhardness of nanocrystalline Ni-Mo-B alloys. *Phys. Solid State* **1998**, *40*, 8–13. [CrossRef]

53. Abrosimova, G.E.; Aronin, A.S.; Kir'janov, Y.V.; Zver'kova, I.I.; Molokanov, V.V.; Alves, H.; Köster, U. The formation, structure and properties of nanocrystalline Ni-Mo-B alloys. *J. Mater. Sci.* **1999**, *34*, 1611–1618. [CrossRef]

54. Hansen, M.; Anderko, K. *Constitution of Binary Alloys*, 2nd ed.; McGraw-Hill Book Company, Inc.: New York, NY, USA; Toronto, ON, Canada; London, UK, 1958.

55. Nakarato, K.; Kawamura, Y.; Tsai, A.P.; Inoue, A. On the growth of nanocrystalline grains in an aluminum-based amorphous alloy. *Appl. Phys. Lett.* **1993**, *63*, 2644–2646. [CrossRef]

56. Greer, A.L. *Nanostructured Materials: Science & Technology*; Chow, G.M., Ed.; NATO ASI Series; Kluwer Academical Publisher: Dordrecht, The Netherlands; London, UK, 1998; p. 457.

57. Inoue, A.; Tomioke, H.; Masumoto, T. Mechanical properties of ductile Fe-Ni-Zr and Fe-Ni-Zr (Nb or Ta) amorphous alloys containing fine crystalline particles. *J. Mater. Sci.* **1983**, *18*, 153–160. [CrossRef]

58. Köster, U.; Herold, U. *Glassy Metals II. Atomic Structure and Dynamics, Electronicv Structure, Magnetic Propertie*; Beck, H., Gűntherodt, H.-J., Eds.; Springer: Berlin, Germany, 1983.

59. Abrosimova, G.E.; Aronin, A.S.; Kir'janov, Y.V. Formation and structure of nanocrystals in an $Al_{86}Ni_{11}Yb_3$ alloy. *Phys. Solid State* **2001**, *43*, 2003–2011. [CrossRef]

60. Köster, U.; Schunemann, M. *Rapidly Solidified Alloys*; Liebermann, H.H., Ed.; Marcel Dekker Inc.: New York, NY, USA, 1993; p. 303.

61. Christian, J.W. *The Theory of Transformation in Metals and Alloys*; Part 1; Pergamon Press: Oxford, UK, 1978; ISBN 9780080542775.

62. Kantrowitz, A. Nucleation in very rapid vapor expansions. *J. Chem. Phys.* **1951**, *19*, 1097–1100. [CrossRef]

63. Greenwood, G.W. *The Mechanism of Phase Transformations in Crystalline Solids*; Inst of Metals: London, UK, 1969; p. 103.

64. Galagher, P.C.J. The influence of alloying, temperature, and related effects on the stacking fault energy. *Metall. Trans.* **1970**, *1*, 2429–2461.

65. Foley, J.C.; Allen, D.R.; Perepezko, J.H. Analysis of nanocrystal development in Al-Y-Fe and Al-Sm glasses. *Scr. Mater.* **1996**, *35*, 655–669. [CrossRef]

66. Hono, K.; Zhang, Y.; Inoue, A.; Sakurai, T. Atom probe studies of nanocrystalline microstructural evolution in some amorphous alloys. *Mater. Trans. JIM* **1995**, *36*, 909–917. [CrossRef]

67. Ham, F.S. Theory of diffusion-limited precipitation. *J. Phys. Chem. Solids.* **1958**, *6*, 335–351. [CrossRef]

68. Allen, D.R.; Folley, J.C.; Perepezko, J.H. Nanocrystal development during primary crystallization of amorphous alloys. *Acta Mater.* **1998**, *46*, 431–440. [CrossRef]

69. Howie, A.; Swann, P.R. Direct measurement of stacking-fault energies from observations of dislocation nodes. *Phil. Mag.* **1961**, *6*, 1215–1226. [CrossRef]

70. Ardell, A.Y. The effect of volume fraction on particle coarsening: Theoretical considerations. *Acta Metal.* **1972**, *20*, 61–71. [CrossRef]

71. Abrosimova, G.E.; Aronin, A.S.; Barkalov, O.I.; Dement'eva, M.M. Formation of the nanostructure in amorphous alloys of Al-Ni-Y system. *Phys. Solid State* **2013**, *55*, 1773–1778. [CrossRef]

72. Kosevich, V.M.; Sokol, A.A. *Crystal Growth from Amorphous Phase*; International School on Crystal Growth: Suzdal, Russia, 1980; p. 161. (In Russian)

73. Walter, J.L.; Bartram, S.R.; Mella, I. Formation and crystallization of alloys with two amorphous phases. *Mat. Sci. Eng.* **1978**, *36*, 193–205. [CrossRef]

74. Vincze, I.; Kemeny, I.; Schaafsma, A.; Lovas, A.; van der Woode, F. *Chemical and Topological Short-Range Order in Metallic Glasses*; KFKI-90 Hungarian Academy of Sciences: Budapest, Hungary, 1980; pp. 11–15.

75. O'Handley, R.C.; Hasegava, R.; Ray, R.; Chou, C.P. Ferromagnetic properties of some new metallic glasses. *Appl. Phys. Lett.* **1976**, *29*, 330–332. [CrossRef]

76. Barkalov, O.I.; Aronin, A.S.; Abrosimova, G.E.; Ponyatovsky, E.G. Formation and structure evolution of the bilk amorpgous $Al_{32}Ge_{68}$ alloy on heating. *J. Non Cryst. Solids* **1996**, *202*, 266–271. [CrossRef]

77. Barkalov, O.I.; Belash, I.T.; Degtyareva, V.F.; Ponyatovsky, E.G. Crystalline and amorphous state of high pressure-treated Al-Ge alloys. *Fiz. Tverd. Tela* **1987**, *29*, 1975–1978.

78. Greer, A.L.; Cheng, Y.Q.; Ma, E. Shear bands in metallic glasses. *Mater. Sci. Eng. R* **2013**, *74*, 71–132. [CrossRef]

79. Li, J.; Wang, Z.L.; Hufnagel, T.C. Characterization of nanometer-scale defects in metallic glasses by quantitative high-resolution transmission electron microscopy. *Phys. Rev. B* **2002**, *65*, 144201. [CrossRef]

80. Glezer, A.; Plotnikova, M.R.; Shalimova, A.V.; Dobatkin, S.V. Megaplastic deformation of amorphous alloys. I. Structure and mechanical properties. *Proc. RAS Phys. Ser.* **2009**, *73*, 1302–1309. (In Russian)

81. Stolpe, M.J.; Kruzic, J.; Busch, R. Evolution of shear bands, free volume and hardness during cold rolling of a Zr-based bulk metallic glass. *Acta Mater.* **2014**, *64*, 231–240. [CrossRef]

82. Spaepen, F. A microscopic mechanism for steady state inhomogeneous flow in metallic glasses. *Acta Metall.* **1977**, *25*, 407–415. [CrossRef]

83. Wang, K.; Fujita, T.; Zeng, Y.Q.; Nishiyama, N.; Inoue, A.; Chen, M.W. Micromechanisms of serrated flow in a $Ni_{50}Pd_{30}P_{20}$ bulk metallic glass with a large compression plasticity. *Acta Mater.* **2008**, *56*, 2834–2842. [CrossRef]

84. Aronin, A.; Abrosimova, G.; Matveev, D.; Rybchenko, O. Structure and Properties Of Nanocrystalline Alloys Prepared By High Pressure Torsion. *Rev. Adv. Mater. Sci.* **2010**, *25*, 52–57.

85. Abrosimova, G.E.; Aronin, A.S.; Dobatkin, S.V.; Kaloshkin, S.D.; Matveev, D.V.; Rybchenko, O.G.; Tatiyanin, E.V.; Zverkova, I.I. The formation of nanocrystalline structure in amorphous Fe-Si-B alloy by severe plastic deformation. *J. Metastable Nanocrystalline Mater. Ser.* **2005**, *24*, 69–72. [CrossRef]

86. Aronin, A.S.; Louzguine-Luzgin, D.V. On nanovoids formation in shear bands of an amorphous Al-based alloy. *Mech. Mater.* **2017**, *113*, 19–23. [CrossRef]

87. Hebert, R.J.; Boucharat, N.; Perepezko, J.H.; Rösner, H.; Wilde, G. Calorimetric and microstructural analysis of deformation induced crystallization reactions in amorphous $Al_{88}Y_7Fe_5$ alloy. *J. Alloy. Compd.* **2007**, *434*, 18–21. [CrossRef]

88. Wilde, G.; Rösner, H. Nanocrystallization in a shear band: An in situ investigation. *Appl. Phys. Lett.* **2011**, *98*, 251904. [CrossRef]

89. Abrosimova, G.; Aronin, A.; Barkalov, O.; Matveev, D.; Rybchenko, O.; Maslov, V.; Tkatch, V. Structural transformations in the $Al_{85}Ni_{6.1}Co_2Gd_6Si_{0.9}$ amorphous alloy during multiple rolling. *Phys. Solid State* **2011**, *53*, 229–233. [CrossRef]

90. Mazzone, G.; Montone, A.; Antisari, M.V. Effect of plastic flow on the kinetics of amorphous phase growth by solid-state reaction in the Ni-Zr system. *Phys. Rev. Lett.* **1990**, *65*, 2019–2023. [CrossRef] [PubMed]

91. Yoshizawa, Y.; Oguma, S.; Yamauchi, K. New Fe-based soft magnetic alloys composed of ultrafine grain structure. *J. Appl. Phys.* **1988**, *64*, 6044–6046. [CrossRef]

92. Allia, P.; Baricco, M.; Knobel, M.; Tiberto, P.; Vinai, F. Nanocrystalline Fe73.5 Cu1Nb3Si13.5B9 obtained by direct-current Joule heating. Magnetic and mechanical properties. *Philos. Mag. B Phys. Condens. Matter.* **1993**, *68*, 853–860. [CrossRef]

93. Ito, N.; Suzuki, K. Improvement of magnetic softness in nanocrystalline soft magnetic materials by rotating magnetic field annealing. *J. Appl. Phys.* **2005**, *97*, 10F503. [CrossRef]

94. Gheiratmand, T.H.; Hosseini, R.M. Fincmet nanocrystalline soft magnetic alloy: Investigation of glass forming ability, crystallization mechanism, production techniques, magnetic softness and the effect of replacing the main constituents by other elements. *J. Magn. Magn. Mater.* **2016**, *408*, 177–192. [CrossRef]

95. Takenaka, K.; Setyawanm, A.D.; Sharma, P.; Nishiyama, N.; Makino, A. Industrialization of nanocrystalline Fe–Si–B–P–Cu alloys for high magnetic flux density cores. *J. Magn. Magn. Mater.* **2016**, *401*, 479–483. [CrossRef]

96. Shuvaeva, E.; Kaloshkin, S.; Churyukanova, M.; Perminov, A.; Khriplivets, I.; Mitra, A.; Panda, A.K.; Roy, R.K.; Premkumar, V.; Zhukova, A. The impact of bending stress on magnetic properties of Finemet type microwires and ribbons. *J. Alloy. Compd.* **2018**, *743*, 388–393. [CrossRef]

97. Mikhalitsyna, E.A.; Kataev, V.A.; Larrañaga, A.; Lepalovskij, V.N.; Kurlyandskaya, G.V. Nanocrystallization in FINEMET-Type $Fe_{73.5}Nb_3Cu_1Si_{13.5}B_9$ and $Fe_{72.5}Nb_{1.5}Mo_2Cu_{1.1}Si_{14.2}B_{8.7}$ thin films. *Materials* **2020**, *13*, 348. [CrossRef]

98. Louzguine, D.V.; Inoue, A. Comparative study of the effect of cold rolling on the structure of Al–RE–Ni–Co (RE = rare-earth metals) amorphous and glassy alloys. *J. Non Cryst. Solids* **2006**, *352*, 3903–3909. [CrossRef]

99. Kovács, Z.; Henits, P.; Hobor, S.; Révész, A. Nanocrystallization process in amorphous alloys during severe plastic deformation and thermal treatments. *Rev. Adv. Mater. Sci.* **2008**, *18*, 593–596.

100. Tkach, V.I.; Rassolov, S.G.; Popov, V.V.; Maksimov, V.V.; Maslov, V.V.; Nosenko, V.K.; Aronin, A.S.; Abrosimova, G.E.; Rybchenko, O.G. Complex crystallization mode of amorphous/nanocrystalline composite $Al_{86}Ni_2Co_{5.8}Gd_{5.7}Si_{0.5}$. *J. Non Cryst. Solids* **2011**, *357*, 1628–1631. [CrossRef]

101. Louzguine-Luzgin, D.V.; Bazlov, A.I.; Ketov, S.V.; Inoue, A. Crystallization behavior of Fe- and Co-based bulk metallic glasses and their glass-forming ability. *Mater. Chem. Phys.* **2015**, *162*, 197–206. [CrossRef]

102. Herold, U.; Kőster, U. Metastabile Phasen in extreme schnell erstarrten Eisen-Bor-Legierungen. *Z Metallk.* **1978**, *69*, 326–332.

103. Khan, Y.; Sostarich, M. Dynamic temperature X-ray diffraction analysis of amorphous $Fe_{80}B_{20}$. *Z Metallk.* **1981**, *72*, 266–268.

104. Abrosimova, G.E.; Aronin, A.S.; Gantmakher, V.F.; Levin, Y.B.; Osherov, M.V. Variation of the electric resistivity of the amorphous Ni-Zr alloy in the initial-stages of crystallization. *Fiz. Tverd. Tela* **1988**, *30*, 1424–1430.

105. Abrosimova, G.E.; Aronin, A.S. Phase transformation in Fe-B alloys at heating. *Metallofizika* **1988**, *10*, 47–52. (In Russian)

106. Abrosimova, G.; Matveev, D.; Pershina, E.; Aronin, A. Effect of treatment conditions on parameters of nanocrystalline structure in Al-based alloys. *Mater. Lett.* **2016**, *183*, 131–134. [CrossRef]

107. Aronin, A.; Matveev, D.; Pershina, E.; Tkatch, V.; Abrosimova, G. The effect of changes in Al-based amorphous phase structure on structure forming upon crystallization. *J. Alloy. Compd.* **2017**, *715*, 176–183. [CrossRef]

108. Aronin, A.; Budchenko, A.; Matveev, D.; Pershina, E.; Tkatch, V.; Abrosimova, G. Nanocrystal formation in light metallic glasses at heating and deformation. *Rev. Adv. Mater. Sci.* **2016**, *46*, 53–69.

109. Boucharat, N.; Hebert, R.; Rösner, H.; Valiev, R.; Wilde, G. Nanocrystallization of amorphous $Al_{88}Y_7Fe_5$ alloy induced by plastic deformation. *Scr. Mater.* **2005**, *53*, 823–828. [CrossRef]

110. Valiev, R.Z.; Alexandrov, I.V. *Bulk Nanocrystalline Metallic Materials*; Akademkniga: Moscow, Russia, 2007; p. 398. (In Russian)

111. Kovacs, Z.; Henits, P.; Zhilyaev, A.P.; Revesz, A. Deformation induced primary crystallization in a thermally non-primary crystallizing amorphous $Al_{85}Ce_8Ni_5Co_2$ alloy. *Scr. Mater.* **2006**, *54*, 1733–1737. [CrossRef]

112. Boucharat, N.; Hebert, R.; Rösner, H.; Valiev, R.; Wilde, G. Synthesis routes for controlling the microstructure in nanostructured $Al_{88}Y_7Fe_5$ alloys. *J. Alloy. Compd.* **2007**, *434*, 252–254. [CrossRef]

113. Aronin, A.; Abrosimova, G.; Matveev, D.; Pershina, E. Nanocrystal formation, structure and magnetic properties of Fe-Si-B amorphous alloy after deformation. *Mater. Lett.* **2013**, *97*, 15–17. [CrossRef]

114. Lewandowski, J.J.; Greer, A.L. Temperature rise at shear bands in metallic glasses. *Nat. Mater.* **2006**, *5*, 15–18. [CrossRef]

115. Csontos, A.A.; Shiflet, G.J. Formation and chemistry of nanocrystalline phases formed during deformation in aluminum-rich metallic glasses. *Nano Struct. Mater.* **1997**, *9*, 281–289. [CrossRef]

116. Georgarakis, K.; Aljerf, M.; Li, Y.; LeMoulec, A.; Charlot, F.; Yavari, A.R.; Chornokhvostenko, K.; Tabachnikova, E.; Evangelakis, G.A.; Miracle, D.B.; et al. Shear band melting and serrated flow in metallic glasses. *Appl. Phys. Lett.* **2008**, *93*, 031907. [CrossRef]

117. Hartley, K.A.; Duffy, J.; Hawley, R.H. Measurement of the temperature profile during shear band formation in steels deforming at high strain rates. *J. Mech. Solids* **1987**, *35*, 283–301. [CrossRef]

118. Li, J.G.; Umemoto, M.; Todaka, Y.; Fujisaku, K.; Tsuchiya, K. The dynamic phase transformation and formation of nanocrystalline structure in SUS304 austenitic stainless steel subjected to high pressure torsion. *Rev. Adv. Mater. Sci.* **2008**, *18*, 577–582.

119. Jiang, W.H.; Atzmon, M. The effect of compression and tension on shear-band structure and nanocrystallization in amorphous $Al_{90}Fe_5Gd_5$: A high-resolution transmission electron microscopy study. *Acta Mater.* **2003**, *51*, 4095–4105. [CrossRef]

120. Kim, J.J.; Choi, Y.; Suresh, S.; Argon, A.S. Nanocrystallization during nanoindentation of a bulk Amorphous metal alloy at room temperature. *Science* **2002**, *295*, 654–656. [CrossRef] [PubMed]

121. Rösner, H.; Peterlechler, M.; Kűbel, C.; Schmidt, V.; Wilde, G. Density changes in shear bands of a metallic glass determined by correlative analytical transmission electron microscopy. *Ultramicroscopy* **2014**, *142*, 1–9. [CrossRef] [PubMed]

122. Schmidt, V.; Rösner, H.; Peterlechler, M.; Wilde, G. Quantitative measurement of density in a shear band of metallic glass monitored along its propagation direction. *Phys. Rev. Lett.* **2015**, *115*, 035501. [CrossRef] [PubMed]

123. Slaughter, S.K.; Kertis, F.; Deda, E.; Gu, X.; Wright, W.J.; Hufnagel, T.C. Shear bands in metallic glasses are not necessarily hot. *APL Mater.* **2014**, *2*, 096110. [CrossRef]

124. Abrosimova, G.; Aronin, A. Nanocrystal formation in Al- and Ti-based amorphous alloys at Deformation. *J. Alloy. Compd.* **2018**, *747*, 26–30. [CrossRef]

125. Yoo, B.G.; Kim, Y.J.; Oh, J.H.; Ramamurty, U.; Jang, J. On the hardness of shear bands in amorphous alloys. *Scr. Mater.* **2009**, *61*, 951–954. [CrossRef]

126. Pan, J.; Chen, Q.; Liu, L.; Li, Y. Softening and dilatation in a single shear band. *Acta Mater.* **2011**, *59*, 5146–5158. [CrossRef]

127. Maaß, R.; Samwer, K.; Arnold, W.; Volkert, C.A. A single shear band in a metallic glass: Local core and wide soft one. *Appl. Phys. Lett.* **2014**, *10*, 17190. [CrossRef]

128. Abrosimova, G.; Aronin, A.; Fokin, D.; Orlova, N.; Postnova, E. The decrease of Young's modulus in shear bands of amorphous $Al_{87}Ni_8Gd_5$ alloy. *Mater. Lett.* **2019**, *252*, 114–116. [CrossRef]

129. Perepezko, J.H. Nucleation-controlled reactions and metastable structures. *Prog. Mater. Sci.* **2004**, *49*, 263–284. [CrossRef]

130. Mu, J.; Fu, H.; Zhu, Z.; Wang, A.; Li, H.; Hu, Z.Q.; Zhang, H. Synthesis and properties of Al-Ni-La bulk metallic glass. *Adv. Eng. Mater.* **2009**, *11*, 530–532. [CrossRef]

131. Yang, B.J.; Yao, J.H.; Chao, Y.S.; Wang, J.Q.; Ma, E. Developing aluminum-based bulk metallic glasses. *Phil. Mag.* **2010**, *90*, 3215–3231. [CrossRef]

132. Du, S.Z.; Li, C.C.; Pang, S.Y.; Leng, J.F.; Geng, H.R. Influences of melt superheat treatment on glass forming ability and properties of $Al_{84}Ni_{10}La_6$ alloy. *Mater. Des.* **2013**, *47*, 358–364. [CrossRef]

133. Chunchu, V.J.; Markandeyulu, G. Magnetoimpedance studies in as quenched $Fe_{73.5}Si_{13.5}B_8CuV_{3-x}AlNb_x$ nanocrystalline ribbons. *Appl. Phys.* **2013**, *113*, 17A321. [CrossRef]

134. Xiang, R.; Zhou, S.; Dong, B.; Zhang, G.; Li, Z.; Wang, Y.; Chang, C. Effect of Co addition on crystallization and magnetic properties of FeSiBPC alloys. *Progr. Nat. Sci. Mater. Intern.* **2014**, *24*, 649–654. [CrossRef]

135. Meng, F.; Tsuchija, K.; Yokoyama, Y. Reversible transition of deformation mode by structural rejuvenation and relaxation in bulk metallic glass. *Appl. Phys. Lett.* **2012**, *101*, 121914. [CrossRef]

136. Tong, Y.; Iwashita, T.; Dmowski, W.; Bei, H.; Yokoyama, Y.; Egami, T. Structural rejuvenation in bulk metallic glasses. *Acta Mater.* **2015**, *86*, 240–246. [CrossRef]

137. Dmowski, W.; Yokoyama, Y.; Chuang, A.; Ren, Y.; Umemoto, M.; Tsuchiya, K.; Inoue, A.; Egami, T. Structural rejuvenation in a bulk metallic glass induced by severe plastic deformation. *Acta Mater.* **2010**, *58*, 429–438. [CrossRef]

138. Tong, Y.; Dmowski, W.; Bei, H.; Yokoyama, Y.; Egami, T. Mechanical rejuvenation in bulk metallic glass induced by thermo-mechanical creep. *Acta Mater.* **2018**, *148*, 384–390. [CrossRef]

139. Guo, W.; Yamada, R.; Saida, J. Rejuvenation and plasticization of metallic glass by deep cryogenic cycling. *Intermetallics* **2018**, *93*, 141–147. [CrossRef]

140. Ketov, S.V.; Sun, Y.H.; Nachum, S.; Lu, Z.; Checchi, A.; Beraldin, A.R.; Bai, H.Y.; Wang, W.H.; Louzguine-Luzgin, D.V.; Carpenter, M.A.; et al. Rejuvenation of metallic glasses by non-affine thermal strain. *Nature* **2015**, *524*, 200–203. [CrossRef] [PubMed]

141. Hufnagel, T.C. Cryogenic rejuvenation. *Nat. Mater.* **2015**, *14*, 87–868. [CrossRef] [PubMed]

142. Bian, X.; Wang, G.; Wang, Q.; Sun, B.A.; Hussain, I.; Zhai, Q.J.; Mattern, N.; Bednarcik, J.; Eckert, J. Cryogenic-temperature-induced structural transformation of a metallic glass. *Mater. Res. Lett.* **2017**, *5*, 284–291. [CrossRef]

143. Guo, W.; Shao, Y.M.; Saida, J.; Zhao, M.; Lu, S.L.; Wu, S.S. Rejuvenation and plasticization of Zr-based bulk metallic glass with various Ta content upon deep cryogenic cycling. *J. Alloy. Compd.* **2019**, *795*, 314–318. [CrossRef]

144. Guo, W.; Saida, J.; Zhao, M.; Lu, S.; Wu, S. Rejuvenation of Zr-based bulk metallic glass matrix composite upon deep cryogenic cycling. *Mater. Lett.* **2019**, *247*, 135–138. [CrossRef]

145. Kang, S.J.; Cao, Q.P.; Liu, J.; Tang, Y.; Wang, X.D.; Zhang, D.X.; Ahn, I.S.; Caron, A.; Jiang, J.Z. Intermediate structural state for maximizing the rejuvenation effect inmetallic glass via thermo-cycling treatment. *J. Alloy. Compd.* **2019**, *795*, 493–500. [CrossRef]

146. Abrosimova, G.; Volkov, N.; Van Tuan, T.; Pershina, E.A.; Aronin, A.S. Cryogenic rejuvenation of Al-based amorphous-nanocrystalline alloys. *Mater. Lett.* **2019**, *240*, 150–152. [CrossRef]

147. Abrosimova, G.; Volkov, N.; Van Tuan, T.; Pershina, E.A.; Aronin, A.S. Amorphous structure rejuvenation under cryogenic treatment of Al-based amorphous-nanocrystalline alloys. *J. Non Cryst. Solids* **2020**, *528*, 119751. [CrossRef]

148. Abrosimova, G.E.; Aronin, A.S. Reversible structure changes in amorphous Fe-B alloys. *Int. J. Rapid Solidif.* **1991**, *6*, 29–40.

149. Abrosimova, G.; Aronin, A.; Ignatieva, E. Phase transformation in the $Ni_{70}Mo_{10}P_{20}$ amorphous alloy at heating. *Phys. Met. Metallogr.* **2003**, *95*, 569–574.

150. Abrosimova, G.; Aronin, A.; Ignatieva, E. A metastable phase forming during crystallization of an amorphous $Ni_{70}Mo_{10}P_{20}$ alloy. *Phys. Solid State* **2006**, *48*, 122–128. [CrossRef]

151. Abrosimova, G.; Aronin, A.; Ignatieva, E. Mechanism of crystallization of the $Ni_{70}Mo_{10}B_{20}$ alloy above the glass transition temperature. *Phys. Solid State* **2006**, *48*, 563–569. [CrossRef]

152. Abrosimova, G.; Aronin, A.; Kabanov, Y.P.; Matveev, D.V.; Molokanov, V.V. Magnetic structure and properties of a bulk $Fe_{72}Al_5P_{10}Ga_2C_6B_4Si_1$ alloy in the amorphous and nanocrystalline states. *Phys. Solid State* **2004**, *46*, 885–890. [CrossRef]

153. Abrosimova, G.; Aronin, A.; Kabanov, Y.P.; Matveev, D.V.; Molokanov, V.V.; Rybchenko, O.G. Effect of thermal treatment on the microstructure and magnetic properties of a bulk amorphous $Fe_{72}Al_5P_{10}Ga_2C_6B_4Si_1$ alloy. *Phys. Solid State* **2004**, *46*, 2232–2237. [CrossRef]

154. Masumoto, T.; Waseda, Y.; Kimura, H.; Inoue, A. Thermal instability and crystallization characteristics of amorphous metal-metalloid system. *Sci. Rep. Res. Inst. Tohoku Univ.* **1976**, *26*, 21–35.

155. Aronin, A.; Ivanov, A.S.; Yakshin, A.E. Increase Of The Near-Surface Crystallization Temperature In A Fe-B-P Amorphous Alloy. *Phys. Solid State* **1991**, *33*, 2527–2532.

 *metals*

Article

Devitrification of $Zr_{55}Cu_{30}Al_{15}Ni_5$ Bulk Metallic Glass under Heating and HPT Deformation

Galina Abrosimova [1,*], Boris Gnesin [1], Dmitry Gunderov [2,3], Alexandra Drozdenko [1], Danila Matveev [1], Bogdan Mironchuk [1,4], Elena Pershina [1], Ilia Sholin [1] and Alexandr Aronin [1]

[1] Institute of Solid State Physics RAS, 142432 Chernogolovka, Russia; gnesin@issp.ac.ru (B.G.); al_krylova@issp.ac.ru (A.D.); matveev@issp.ac.ru (D.M.); milana12032011@mail.ru (B.M.); pershina@issp.ac.ru (E.P.); sholin@issp.ac.ru (I.S.); Aronin@issp.ac.ru (A.A.)
[2] Ufa State Aviation Technical University, 12 K. Marx str., 450008 Ufa, Russia; dimagun@mail.ru
[3] Institute of Molecule and Crystal Physics, Ufa Federal Research Center RAS, P. October 151 Ufa, Russia
[4] National Research University Higher School of Economics, 101000 Moscow, Russia
* Correspondence: gea@issp.ac.ru; Tel.: +7-4-965-228-462

Received: 9 September 2020; Accepted: 2 October 2020; Published: 5 October 2020

Abstract: The nanocrystal formation in $Zr_{55}Cu_{30}Al_{15}Ni_5$ bulk metallic glass was studied under heat treatment and deformation. The activation energy of crystallization under heating is 278 kJ/mol. Different crystalline phases were found to be formed during crystallization under heating and deformation. At the first crystallization stage, the metastable phase with a hexagonal structure (lattice of space group $P6_3/mmc$ with the parameters a = 8.66 Å, c = 14.99 Å) is formed under heat treatment. When the temperature rises, the metastable phase decays with the formation of stable crystalline phases. The crystalline Zr_2Cu phase with the lattice of space group Fd3m is formed during crystallization under the action of deformation. It was determined that during deformation nanocrystals are formed primarily in the subsurface regions of the samples.

Keywords: metallic glass; nanostructure; HPT deformation; metastable phases

1. Introduction

The first metallic glass was produced in 1960 [1]. Since this moment, these materials have been provoking great interest as both non-crystalline metallic materials and the basis for the creation of composite amorphous-nanocrystalline materials. A special group of metallic glasses is bulk amorphous alloys, many of which have good mechanical properties. Among these alloys are high-strength Zr-based bulk amorphous alloys, which, in particular, can be used for medical applications such as struts for cardiovascular stents [2,3]. Zr-based bulk alloys were studied in a number of works [4–6]. The processes of crystallization of Zr-based bulk amorphous alloys were investigated mainly under heat treatment. At an initial stage of the devitrification of these alloys, the formation of metastable crystalline [7,8] and quasi-crystalline [9,10] phases was observed. Under heating or annealing, after the completion of the first crystallization stage alloys have an amorphous-nanocrystalline structure, with the fraction of the crystalline phase depending on heat treatment conditions. Another method of impact on the structure of bulk metallic glasses is severe plastic deformation. One of its main methods is high-pressure torsion. This action also leads to crystal formation in the amorphous phase, with crystal formation starting in the regions of plastic deformation localization, i.e., shear bands or their vicinity. Crystal formation in the places of plastic deformation localization is caused by an increase in the free volume fraction and, correspondingly, by enhanced values of diffusion coefficients in these regions [11–13]. The fraction of the nanocrystalline phase under plastic deformation also depends on treatment conditions, i.e., the value of applied pressure, the rate, duration, and temperature of deformation. In turn, the properties of a produced material depend on the formed structure and the fraction of the crystalline phase. Plastic deformation is realized by the formation and propagation of shear bands. At that stage, nanocrystals being present in the structure can have an inhibitory action on the propagation of the bands [14].

In some studies of the processes of metallic glass crystallization, it was shown that the structure formed under crystallization depends significantly on the conditions of production of an amorphous alloy [15], as well as on the conditions of treatments (heat or deformation ones) of the amorphous structure both in the limit of an amorphous state and at early crystallization stages [15,16]. However, despite active studies of the processes of the devitrification of Zr-based alloys, treatment parameters necessary for the formation of an amorphous-nanocrystalline structure remain unknown. Therefore, the present work aims at carrying out a comparative study of devitrification processes of $Zr_{55}Cu_{30}Al_{15}Ni_5$ bulk alloy under heat treatment and deformation.

2. Materials and Methods

The samples of $Zr_{55}Cu_{30}Al_{15}Ni_5$ amorphous alloy were produced by melting and quenching into a copper mold. They were rods with a diameter of 8 mm. The thermal analysis of the alloy was performed by differential scanning calorimetry (DSC) (Perkin-Elmer DSC-7). The $Zr_{55}Cu_{30}Al_{15}Ni_5$ bulk amorphous alloy was heated to the preset temperatures (up to 848 K), then cooled to room temperature in the calorimeter with Ar flow. The use of an argon atmosphere avoided oxidation of the sample. The heating rates (β) were 5, 10, 20, and 40 K/min. The kinetic characteristics of the crystallization reaction were determined by the obtained series of DSC curves. The error in the measurement of activation energy (E_a) and temperature was 1.9 kJ/mol and 3 K, respectively.

The samples were deformed by high-pressure torsion (HPT). The samples of an initial alloy were cut into disks with a thickness of 0.5 mm and polished before deformation. They were deformed at a rate of 1 rotation per minute, with deformation at 1, 5, and 10 rotations being used. The deformation was carried out at a pressure of 6 GPa at room temperature. The deformation degree was estimated by the formula:

$$e = ln\left(1 + \left(\frac{\varphi \cdot r}{h}\right)^2\right)^{0.5} + ln\left(\frac{h_0}{h}\right) \tag{1}$$

where: r is the radius of a sample, u is the angle of the punch rotation, h_0 is the thickness of an initial sample, h is the thickness of a deformed sample [17]. Thus, e = 4.8, 6.4, 7.1 for 1, 5, and 10 rotations, respectively. The deformation degree was determined for the middle of the sample radius. The diameter of the sample was 8 mm. All the subsequent measurements were performed for a sample region which was in the middle of the deformed sample radius. The deformation of the samples was carried out at room temperature. No oxide layer was found on the surface after deformation. The structural studies were carried out by X-ray diffraction (using Co Kα and Mo Kα radiations), high-resolution transmission electron microscopy (HREM), scanning electron microscopy (SEM), and X-ray microanalysis (EDS). A focused ion beam (FIB) was used to prepare electron microscope foils from certain regions of the deformed samples.

3. Resultsand Discussion

The structure of the samples after the production was amorphous. Figure 1 shows an X-ray diffraction pattern of the alloy. It contains only broad diffuse halos from the amorphous phase. No reflections from the crystalline phases are observed.

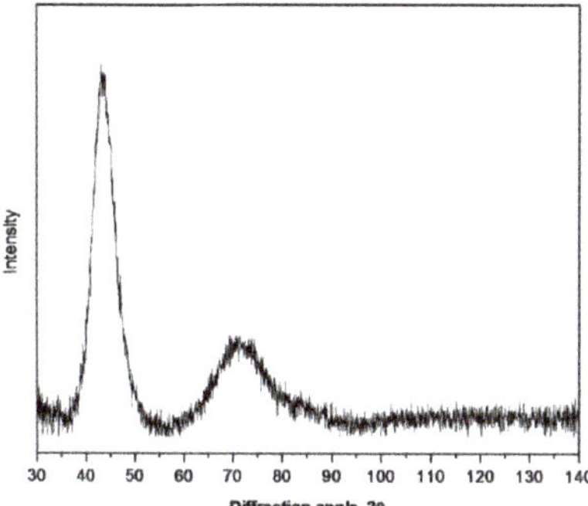

Figure 1. X-ray diffraction pattern of glassy $Zr_{55}Cu_{30}Al_{15}Ni_5$ sample.

3.1. Analysis of Devitrification under Heating

The amorphous alloy crystallizes under heating. DSC curves for the $Zr_{55}Cu_{30}Al_{10}Ni_5$ alloy under investigation are presented in Figure 2. One can see in Figure 2 that the temperatures of phase transitions occurring in the samples depend on the heating rate. The crystallization temperature rises as the heating rate increases. Table 1 provides crystallization temperatures depending on the heating rate.

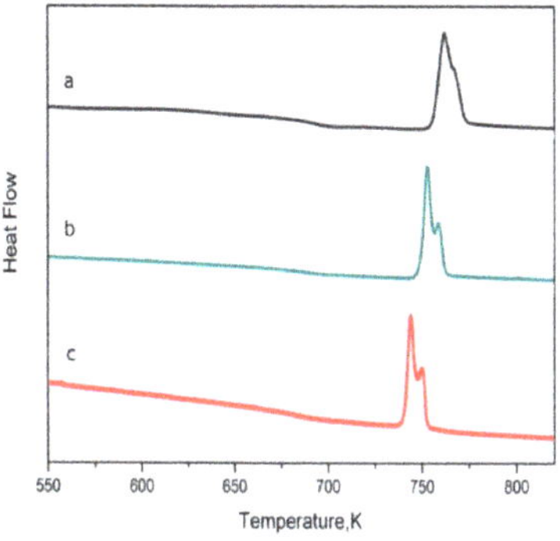

Figure 2. Differential scanning calorimetry (DSC) curves of $Zr_{55}Cu_{30}Al_{10}Ni_5$ alloy heated to 848 K at a constant rate of 5 (**a**), 10 (**b**) and 20 (**c**) K/min.

Table 1. Crystallization temperatures for different heating rates.

Heating Rate β, K/min	Temperature of Crystallization Start T_s, K
5	730
10	740
20	755

The feature of the first exothermic peak is its double shape (Figure 2). This thermogram is typical of alloys of Zr-Cu-Al-Ni system. The material remained amorphous under heating below the crystallization temperature: no indication of the presence of phase transitions was observed in the thermograms of the corresponding samples. The temperatures of crystallization start (start of the first crystallization stage) of an initial $Zr_{55}Cu_{30}Al_{10}Ni_5$ alloy are $T_x \sim 730$ K, 740 K, and 755 K for a heating rate of 5, 10, and 20 K/min, respectively (Table 1).

Based on the analysis of the shift of DSC curves depending on the heating rate, the activation energy of crystallization (E_a) was determined using the Kissinger method. E_a was determined in the temperature range corresponding to the first peak in the curves since this peak corresponds to the process of primary crystallization of this alloy. Figure 3 shows the corresponding Kissinger plot for the determination of the activation energy of $Zr_{55}Cu_{30}Al_{10}Ni_5$ alloys by a series of DSC scans at heating rates of 5, 10, and 20 K/min.

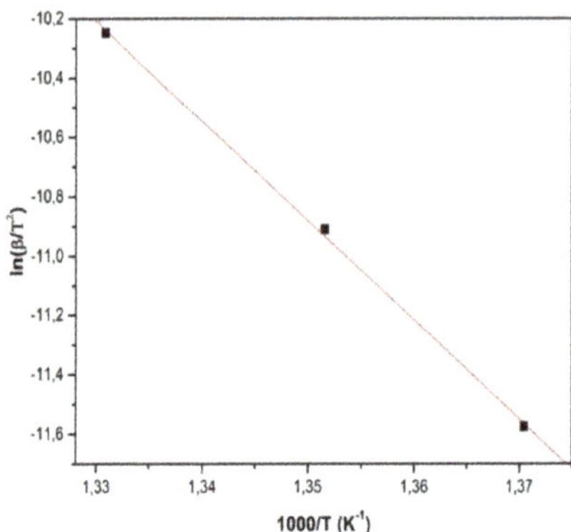

Figure 3. The Kissinger plot of $Zr_{55}Cu_{30}Al_{10}Ni_5$.

The activation energy of crystallization for the alloy, determined by this plot, was 278 kJ/mol. This value of activation energy agrees with the known values for alloys of similar composition. The activation energy of crystallization for this alloy, determined in other work, is in the range between 230 and 315 kJ/mol [18–21].

In the temperature range under study, DSC curves have a peak with a complex shape. This peak shape indicates the successive formation of several crystalline phases.

The X-ray diffraction patterns of heated samples are presented in Figure 4. The temperature of 738 K corresponds to the start of the DSC curve peak. The temperature of 750 K is between the peaks forming the peak with a complex shape. It is clear that the complex shape of the peak is related to the successive formation of several crystalline phases. This treatment was performed to produce a sample with a great amount of primary formed phase without a significant amount of the second phase (assuming that only one, not several phases are formed at the first stage).

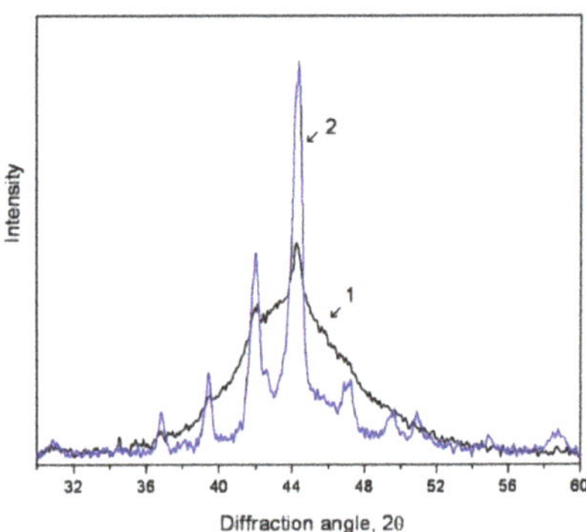

Figure 4. X-ray diffraction patterns of the samples heated in the calorimeter to 738 (1) and 750 (2) K (Co Kα).

Curve 1 (black) corresponds to the sample heated in the calorimeter before the start of the first peak (738 K). Curve 2 (blue) corresponds to the sample after the completion of the first peak in the DSC curve. According to the literature [22,23], the metastable phase is formed under crystallization which then transits to the equilibrium Zr_2Cu phase. The authors of the works above assumed that the metastable phase has a lattice of the distorted tetragonal Zr_2Ni phase with space group I4/mcm.

The analysis of diffraction reflections in Figure 4 showed that under heat treatment of the alloy under study, the metastable phase is formed at the first crystallization stage. Its structure can be described by a hexagonal lattice with space group $P6_3$/mmc with the parameters a = 8.66 Å, c = 14.99 Å. When the temperature rises, the metastable phase decays with the formation of stable crystalline phases (Figure 5).

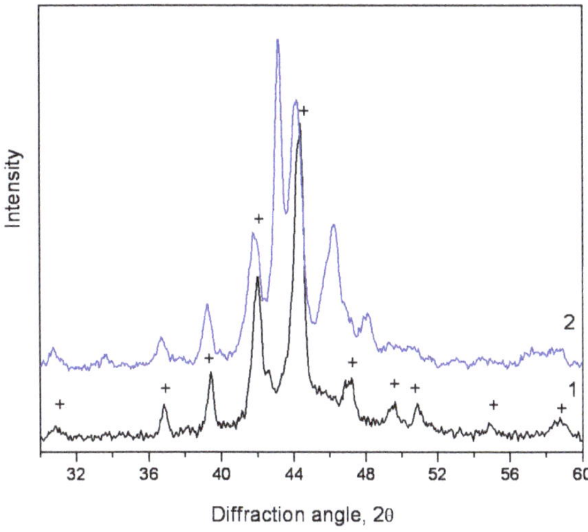

Figure 5. X-ray diffraction patterns of the alloy after annealing at 750 (1) and 873 K (2) (Co Kα).

One can see in Figure 5 that additional lines appear along with reflections from the metastable phase, marked with crosses. Besides the partially conserved metastable phase, the observed diffraction pattern indicates the possible presence of the following well-known crystalline phases: Zr_2Cu with the lattice of space group I4/mmm with the parameters a = 3.22 Å, c = 11.18 Å, Zr_5Al_3 with the lattice of space group $P6_3$/mmc (a = 8.18 Å, c = 5.70 Å) or I4/mcm (a = 11.04 Å, c = 5.39 Å), or Zr_2Al with the lattice of space group I4/mmm (a = 6.853 Å, c = 5.50 Å). At that, a small amount of the amorphous phase remains in the sample, too. All these phases are found in the alloys of the studied system. Since nanocrystals are formed at initial stages of the crystallization of an amorphous structure, broad diffraction lines are present in the X-ray diffraction patterns; in some cases, these lines overlap each other, and the accuracy of determination of phase composition is low at this stage. It is important to note that only one metastable phase is formed at an initial stage of the decay of $Zr_{55}Cu_{30}Al_{15}Ni_5$ bulk amorphous alloy. Consequently, this alloy crystallizes by the primary mechanism.

3.2. Crystallization of the Alloy during Deformation

Devitrification of the amorphous phase also occurs under the deformation of the alloy by high-pressure torsion (HPT). The fraction of the crystalline phase increases as the deformation degree increases. Figure 6 demonstrates X-ray diffraction patterns of the alloy after deformation. No crystalline phases are observed in the sample deformed by 4.8 (one rotation). Reflections from the crystalline phases arise in the samples deformed by 6.4 and 7.1. The intensity of reflections from crystals remains low. Figure 7 shows experimental curve (1), diffuse halo from the amorphous phase (3), set of diffraction peaks, and summation curve (2). The observed seven reflections agree well with the known crystalline Zr_2Cu phase with the lattice of space group Fd3m.

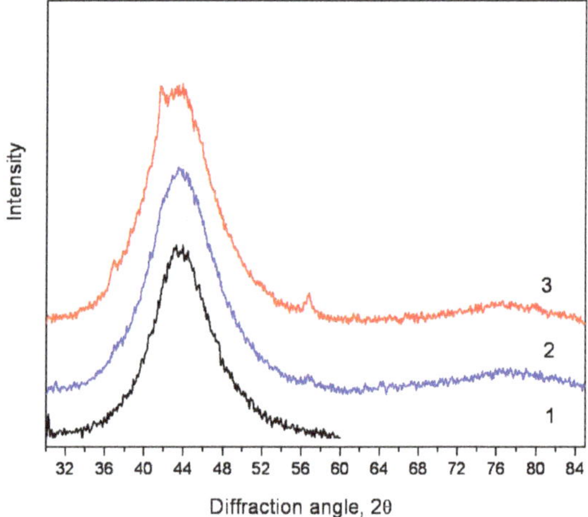

Figure 6. X-ray diffraction patterns of initial (1) and deformed samples of $Zr_{55}Cu_{30}Al_{10}Ni_5$ alloy (2–1 rotation, 3–10 rotations) (Co Kα).

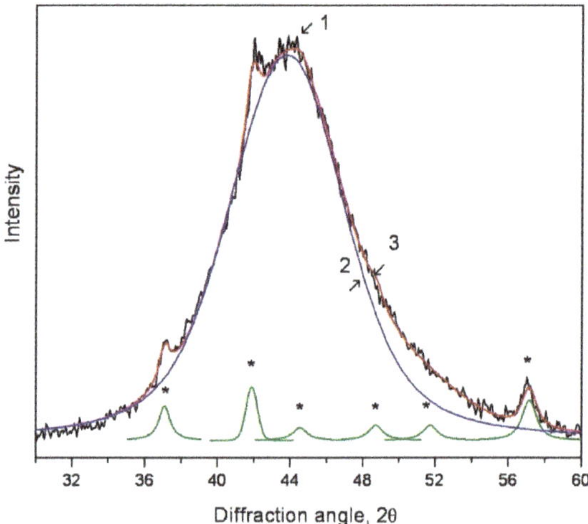

Figure 7. Result of the expansion of the X-ray diffraction pattern of the sample after deformation at 5 rotations: 1(black)–experimental spectrum, 2 (red)–summation curve, 3 (blue)–diffuse halo from the amorphous phase (reflections corresponding to the crystalline phase (green) are marked with asterisks).

Thus, the crystalline phases formed under heating and deformation turn out to be different. For comparison, Figure 8 demonstrates X-ray diffraction patterns (the region of the most intense reflections) of the samples after HPT (1) and heating in the calorimeter (2) to 750 K (the temperature corresponding to the completion of the first DSC peak). For illustration purposes, the positions of diffraction reflections from the Zr_2Cu phase are marked with vertical lines. One can see in the figure that at an initial crystallization stage the phase compositions of the deformed and heated samples are different. There are some additional reflections (for example, intense lines corresponding to angles of ~31.1, 39.5, 47 degrees, etc.) in the X-ray diffraction pattern; other lines (~36.9, 42 degrees) are shifted.

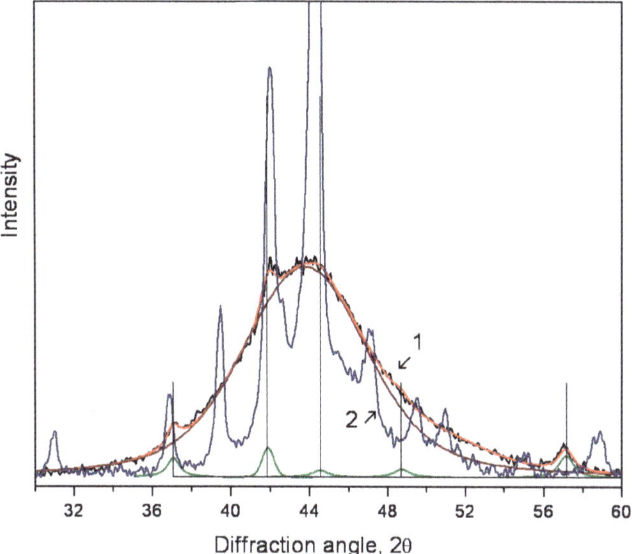

Figure 8. X-ray diffraction patterns of the samples after high-pressure torsion (HPT) (1) and heating in the calorimeter (2, wine) to the temperature corresponding to the completion of the first crystallization stage (Co Kα). Decomposition of curve 1: 1(black)–experimental spectrum, 2 (red)–summation curve, 3 (blue)–diffuse halo from the amorphous phase (reflections corresponding to the crystalline phase (green) are marked with asterisks).

Note that the formation of different structures under the heating and deformation of the amorphous phase was observed earlier in metallic glasses of an Fe-B system. In amorphous alloys of an Fe-B system of eutectic and hypereutectic compositions, crystallization resulted in the formation of eutectic colonies consisting of α-Fe and Fe_3B. If crystallization occurred during deformation, only nanocrystals of α-Fe(Si) solid solution were formed [24].

The peaks corresponding to the crystalline phase, which was formed under deformation, are broad that conforms to nanocrystals. However, the low intensity of these reflections does not allow the correct determination of their size. Since deformation under HPT is non-uniformly distributed over the sample section, one can assume that the crystalline phases are formed primarily in a subsurface region which is deformed more strongly under HPT. In this case, their distribution over the sample section is non-uniform, and their largest amount is near the surface. To check this assumption, the X-ray diffraction patterns of the deformed sample (e = 6.4) were recorded using harder radiation (Mo). According to the performed calculations, the depth of penetration of X-ray Mo Kα radiation into the regions of wave vectors corresponding to the diffuse maximum is 15 μm. The corresponding value for Co Kα radiation is about 5 μm. Figure 9 illustrates a section of the X-ray diffraction pattern in the region of a diffuse halo for Mo Kα radiation.

There are no reflections from the crystalline phases in this X-ray diffraction pattern; only a diffuse halo from the amorphous phase is present. It is obvious that the fundamental difference between the X-ray diffraction patterns recorded in Co and Mo radiations (Figures 7 and 9), is related to the different depths of X-ray beam penetration into the sample. When using Mo Kα radiation, a thicker layer of the sample takes part in scattering (due to different depths of X-ray beam penetration into the sample). As was mentioned above, the depth of X-ray beam penetration in the region of the main diffuse maximum is 5 and 15 μm for Co and Mo radiations, respectively. If the crystalline phases are formed primarily in subsurface regions, their fraction in the volume of a material, which is analyzed using Mo Kα radiation, will be significantly less. In this case, method sensitivity can be not enough to detect them. In principle, such a non-uniform distribution in the sample section distribution of phases formed under HPT was observed [25,26].

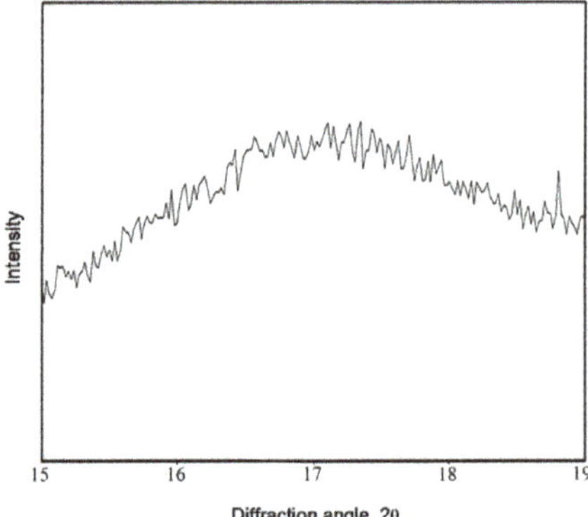

Figure 9. X-ray diffraction pattern of the deformed sample (e = 6.4), (Mo Kα).

To obtain more information on the morphology and structure of the deformed sample, an electron microscope foil was prepared by the method of a focused ion beam. The electron microscope foil was prepared from a subsurface region of the deformed sample. Figure 10 shows a sample microstructure after deformation. In the image, one can see an amorphous structure, to which the mazy contrast corresponds, and ordered regions in it which correspond to nanocrystals. The size of these regions is 1–3 nm. The number of nanocrystals is huge. The images of some (by no means all) nanocrystals are marked with a box. The results of the fast Fourier transformation (FFT) of the marked regions are provided under the image of the structure. Figure numbers at the bottom correspond to the numbers of the marked regions. Reflections corresponding to the crystals can be seen along with the diffuse halo. For comparison, the HREM image of an as-cast amorphous sample is shown in Figure 11. In this image, only mazy contrast is seen. Thus, the data of electron microscopy agree with the results of X-ray studies on nanocrystal formation in the subsurface regions of the$Zr_{55}Cu_{30}Al_{10}Ni_5$ sample under HPT.

Nowadays, the reasons for the formation of different crystalline phases under heating and deformation are not clear. As we stated above, the formation of different structures under the heating and deformation of the amorphous phase was observed in metallic glasses of an Fe-B system (the formation of α-Fe and Fe_3B under heating and of only α-Fe(Si) solid solution under deformation [24]). A similar situation was observed also in alloys of Fe-Zr system [27] where the formation of α-Fe and Fe_3Zr crystals was observed under heating, and only α-Fe crystals were formed under mechanical alloying. It is natural that under heating and deformation the amorphous phase crystallizes under different conditions. Nucleating crystals grow in a homogeneous amorphous matrix under heating, while stresses arise in the amorphous phase under deformation, which are inhomogeneously distributed over a sample. The deformation of amorphous alloys at low temperatures (significantly below the glass transition temperature) is localized and is carried out by the formation and propagation of shear bands. A lot of works [5,11,12,14,23,24,28–31] are devoted to the study of the processes of shear band formation and crystallization under deformation. Nanocrystal formation in shear bands and their vicinity is caused by an enhanced value of the diffusion coefficient in these regions. The reasons for diffusion acceleration are usually related to either a local significant but short-term (~30 ps) temperature rise in the region of deformation localization [32,33] or a decrease in the material density (an increase in the free volume fraction) in a shear band [34,35]. In a number of works, for example, in [30], it is shown that not only shear bands but also compressed and extended regions are formed under deformation. The authors of [30] explained the formation of these regions in the following way. A significant amount of excess free volume is concentrated in these regions during the nucleation of shear bands [36,37].

This is caused by a stretching effect. As a result, viscosity decreases significantly in shear bands and reaches the values typical of a supercooled liquid [38]. Thus, in addition to shear bands, regions are formed under deformation, which include tens of atoms and are characterized by their collective motion. These regions were called shear transformation zones (STZ) [39,40]. The accumulation of excess free volume results in viscous flow. However, since there is an undeformed amorphous matrix around shear bands and STZ, their neighboring regions turn to be under the action of compressive stresses. It was demonstrated in [30] that the rate of homogeneous nanocrystal nucleation in the compressed regions of an amorphous matrix is significantly higher than that in the extended regions. As we stated above, in our work rolling deformation was carried out at room temperature; under rolling, the deformation of subsurface regions is higher than that deeper in a sample. It is these regions where the formation of a larger number of nanocrystals was observed. The nanocrystals formed have the sizes of several nanometers, and their number is large (Figure 10). This corresponds to a high nucleation rate. The formation of small nanocrystals is typical of deformation-induced nanocrystallization [15]. Such a small size of nanocrystals may be related to the fact that they are located close to STZ. Since the STZ size does not exceed several nanometers, the size of compression regions caused by these regions has probably the same scale. Far from STZ, the characteristics of an amorphous matrix change, and the conditions favorable for crystal nucleation and growth disappear. High STZ concentration provides the nucleation of a large number of nanocrystals in compression regions. Such nucleation has similarities with heterogeneous nucleation [41,42], and in this case compression regions are the places of facilitated crystal nucleation.

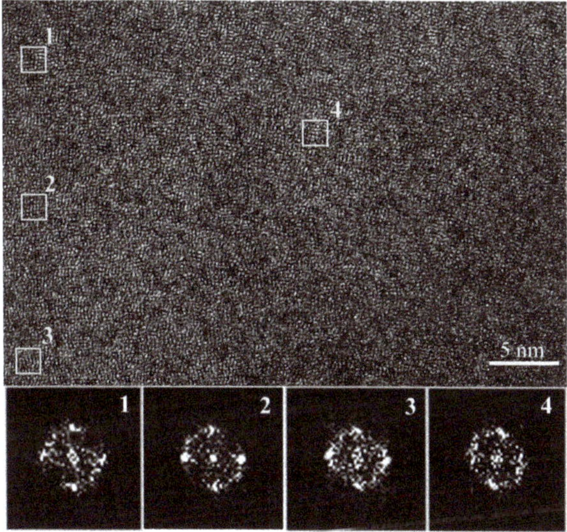

Figure 10. High-resolution transmission electron microscopy (HREM) image of the deformed structure. Bottom of the figure: results of the fast Fourier transformation (FFT) of the marked regions. Figure numbers at the bottom correspond to the numbers of the marked regions.

It was shown earlier that different crystalline phases are formed under the heating and deformation of the investigated alloy. The Zr_2Cu phase with a cubic lattice is the first to be formed under deformation. The formation of a phase of this type (big cube) under deformation of different types was observed earlier, for example, in [28,43], so this is not surprising. It is surprising that the other phase—the metastable phase with a hexagonal lattice—is formed under heating. In principle, this difference may be related to different atomic mobility in the heated and deformed samples. One can assume that due to lower atomic mobility in the sample deformed at room temperature, the composition of the formed crystals is close to that of an amorphous matrix since long-range atom shifts are hindered (polymorphous crystallization-type transformation). A different situation is observed under heating. The crystallization temperature of the alloy is quite high, and atomic mobility is significantly higher at this temperature. Under such conditions, diffusion paths will be significantly longer. In fact, nanocrystals are formed from the state of a supercooled liquid; there are no compression regions that facilitate crystal nucleation. It is natural that the crystallization mechanism should change. One can assume in this case that the metastable hexagonal phase is formed by the primary crystallization mechanism, and its composition differs from that of an amorphous matrix. The formation of several crystalline phases with different compositions at the consequent crystallization stages is the evidence of this assumption. These issues require further research.

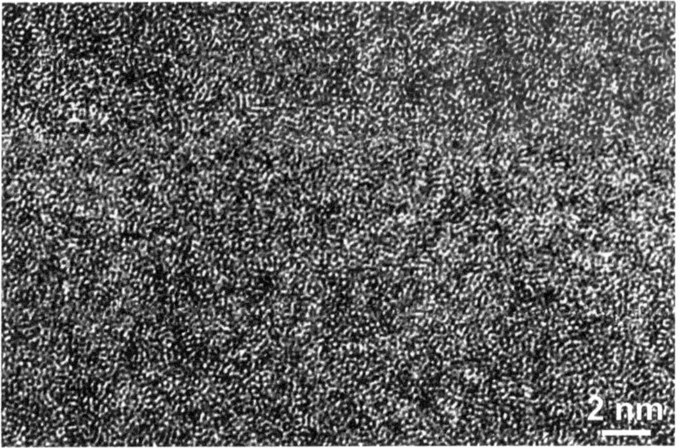

Figure 11. HREM image of an as-prepared amorphous sample.

4. Conclusions

A comparative study of the processes of devitrification of $Zr_{55}Cu_{30}Al_{15}Ni_5$ bulk alloy under heat treatment and deformation has been carried out.

The activation energy of crystallization, determined by the Kissinger method, is 278 kJ/mol.

It has been shown that different crystalline phases are formed during crystallization under heating and deformation. At the first crystallization stage, the metastable phase with a hexagonal structure (lattice of space group $P6_3/mmc$ with the parameters a = 8.66 Å, c = 14.99 Å) is formed under heat treatment. When the temperature rises, the metastable phase decays with the formation of stable crystalline phases. The well-known crystalline Zr_2Cu phase with the lattice of space group Fd3m is formed during crystallization under the action of deformation.

It has been determined that during deformation, nanocrystals are formed primarily in the subsurface regions of the samples. The nanocrystal size is several nanometers.

Author Contributions: Conceptualization, A.A.; sample preparation, D.G.; X-ray diffraction experiments, G.A. and B.G.; deformation, heat treatment, A.D. and B.M.; electron microscopy study, A.A., E.P. and D.M.; DTA experiments, I.S.; all authors analyzed the data, discussed the results and wrote the paper. All authors have read and agreed to the published version of the manuscript.

Funding: Russian foundation for Basic Research (partially).

Acknowledgments: The work was partially supported by RFBR (grants 19-03-00355 and 20-08-00497).

References

1. Duwez, P.; Willens, R.H.; Klement, W. Continuous Series of Metastable Solid Solutions in Silver-Copper Alloys. *J. Appl. Phys.* **1960**, *31*, 1136. [CrossRef]
2. Luo, Y.; Ke, H.; Zeng, R.; Liu, X.; Luo, J.; Zhang, P. Crystallization behavior of $Zr_{60}Cu_{20}Fe_{10}Al_{10}$ amorphous alloy. *J. Non-Cryst. Solids* **2020**, *528*, 119728. [CrossRef]
3. Li, H.; Zheng, Y. Recent advances in bulk metallic glasses for biomedical applications. *Acta Biomater.* **2016**, *36*, 1–20. [CrossRef] [PubMed]
4. Stoica, M.; Das, J.; Bednarcik, J.; Franz, H.; Mattern, N.; Wang, W.H.; Eckert, J. Strain distribution in $Zr_{64.13}Cu_{15.75}Ni10._{12}Al_{10}$ bulk metallic glass investigated byin situtensile tests under synchrotron radiation. *J. Appl. Phys.* **2008**, *104*, 013522. [CrossRef]
5. Stolpe, M.; Kruzic, J.; Busch, R. Evolution of shear bands, free volume and hardness during cold rolling of a Zr-based bulk metallic glass. *Acta Mater.* **2014**, *64*, 231–240. [CrossRef]
6. Guo, W.; Shao, Y.; Saida, J.; Zhao, M.; Lü, S.; Wu, S. Rejuvenation and plasticization of Zr-based bulk metallic glass with various Ta content upon deep cryogenic cycling. *J. Alloys Compd.* **2019**, *795*, 314–318. [CrossRef]
7. Abrosimova, G.E.; Aronin, A.S.; Kir'janov, Y.V.; Matveev, D.V.; Zver'kova, I.I.; Molokanov, V.V.; Pan, S.; Slipenyuk, A. The structure and mechanical properties of bulk $Zr_{50}Ti_{16.5}Cu_{14}Ni_{18.5}$ metallic glass. *J. Mater. Sci.* **2001**, *36*, 3933–3939. [CrossRef]
8. Abrosimova, G.; Aronin, A.; Matveev, D.; Molokanov, V. Formation and structure of nanocrystals in bulk стеклеZr$_{50}$Ti$_{16}$Cu$_{15}$Ni$_{19}$ metallic glass. *Phys. Solid State* **2004**, *46*, 2191–2195. [CrossRef]
9. Saida, J.; Kato, H.; Inoue, A. Primary precipitation of icosahedral quasicrystal with rearrangement of constitutional elements in Zr65Al7.5Cu27.5 glassy alloy with low oxygen impurity. *J. Mater. Res.* **2005**, *20*, 303–306. [CrossRef]
10. Zhao, X.; Pang, S.; Ma, C.; Zhang, T. Precipitation of Icosahedral Phase in Zr-Ni-Nb-Cu-Al Metallic Glasses. *Mater. Trans.* **2009**, *50*, 1838–1842. [CrossRef]
11. Wilde, G.; Rösner, H. Nanocrystallization in a shear band: An in situ investigation. *Appl. Phys. Lett.* **2011**, *98*, 251904. [CrossRef]
12. Abrosimova, G.; Aronin, A.; Barkalov, O.; Matveev, D.; Rybchenko, O.; Maslov, V.; Tkach, V. Structural transformations in the Al85Ni6.1Co2Gd6Si0.9 amorphous alloy during multiple rolling. *Phys. Solid State* **2011**, *53*, 229–233. [CrossRef]
13. Aronin, A.; Louzguine-Luzgin, D.V. On nanovoids formation in shear bands of an amorphous Al-based alloy. *Mech. Mater.* **2017**, *113*, 19–23. [CrossRef]
14. Glezer, A.M.; Permyakova, I.E.; Manaenkov, S.E. Plasticizing effect in the transition from an amorphous state to a nanocrystalline state. *Dokl. Phys.* **2008**, *53*, 8–10. [CrossRef]
15. Abrosimova, G.; Matveev, D.; Pershina, E.; Aronin, A. Effect of treatment conditions on parameters of nanocrystalline structure in Al-based alloys. *Mater. Lett.* **2016**, *183*, 131–134. [CrossRef]
16. Aronin, A.; Matveev, D.; Pershina, E.; Tkatch, V.; Abrosimova, G. The effect of changes in Al-based amorphous phase structure on structure forming upon crystallization. *J. Alloys Compd.* **2017**, *715*, 176–183. [CrossRef]
17. Zhilyaev, A.; Langdon, T. Using high-pressure torsion for metal processing: Fundamentals and applications. *Prog. Mater. Sci.* **2008**, *53*, 893–979. [CrossRef]
18. Liu, L.; Wu, Z.; Zhang, J. Crystallization kinetics of $Zr_{55}Cu_{30}Al_{10}Ni_5$ bulk amorphous alloy *J. Alloys Compd.* **2002**, *339*, 90–95. [CrossRef]
19. Tariq, N.; Iqbal, M.; Shaikh, M.; Akhter, J.I.; Ahmad, M.; Ali, G.; Xu, M. Evolution of microstructure and non-equilibrium phases in electron beam treated $Zr_{55}Cu_{30}Al_{10}Ni_5$ bulk amorphous alloy. *J. Alloys Compd.* **2008**, *460*, 258–262. [CrossRef]
20. Tao, P.J.; Yang, Y.Z.; Bai, X.J.; Mu, Z.X.; Li, G. Non-Isothermal Crystallization Behavior in $Zr_{55}Cu_{30}Ni_5Al_{10}$ Bulk Metallic Glass. *Adv. Mater. Res.* **2010**, *146*, 560–564. [CrossRef]
21. Gao, Y.-L.; Shen, J.; Sun, J.-F.; Wang, G.; Xing, D.-W.; Xian, H.-Z.; Zhou, B.-D. Crystallization behavior of ZrAlNiCu bulk metallic glass with wide supercooled liquid region. *Mater. Lett.* **2003**, *57*, 1894–1898. [CrossRef]
22. Yavari, A.; Le Moulec, A.; Botta, W.J.; Inoue, A.; Rejmankova, P.; Kvick, A. In Situ crystallization of $Zr_{55}Cu_{30}Al_{10}Ni_5$ bulk glass forming from the glassy and undercooled liquid states using synchrotron radiation. *J. Non-Cryst. Solids* **1999**, *247*, 31–34. [CrossRef]

23. Yavari, A.; Le Moulec, A.; Inoue, A.; Botta, W.J.; Vaughan, G.; Kvick, A. Metastable phases in Zr-based bulk glass-forming alloys detected using a synchrotron beam in transmission. *Mater. Sci. Eng. A* **2001**, *304–306*, 34–38. [CrossRef]

24. Aronin, A.; Abrosimova, G.; Matveev, D.; Rybchenko, O. Structure and properties of nanocrystalline alloys prepared by high pressure torsion. *Rev. Adv. Mater. Sci.* **2010**, *25*, 52–57.

25. Ubyivovk, E.V.; Boltynjuk, E.; Gunderov, D.; Churakova, A.A.; Kilmametov, A.; Valiev, R. HPT-induced shear banding and nanoclustering in a TiNiCu amorphous alloy. *Mater. Lett.* **2017**, *209*, 327–329. [CrossRef]

26. Abrosimova, G.; Aronin, A. Nanocrystal formation in Al- and Ti-based amorphous alloys at deformation. *J. Alloys Compd.* **2018**, *747*, 26–30. [CrossRef]

27. Trudeau, M.L. Deformation induced crystallization due ti instability in amorphous FeZr alloys. *Appl. Phys. Lett.* **1994**, *64*, 3661–3663, doiorg/101063/1111953. [CrossRef]

28. El-Eskandarany, M.S.; Saida, J.; Inoue, A. Room-temperature mechanical induced solid-state devitrification od glassy $Zr_{65}Al_{7.5}Ni_{10}Cu_{12.5}Pd_5$ alloy. *Acta. Mater.* **2003**, *51*, 4519–4532. [CrossRef]

29. Yavari, A.R.; Georgarakis, K.; Antonowicz, J.; Stoica, V.; Nishiyama, N.; Vaughan, G.; Chen, M.; Pons, M. Crystallization during Bending of a Pd-based metallic glass detected by X-ray microscopy. *Phys. Rev. Lett.* **2012**, *109*, 085501. [CrossRef]

30. Yan, Z.H.; Song, K.; Hu, Y.; Dai, F.; Chu, Z.; Eckert, J. Localized crystallization in shear bands of a metallic glass. *Sci. Rep.* **2016**, *6*, 19358. [CrossRef]

31. Abrosimova, G.; Aronin, A.; Fokin, D.; Orlova, N.; Postnova, E. The decrase of Young's modulus in shear bands of amorphous $Al_{87}Ni_8Gd_5$ alloy. *Mater. Lett.* **2019**, *252*, 114–116. [CrossRef]

32. Lewandowski, J.J.; Greer, A.L. Temperature rise at shear bands in metallic glasses. *Nat. Mater.* **2006**, *5*, 15–18. [CrossRef]

33. Georgarakis, K.; Aljerf, M.; Li, Y.; LeMoulec, A.; Charlot, F.; Yavari, A.R.; Chornokhvostenko, K.; Tabachnikova, E.; Evangelakis, G.A.; Miracle, D.B.; et al. Shear band melting and serrated flow in metallic glasses. *App. Phys. Lett.* **2008**, *93*, 031907. [CrossRef]

34. Schmidt, V.; Rösner, H.; Peterlechler, M.; Wilde, G. Quantitative Measurement of Density in a Shear Band of Metallic Glass Monitored Along its Propagation Direction. *Phys. Rev. Lett.* **2015**, *115*, 035501. [CrossRef]

35. Greer, A.L.; Cheng, Y.Q.; Ma, E. Shear bands in metallic glasses. *Mater. Sci. Eng.* **2013**, *74*, 71–132. [CrossRef]

36. Flores, K.M.; Suh, D.; Dauskardt, R.H. Characterization of free volume in a bulk metallic glass using positron annihilation spectroscopy. *J. Mater. Res.* **2002**, *17*, 1153–1161. [CrossRef]

37. Kanungo, B.P.; Gladeb, S.C.; Asoka-Kumarb, P.; Floresa, K.M. Characterization of free volume changes associated with shear band formation in Zr- and Cu-based bulk metallic glasses. *Intermetallics* **2004**, *12*, 1073–1080. [CrossRef]

38. Huo, L.S.; Ma, J.; Ke, H.B.; Bai, H.Y.; Zhao, D.Q.; Wang, W.H. The deformation units in metallic glasses revealed by stress-induced localized glass transition. *J. Appl. Phys.* **2012**, *111*, 113522. [CrossRef]

39. Tang, X.P.; Geyer, U.; Busch, R.; Johnson, W.L.; Wu, Y. Diffusion mechanisms in metallic supercooled liquids and glasses. *Nature* **1999**, *402*, 160–162. [CrossRef]

40. Bokeloh, J.; Divinski, S.V.; Reglitz, G.; Wilde, G. Tracer measurements of atomic diffusion inside shear bands of a bulk metallic glass. *Phys. Rev. Lett.* **2011**, *107*, 235503. [CrossRef]

41. Foley, J.C.; Allen, D.R.; Perepezko, J.H. Analysis of nanocrystal development in Al-Y-Fe and Al-Sm glasses. *Scr. Mater.* **1996**, *35*, 655–669. [CrossRef]

42. Abrosimova, G.E.; Aronin, A.S. Effect of the Concentration of a rare-earth component on the parameters of the nanocrystalline structure in aluminum-based alloys. *Phys. Solid State* **2009**, *51*, 1765–1771. [CrossRef]

43. El-Eskandarany, M.S.; Inoue, A. Solid-state crystalline-glassy cyclic phase transformations of mechanically alloyed $Cu_{33}Zr_{67}$ powders. *Met. Trans. A* **2002**, *33*, 135. [CrossRef]

Article

Amorphous-Nanocrystalline Composites Prepared by High-Pressure Torsion

Inga Permyakova [1,*] and Alex Glezer [1,2]

[1] Scientific Center of Metal Science and Metal Physics, Bardin Central Research Institute of Ferrous Metallurgy, 105005 Moscow, Russia; a.glezer@mail.ru

[2] Department of Physical Materials, National University of Science and Technology "MISIS", 119049 Moscow, Russia

* Correspondence: inga_perm@mail.ru; Tel.: +7-495-777-93-50

Received: 23 February 2020; Accepted: 14 April 2020; Published: 15 April 2020

Abstract: This article presents systematic studies of the preparation method and the specific features of the changes in the structure and properties of amorphous-nanocrystalline composites formed from melt-quenched ribbons of iron- and cobalt-based amorphous alloys and the Cu-Nb crystalline nanolaminates by severe plastic deformation by torsion in the Bridgeman chamber at high quasi-hydrostatic pressure.

Keywords: composite; amorphous alloy; nanolaminate; severe plastic deformation; high-pressure torsion; hardness; crack resistance

---

1. Introduction

The development of the basic principles for creating new composite materials is undoubtedly an important task of modern materials science. The demand for such materials is due to the possibility of combining enhanced mechanical properties (strength, wear resistance, crack resistance, stiffness, heat resistance and fatigue limit) [1–3].

Along with natural composites (mollusk shells, bones, wood), there are man-made artificial composites based on polymer, metal and ceramic matrices reinforced with fibers or filled with disperse particles. Such variety allows one to permanently expand the application field for composite materials.

Comprehensive studies have shown that enhanced mechanical properties can be achieved in multilayer composite systems consisting of amorphous and crystalline materials [4–9]. In addition, metallic materials with discrete structure constituents of nanoscale range can combine increased ductility with sufficiently high strength characteristics. This differs amorphous-nanocrystalline composites and advanced nanocrystalline materials from conventional structural materials produced by conventional technologies. Moreover, the controlled transformation of amorphous into nanocrystalline state makes it possible to successfully manage the functional properties of amorphous-nanocrystalline composites [10–13]. There is a well-known Ulitovsky-Taylor method for manufacturing one-dimensional composite material consisting of a high-strength magnetically soft metal base with an amorphous and/or nanocrystalline structure and an outer glass shell [14,15]. Another way to obtain amorphous-nanocrystalline composites is to decrease the critical rate of melt cooling upon manufacture of amorphous alloys (AA) by melt spinning method [16]. Heat treatment of AA at controlled temperature and time parameters initiates the nucleation and growth of nanocrystals, i.e., the formation of a composite material with an amorphous matrix [17–19]. It is possible to modify the AA surface by laser for the manufacture of "sandwich" amorphous-nanocrystalline composites and gradient structures with amorphous-nanocrystalline components [20–22]. The composite consisting of an amorphous phase with a nanocrystalline filler can surpass fully amorphous or fully crystalline analogs in the combination of properties [23–26].

As is known, high-pressure torsion (HPT) allows one to create new structure states by the consolidation of small fractions, due to the occurrence of the "crystalline-amorphous" phase transitions in the material [27–29]. In the framework of the present work, the idea arose to use the possibilities of severe plastic deformation (SPD) in a Bridgman chamber for the preparation of layered amorphous-nanocrystalline composites. We used two fundamentally different methodological techniques (Figure 1):

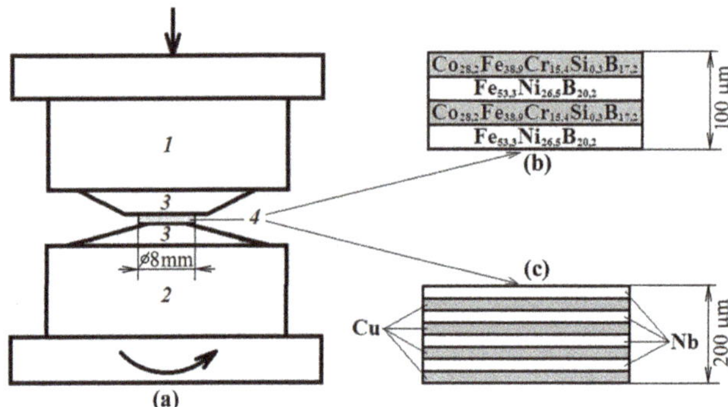

Figure 1. (**a**) Severe plastic deformation (SPD) scheme for the composite preparation: Bridgman chamber consisting of immobile *1* and movable *2* anvils, carbide inserts *3*, sample *4* of alternating amorphous alloys (AA) ribbons (**b**) or the Cu-Nb nanolaminate (**c**).

(1) HPT consolidation of AA melt spun ribbons differing in chemical compositions and mechanical properties [30];

(2) the use of Bridgman chamber for deformation processing of nanocrystalline Cu-Nb laminates prepared by multiple pack rolling (MPR) [31].

AA, due to the presence of a homogeneous structure and the absence of defects (dislocations and grain boundaries), demonstrate a higher level of mechanical properties that exceeds the level of properties achieved in the crystalline alloys. However, these materials have a serious flaw—the absence of tensile plasticity and low plastic deformation under compressive loads—which makes them prone to brittle fracture, and, accordingly, greatly limits their possible application. Structure changing during crystallization is an important aspect of AA research, since one of the ways to increase the ductility of AA is the formation of composite structure "glass-crystal".

Ternary AA of the Fe-Ni-B system (for example, $Fe_{58}Ni_{25}B_{17}$, $Fe_{53.3}Ni_{26.5}B_{20.2}$, $Fe_{50}Ni_{33}B_{17}$) are model alloys. External actions on the Fe-Ni-B AA cause precipitation of a Fe-Ni nanocrystalline phase, which can vary in crystal lattice types (BCC, FCC) depending on the ratio between the iron and nickel concentrations. Thus, it possible to establish the effect of the type crystal lattice of nanocrystals on the mechanical behavior of materials with an amorphous-nanocrystalline structure. In addition, partial crystallization can favor changes in their soft magnetic characteristics.

The Co-Fe-Cr-Si-B AA is related to corrosion resistant materials and exhibits a high electrical resistivity, low magnetization-reversal loss over a wide frequency range, low coercive force and resistance to impacts and vibrations. The high-cobalt amorphous alloys are characterized by near-zero saturation magnetostriction ($\lambda_S \leq 10^{-7}$) and very high magnetic permeability. For this reason, such AA show promise as materials for magnetic shields [23].

Multilayer Cu-Nb laminates with nanoscale layer thicknesses are typical representatives of nanostructured composite materials with a unique combination of properties: good ductility, high electrical conductivity of copper and superconductivity of niobium. The combination of the copper-niobium system is demanded, and is actively used in the manufacture of microwires in resonant power transmission systems, inductors for magnetic pulse stamping and welding, foils in electronics for flexible printed circuits; in large magnetic systems at 50–100 T and in high-field cryogenic synchronizers of industrial frequency.

2. Materials and Methods

In the first case, the $Fe_{53.3}Ni_{26.5}B_{20.2}$ and $Co_{28.2}Fe_{38.9}Cr_{15.4}Si_{0.3}B_{17.2}$ (at. %) melt quenched AA ribbons 25 μm thick were taken as objects of the study and composite constituents (Figure 1b). The total thickness of the initial sample (before HPT), consisting of 4 alternating layers of ribbons of AA, was 100 μm, respectively.

In this case, SPD was performed by HPT (P = 6 GPa, v = 1 rpm) in a Bridgman chamber to different degrees of deformation preset by varying the number of revolutions (N) of rotating anvil from 1/2 to 9. Before HPT, the AA ribbons were cut into fragments 1 cm × 1 cm in size, and then the fragments were piled in groups of four and deformed in the Bridgman chamber to a given number of revolutions. Ethanol was applied to the degreasing of the ribbons surface before the HPT. In such a way, the deformation-induced disk samples were formed from each alloy and similar composite samples were formed from alternating $Fe_{53.3}Ni_{26.5}B_{20.2}$ and $Co_{28.2}Fe_{38.9}Cr_{15.4}Si_{0.3}B_{17.2}$ AA layers. The amorphous and crystalline phases in the alloys and composites were identified by transmission electron microscopy (TEM) with a JEM 1400 microscope (Jeol Ltd., Tokyo, Japan) at an accelerating voltage of 120 kV and by X-ray diffraction (XRD) analysis with an Ultima IV multifunctional diffractometer (Rigaku Corp, Tokyo, Japan) with CoK_α radiation. The microhardness of the disk samples was measured at 1/2 radius by indentation with a Vickers pyramid using a MHT-3M microhardness tester (Lomo, St. Petersburg, Russia) at a load of 0.40 N by a standard technique. K_{1c} values were calculated by the formula:

$$K_{1c} = A(E/HV)^{1/2}P/C^{3/2} \tag{1}$$

where A = 0.016 is the calibration coefficient of proportionality for thin ribbons of amorphous alloys; E is Young's modulus measured by dynamic indentation methods; HV is Vickers microhardness; P is the critical load for the appearance of radial cracks in the process of local loading of samples of amorphous alloys; C is the average length of cracks [32,33]. The indentation of amorphous alloys was carried out only in the plane of the sample.

The magnetic properties were measured at room temperature in fields of up to 20 kOe with a VSM 250 vibrating-sample magnetometer (Xiamen Dexing Magnet Tech. Co., Ltd., Xiamen, China). Measured hysteresis loops were used to determine the saturation magnetization (σ_s) and the coercive force (H_c).

In the second case, the initial Cu-Nb nanolaminates (Figure 1c) were produced by a series of the following sequential operations making up technological cycle of the MPR process [34,35]: assembling a pack with a given number of layers, rolling of the pack in vacuum at a temperature of 750–800 °C, and then cold rolling in air to a thickness equal to that of a single initial layer of the composite. This procedure is more efficient than that reported in [36,37]. In addition, the holding time of the compacted pack at high temperature within this technological scheme is much shorter than that is in the case of diffusion welding. The initial plates of 50 mm × 100 mm in area and 0.35 mm thick were assembled into a pack of 32 alternating copper and niobium layers. The total degrees of reduction were 40% upon vacuum rolling and 10% upon cold rolling.

The prepared nanolaminates were subjected to HPT in Bridgman anvils at P = 4 GPa to 1/2–4 revolutions. Before HPT, the samples were cut to fragments of 1 cm × 1 cm in area. The thickness of the Cu-Nb laminate samples (before HPT) was 200 μm (Figure 1c).

The structure changes in the samples were examined by the methods of transmission electron microscopy (TEM) and scanning electron microscopy (SEM) with a JEM 2100 microscope (Jeol Ltd., Tokyo, Japan) equipped with a BSE detector. The chemical composition of the elements and their distribution in the specimens subjected to SPD were determined by the methods of bright-field and dark-field scanning transmission electron microscopy (BF-STEM/DF-STEM) and energy dispersive

X-ray spectroscopy (EDS) with a JEM ARM-200F cold-emission microscope (Jeol Ltd., Tokyo, Japan) equipped with a CENTURIO EDX E-Max EDS detector (Jeol Ltd., Tokyo, Japan). The foils for the high-resolution electron-microscopic examinations were prepared from cross-section samples by a standard procedure [38]. The XRD spectra were obtained with an Ultima IV multifunction diffractometer (Rigaku Corp, Tokyo, Japan), using copper emission and a K_β filter (Ni). The Vickers hardness was measured in the half-radius region of consolidated disk-shaped specimens using a standard procedure with a microhardness tester in three dimensions (3D). For this purpose, the disk-shaped nanocomposite samples were cut in four equal segments, and their flat surfaces and two orthogonally related butt-ends after polishing were subjected to indentation.

3. Results and Discussion

3.1. Amorphous-Nanocrystalline Composites Prepared by Consolidation of Two Amorphous Alloys upon HPT

At the initial SPD stages, the consolidation of AA ribbon samples of the same composition, $Fe_{53.3}Ni_{26.5}B_{20.2}$ or $Co_{28.2}Fe_{38.9}Cr_{15.4}Si_{0.3}B_{17.2}$, is difficult because of weak adhesion between the layers. Monolithic samples without delamination into individual ribbon components were obtained only after HPT to $N = 3$ revolutions. By contrast, the composites consisting of alternating $Fe_{53.3}Ni_{26.5}B_{20.2}$ and $Co_{28.2}Fe_{38.9}Cr_{15.4}Si_{0.3}B_{17.2}$ AA ribbons exhibit adequate adhesiveness of the layers starting from $N = 1$.

According to the XRD data for the $Fe_{53.3}Ni_{26.5}B_{20.2}$ AA, no clear changes (except for a decrease in the intensity of peak corresponding to the first halo) in the XRD patterns are observed as the number of revolution increases (Figure 2a). The $Co_{28.2}Fe_{38.9}Cr_{15.4}Si_{0.3}B_{17.2}$ AA subjected to HPT to $N = 2$ exhibits the onset of crystallization (Figure 2b), where the volume fraction of crystalline phase is $V_{cr} \approx 10\%$. The crystallite peaks disappear at $N = 3$ and appear again at $N = 5$, at which $V_{cr} \approx 16\%$.

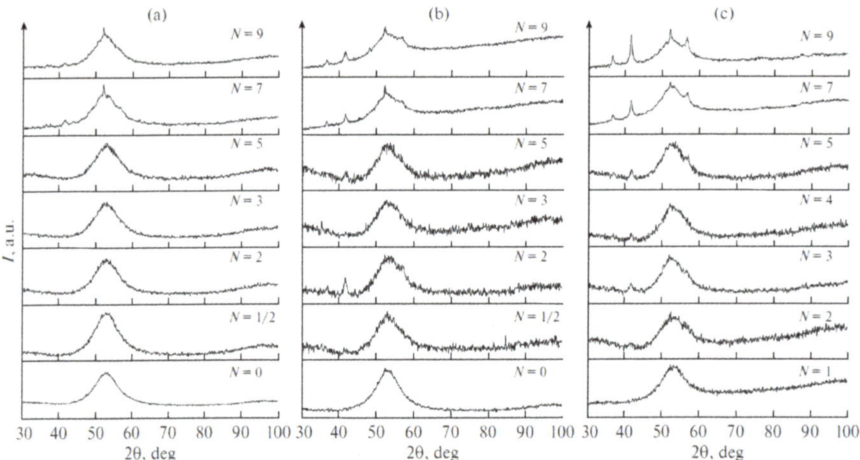

Figure 2. X-ray diffraction (XRD) patterns of multilayer AA samples subjected to high-pressure torsion (HPT): (**a**) $Fe_{53.3}Ni_{26.5}B_{20.2}$, (**b**) $Co_{28.2}Fe_{38.9}Cr_{15.4}Si_{0.3}B_{17.2}$ and (**c**) composite material.

Figure 2c shows the XRD patterns for the composites subjected to HPT. At $N = 2$, the peak intensity corresponding to the first halo substantially decreases. With an increasing degree of deformation, the composites undergo a partial crystallization. The volume fraction of crystalline phase decreases at $N = 4$; $V_{cr} \approx 18\%$, 9% and 19% at $N = 3$, 4 and 5, respectively. Thus, up to $N = 5$, we failed to transfer the alloys and the composites into a completely nanocrystalline state. More intense crystallization was detected at higher degrees of deformation ($N = 7$–9) (Figure 2).

To refine the phase transformations, and to identify the formed crystal structures, we supplemented the XRD data by TEM examination results. The TEM data confirmed that the $Fe_{53.3}Ni_{26.5}B_{20.2}$ and $Co_{28.2}Fe_{38.9}Cr_{15.4}Si_{0.3}B_{17.2}$ AA remain completely amorphous at low degree of deformation.

First nanocrystals appear in the amorphous phase in the $Fe_{53.3}Ni_{26.5}B_{20.2}$ AA at $N = 3$ at the edge of the disk samples (Figure 3a). Such crystals correspond to α-Fe (Im3m, $a = b = c = 0.2857$ nm). At the late stages of deformation, the volume fraction of the crystalline phase substantially increases (Figure 3b), and the Fe_2B (I4/mcm, $a = b = 0.5099$ nm, $c = 0.4240$ nm) and FeB (Pnma, $a = 0.4053$ nm, $b = 0.5495$ nm, $c = 0.2946$ nm) boride precipitates are also observed. The nanocrystallite size is 10–40 nm.

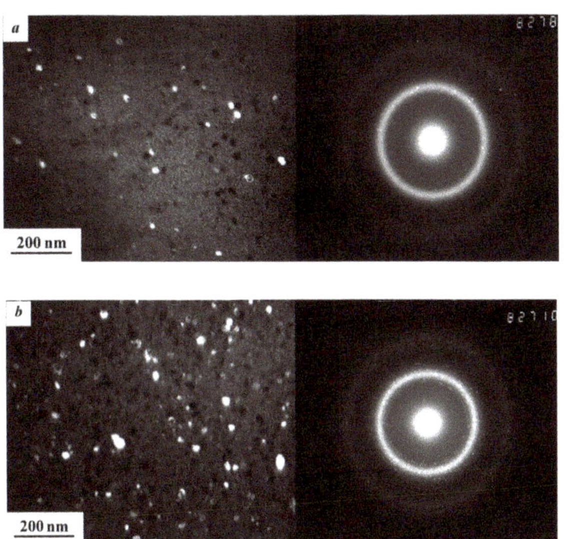

Figure 3. Dark-field images of the of $Fe_{53.3}Ni_{26.5}B_{20.2}$ AA structure subjected to HPT to $N = 3$ (**a**) and $N = 9$ (**b**) and the corresponding selected area electron diffraction (SAED) patterns.

The first evidences of crystallization in the $Co_{28.2}Fe_{38.9}Cr_{15.4}Si_{0.3}B_{17.2}$ AA are observed at $N = 2$; local signs are also observed at the edge of the samples (Figure 4). The extended and coarse-grained phases precipitated from the amorphous matrix were identified as α-Fe with BCC crystal lattice (Figure 4a) and α-Co with HCP lattice (Figure 4b).

With increasing degree of deformation ($N \geq 3$), V_{cr} of fine crystalline phase in $Co_{28.2}Fe_{38.9}Cr_{15.4}Si_{0.3}B_{17.2}$ AA increases from the center of the sample to its periphery. Deformation to $N \geq 7$ leads to the formation of homogeneous structure throughout the sample. The nanocrystal size (5–25 nm) in the $Co_{28.2}Fe_{38.9}Cr_{15.4}Si_{0.3}B_{17.2}$ AA is unchanged upon deformation to $N = 4$–9. The crystalline particles correspond to α-Co (P6$_4$/mmc(A3), $a = b = 0.2514$ nm, $c = 0.4105$ nm) and the Fe_2B and FeB borides. The deformation-induced composites prepared from the Fe-Ni-B and Co-Fe-Cr-Si-B AA layer-by-layer packets at $N = 2$ exhibit the following specific feature, which is similar to that of the $Co_{28.2}Fe_{38.9}Cr_{15.4}Si_{0.3}B_{17.2}$ AA: coarse-grained crystalline areas are observed at 1/2 radius of the disk samples. Similar grains are formed in the case where the crystallization of the $Co_{28.2}Fe_{38.9}Cr_{15.4}Si_{0.3}B_{17.2}$ AA occurs upon high-temperature annealing.

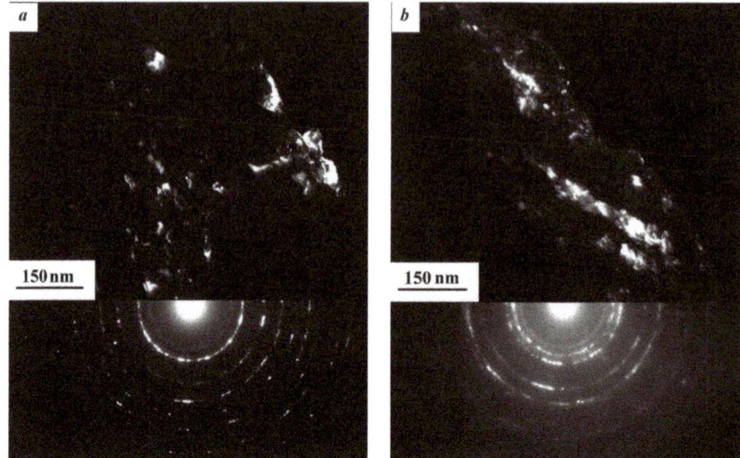

Figure 4. Dark-field images (TEM) and SAED patterns of α-Fe (**a**) and α-Co (**b**) in the $Co_{28.2}Fe_{38.9}Cr_{15.4}Si_{0.3}B_{17.2}$ AA subjected to HPT to $N = 2$.

The shear bands (SB) observed in the composites formed by HPT up to $N = 5$ exhibit the occurrence of highly localized plastic deformation (Figure 5). The SB become thinner with increasing degree of deformation.

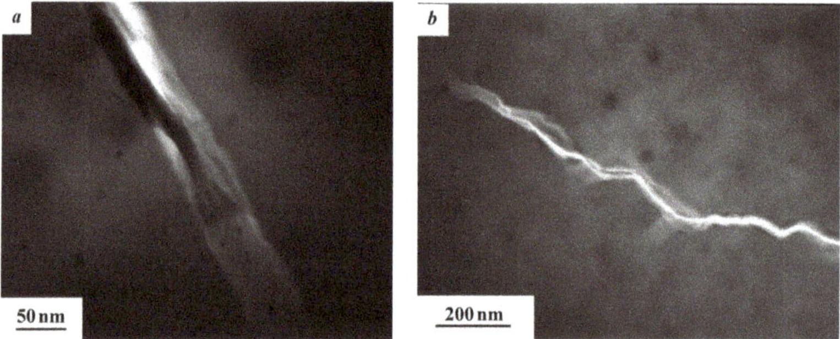

Figure 5. Shear bands in the composites formed from AA and subjected to HPT at $N = 3$ (**a**) and $N = 5$ (**b**).

Because of the small size of the nanocrystals formed upon HPT, the priority mechanisms of their interaction with SB (in accordance with the classification proposed in earlier works [39,40]) are suggested to be the "absorption" mechanism, at which moving SB absorbs nanoparticles without changing the motion path in the amorphous matrix (Figure 5a), and the "accommodation" mechanism, at which meeting of SB with an elastically stressed nanocrystal initiates the exit of one or several secondary SB to the amorphous matrix (Figure 5b).

In composites, an increase in the degree of deformation increases the volume fraction of the nanocrystalline phase and decreases the nanocrystal size (Figure 6).

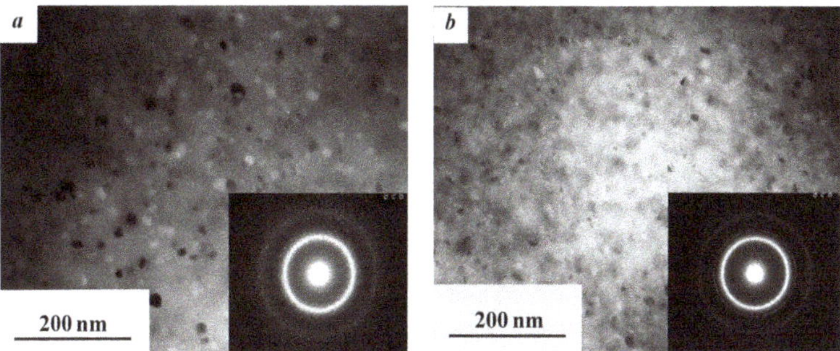

Figure 6. Bright-field images (TEM) of the composite structure after HPT to $N = 3$ (**a**) and $N = 7$ (**b**) and the corresponding SAED patterns.

At $N = 9$, the composite passes into a completely nanocrystalline state (Figure 7). The precipitated phases were identified as BCC α-Fe, HCP α-Co and the Fe_2B and Co_2B borides.

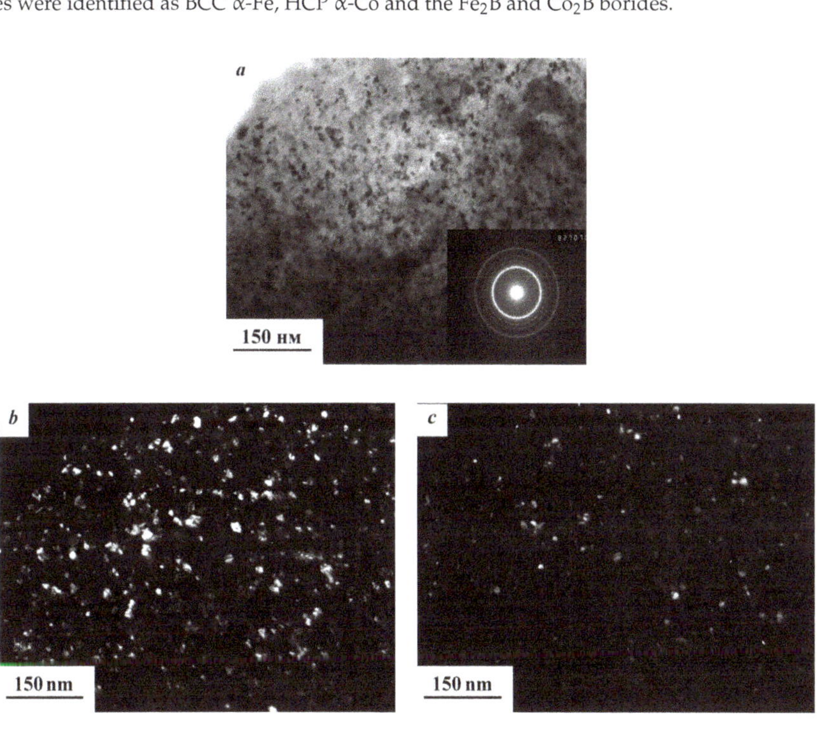

Figure 7. Bright-field image (TEM) of the composite after HPT to $N = 9$ and the corresponding SAED pattern (**a**); and dark-field images (TEM) from the 1st (**b**) and 4th circles (**c**) of SAED pattern.

After small deformation to $N = 2$, the microhardness HV of the composite decreases by 16% (Figure 8). The minimum HV of the $Fe_{53.3}Ni_{26.5}B_{20.2}$ AA also corresponds to $N = 2$. The microhardness of the $Co_{28.2}Fe_{38.9}Cr_{15.4}Si_{0.3}B_{17.2}$ AA decreases progressively over a wider strain range, and HV_{min} is reached at $N = 5$ (Figure 8). This effect can be related to the changes occurring upon HPT in the topological and chemical short-range orders before crystallization. Moreover, HPT initiates the formation of numerous SB in the amorphous matrix [41]. As is known, the presence of SB decreases the strength of AA and substantially facilitates plastic flow. We should also note an important role of migration processes and the redistribution of excess free volume regions near and inside SB.

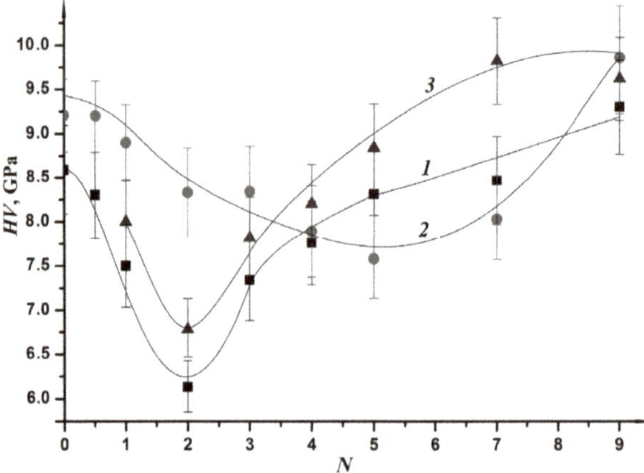

Figure 8. Microhardness of the materials as a function of number of revolutions upon HPT: 1–Fe$_{53.3}$Ni$_{26.5}$B$_{20.2}$ AA; 2–Co$_{28.2}$Fe$_{38.9}$Cr$_{15.4}$Si$_{0.3}$B$_{17.2}$ AA; 3–composite formed from different AA.

Figure 8 shows a very interesting and important effect related to the fact that the average microhardness (strength) of the composite obtained by HPT to $N > 4$ (curve 3) exceeds microhardnesses of the individual components (curves 1 and 2), from which it is formed. If we follow the additivity rule, then microhardness of the composite should be between microhardnesses of its components. The observed synergic effect can be associated with the interdiffusion of the components of amorphous precursors upon HPT, and the formation of a new composition of amorphous or nanocrystalline state in the region of the interface between two amorphous layers. Such a new state can be anomalously hard in a fairly extended region of the contacting initial phases.

Preliminary studies demonstrated a plasticizing effect of heat treatment on the AA under consideration. The K_{1c} parameter of the Fe$_{53.3}$Ni$_{26.5}$B$_{20.2}$ and Co$_{28.2}$Fe$_{38.9}$Cr$_{15.4}$Si$_{0.3}$B$_{17.2}$ AA increases at temperatures ranging from 425 °C to 445 °C and from 425 °C to 485 °C, respectively. With allowance for this result, the behavior of the K_{1c} parameter after HPT was studied for both AA separately and for deformation-induced composites based on them (Figure 9). For plastic deformation values up to $N = 2$ revolutions, K_{1c} may depend on the orientation of layers. At high degrees of plastic deformation by HPT, intensive mixing of the layers occurs and the K_{1c} anisotropy will be leveled.

At low degrees of deformation, K_{1c} of the Fe$_{53.3}$Ni$_{26.5}$B$_{20.2}$ AA, Co$_{28.2}$Fe$_{38.9}$Cr$_{15.4}$Si$_{0.3}$B$_{17.2}$ AA, and the composite increases by 5.6% with a maximum at $N = 2$, by 6% with a maximum at $N = 3$–4, and by 7.3% with a maximum at $N = 4$, respectively. Then, the crack resistance decreases for the composite and its individual components approximately by a factor of 1.3–1.4, relative to the initial levels. However, the comparison of K_{1c} at $N = 7$–9 (Figure 9) and K_{1c} at high temperatures corresponding to the complete crystallization of the initially amorphous alloys shows that the crack resistance of the nanocrystalline structure obtained by SPD is higher than that of the coarse-grained structure initiated by annealing. In addition, even at high degrees of deformation ($N = 7$–9), K_{1c} of the samples remain comparable with those of zirconia "ceramic steel" or laminar (layered) alumina-zirconia laminar composites, characterized by K_{1c} ranging between 9 and 15 MPa·m$^{1/2}$ [42].

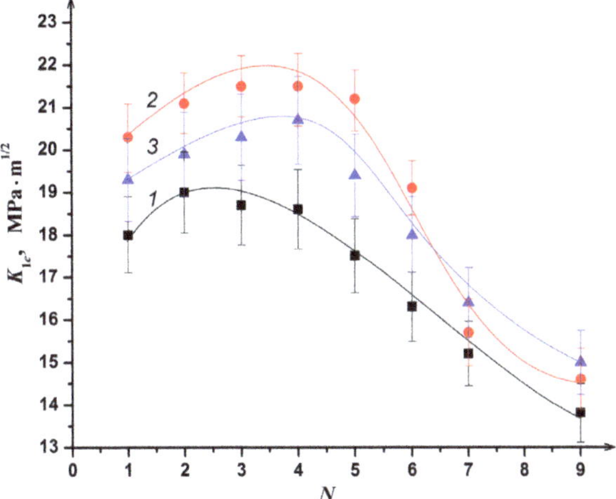

Figure 9. Crack resistance behavior of the materials subjected to HPT: *1*–Fe$_{53.3}$Ni$_{26.5}$B$_{20.2}$ AA; *2*–Co$_{28.2}$Fe$_{38.9}$Cr$_{15.4}$Si$_{0.3}$B$_{17.2}$ AA; *3*–composite.

The response of the magnetic characteristics of the materials after SPD has been studied. No significant effect of HPT on the specific saturation magnetization σ_s was noted for the Fe$_{53.3}$Ni$_{26.5}$B$_{20.2}$ and Co$_{28.2}$Fe$_{38.9}$Cr$_{15.4}$Si$_{0.3}$B$_{17.2}$ AA, and for the composite based on the AA. The change in σ_s does not exceed 2–3% in the entire strain range. Coercive force H_c as a more structure-sensitive characteristic nonmonotonically changes with increasing degree of deformation and exhibits a distinct maximum at $N = 1$ (Figure 10). For the Fe$_{53.3}$Ni$_{26.5}$B$_{20.2}$ and Co$_{28.2}$Fe$_{38.9}$Cr$_{15.4}$Si$_{0.3}$B$_{17.2}$ AA, the maximum H_c is higher than that of the initial state by a factor of 1.5–2. Further, H_c decreases to the levels slightly exceeding the initial ones and becomes stabilized with the retention of the soft magnetic state of the alloys (Figure 10).

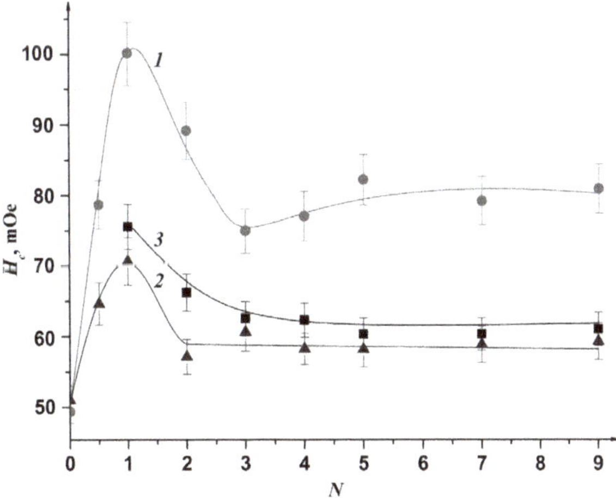

Figure 10. Coercive force as a function of the degree of deformation upon HPT: *1*–Fe$_{53.3}$Ni$_{26.5}$B$_{20.2}$ AA; *2*–Co$_{28.2}$Fe$_{38.9}$Cr$_{15.4}$Si$_{0.3}$B$_{17.2}$ AA; *3*–composite.

The jumps of the coercive force at the initial stages of deformation can be associated with the processes of the delamination of the amorphous matrix, further clustering and, as a consequence, the accompanying changes in the exchange interaction between the ferromagnetic components. In addition, with increasing degree of deformation, the resulting SB can branch, due to the frontal formation of nanocrystals [23]. As a result, the additional local stresses generated in the structure of deformation-induced composites can affect the growth of the coercive force.

Thus, it is shown that SPD can be used as an effective method to create composites with improved properties by consolidating different AA ribbons.

3.2. Amorphous-Nanocrystalline Composites Prepared by HPT of Cu-Nb Nanolaminates

Multilayer nanolaminates consisting of non-miscible or partially mutually soluble metals with nanometer (about 10 nm) thickness of the layers of each component were first obtained by the method of magnetron deposition for the Cu-Nb system [43]. Currently, such composites are obtained by MPR of copper and niobium sheets [34,35,44,45]. The Cu-Nb nanolaminates are of interest for their special physical properties, such as high plasticity and electrical conductivity of copper combined with superconductivity of niobium.

The elasto-plastic properties (Young's modulus, amplitude-independent decremen, and stress-strain diagram in the range of microplastic deformation) of the Cu-Nb nanolaminate were comprehensively studied in [46]. The analysis of the data on E, δ and σ_s before and after the action of high hydrostatic pressure (1 GPa) allowed us to assume that the nanolaminate samples contain discontinuities, which can be formed at the interfaces between the Cu and Nb layers upon MPR.

The Cu/Nb nanolaminates under the effect of severe deformation were studied in more detail in [47]. The steady-state solubility of Nb in Cu and Cu in Nb were found to reach ~1.5 at. % and ~10 at. %, respectively, in near-boundary areas, but no phase transformations, including amorphization, occur in the nanolaminate.

The energy anisotropy of the interface in a system of non-miscible elements was first studied in [48–50] by the example of molecular-dynamics simulation of Cu/Nb bicrystals. It was established that the mechanism of niobium dissolution in the copper matrix involves the formation of niobium clusters coherent with the matrix. Such coherent dissolution of non-miscible elements on the interface was noted in the literature for the Cu-Fe system upon mechanical alloying of components [51].

The observed [48–50] phenomenon of niobium cluster dissolution near the Cu/Nb interface of finite curvature is proposed as a physical explanation of the amorphization of such interface. Because of the lattice distortion by clusters, a sharp TEM image of the projection of close packed copper atom rows in the interface region disappears, and such regions are recognized as amorphous.

Our experiments have shown that the nanolaminates fabricated by MPR consist of about ten thousand Cu and Nb layers, which can vary in thickness from 100 to 300 nm. The electron microscopic data confirm the nanolayering of the Cu-Nb composite, as is illustrated in Figure 11. The layers have a wavy shape and sharp boundaries. In the image (Figure 11a) taken in backscattered electrons, the darker and lighter strips correspond to copper and niobium layers, respectively. Figure 11b shows the EDS data on multilayer mapping of elemental composition.

A noticeable thinning of nanolaminate layers is observed after SPD by HPT. After HPT to a degree of deformation of $N = 2$ turns, the layer thickness decreases approximately by a factor of two (Figure 12a), relative to the initial average layer thickness of ~200 nm (Figure 11). At higher degree of deformation ($N = 4$), the effect of layer thinning is amplified, and their integrity is destroyed (Figure 12b).

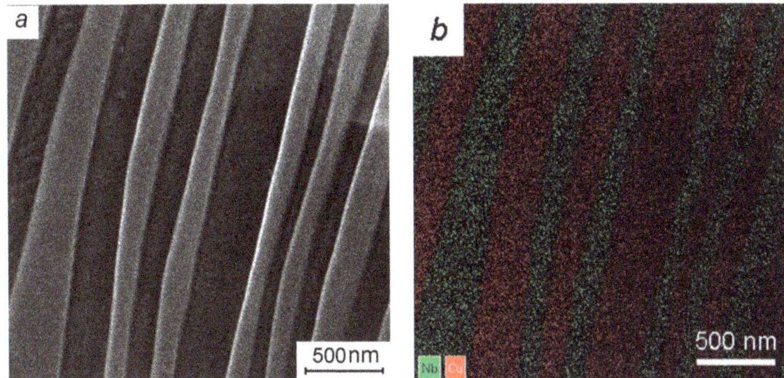

Figure 11. Structure of the cross section of Cu-Nb composite as-fabricated by multiple pack rolling (MPR) (N = 0): bright-field scanning transmission electron microscopy (BF-STEM) image (**a**), energy dispersive X-ray spectroscopy (EDS) mapping (**b**).

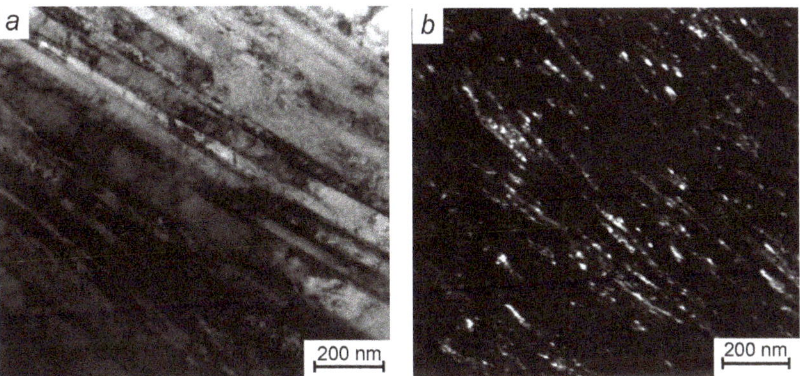

Figure 12. Planar microstructure of Cu-Nb nanolaminate subjected to HPT: N = 2, bright-field TEM image (**a**); N = 4, dark-field TEM image (**b**).

After HPT to N = 4, the areas are observed, in which the layers are mixed, and the structure is substantially refined, but the grains retain their initial preferred orientation. The average size of crystalline grains decreases to tens of nanometers. The ring reflections of the FCC copper structure and the BCC niobium structure are clearly observed in the SAED pattern.

Table 1 shows the calculations of the true deformation value upon the HPT of Cu-Nb-nanolaminates in the middle of the sample radius, corresponding to the number of revolutions according to the formula:

$$e \approx \ln\left(\frac{2\pi r N h_0}{h^2}\right) \tag{2}$$

where r is the radius of the sample; h_0 is the initial thickness of the sample; h is the final thickness of the sample; N is the number of full revolutions of the movable anvil.

Table 1. Geometric parameters of Cu-Nb nanolaminates befor and after HPT, calculated logarithmic true strain.

Layer Thickness, nm	Number of Full Revolutions	Logarithmic True Strain, e
200	0	-
180	1/2	4.61
130	1	5.42
100	2	6.33
50	3	6.54
20	4	6.83

If we conduct a comparative analysis of our Table 1 with data by Beyerlein I.J. et.al. [44], in our case, as a result of the HPT in the Bridgman anvils, to obtain nanolaminates with a layer thickness of 20 nm, the logarithmic true strain e = 6.83 is required, while accumulative roll bonding requires much large value: e = 11.51. However, it should be noted that such a comparison seems incorrect to us. The fact is that in our studies, we subjected to HPT the finished Cu-Nb composites obtained previously by the MPR method, which made it possible to reduce the layer thickness to 200 nm due to multi-stage accumulation of deformation. HPT is a subsequent and completely different SPD technology, which has a different loading MPR scheme, with a different stress condition of the sample, contributing to a further decrease in the layer thickness up to 20 nm. Thus, we are dealing with two stages of material processing. The true deformations of each of them are different from each other and are not subject to simple summation.

Typical high-resolution (HR) TEM images of the atomic structure of the samples after HPT are shown in Figure 13a–c. Also, the fast Fourier transform (FFT) of the HR TEM image was calculated (Figure 13d). In addition to the crystalline grains of 10–30 nm in size, areas with a characteristic contrast are observed in Figure 13a,b, which unambiguously indicates the appearance of amorphous phase of up to ~100 nm in size in the structure, after deformation by HPT to N = 4. The volume fraction of such areas does not exceed 5–10%. As a rule, they are elongated along the interface between Cu and Nb crystals. It can be seen that the crystalline regions *1,2,5,6* in Figure 13a,b according to the FFT analysis, have discrete point reflections from Cu or Nb, while the amorphous regions *3,4,7,8* are continuous rings—a diffuse halo in FFT diffractograms (Figure 13d). Figure 13c demonstrates a periodic banded contrast of the crystalline grain, which is in the reflecting position. This contrast represents the projections of the {110} planes onto the image plane. The direction of the incident electron beam coincides with the [111] direction. The BCC niobium and FCC copper structures have the same symmetry of atom projection onto the (111) plane. Therefore, it is difficult to determine what crystalline grain is in the observation field by the HR TEM image.

The X-ray mapping of element distributions (Cu, Nb) is carried out by the EDS method for the samples of Cu-Nb nanolaminate subjected to HPT (Figure 14). We note that the structure of the sample after HPT is formed by oriented alternating Cu and Nb grains, which are clearly seen in the BSE image and element distribution maps (Figure 14). The transverse thickness of the elongated grains varies from 4 to 40 nm. Thus, the mechanical treatment of the samples by HPT leads to the structure refinement almost by an order of magnitude. However, the general directionality (orientation) of the structure components is retained. Note that some niobium grains on the surface of the samples are enriched with oxygen.

In addition, substantial "mixing" is observed in copper and niobium layers. The boundaries of the layers lose their sharpness (Figure 15). As a consequence, regions with equal contents of the elements appear in the structure. The concept of "layer" is losing its original meaning. EDS mapping demonstrates the distribution of elements in the sample at a purely qualitative level.

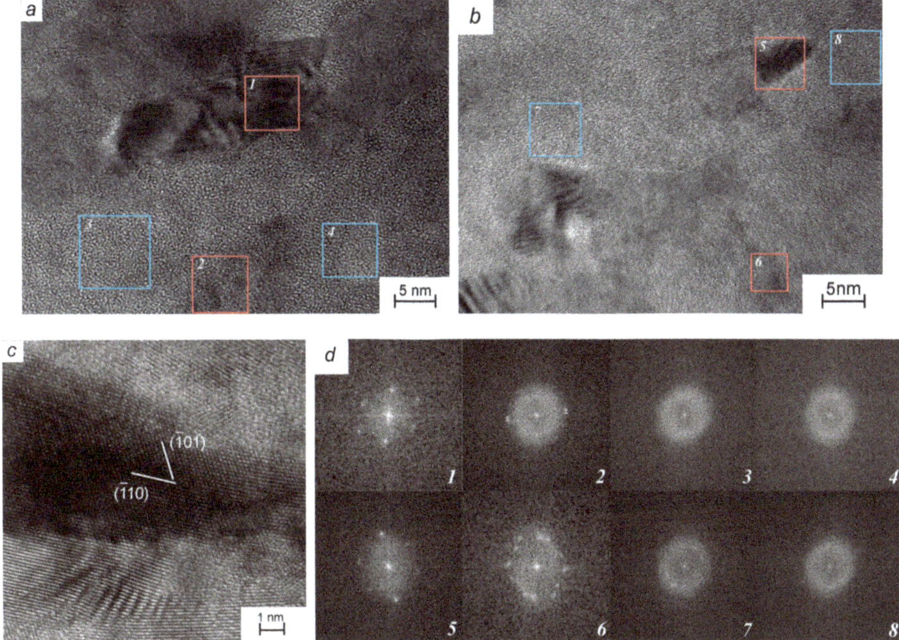

Figure 13. HR TEM images of the structure of Cu-Nb nanolaminate after HPT to $N = 4$ (**a**,**b**,**c**) and fast Fourier transform (FFT) diffractograms (**d**) of the outlined areas *1–8* of (**a**),(**b**), respectively.

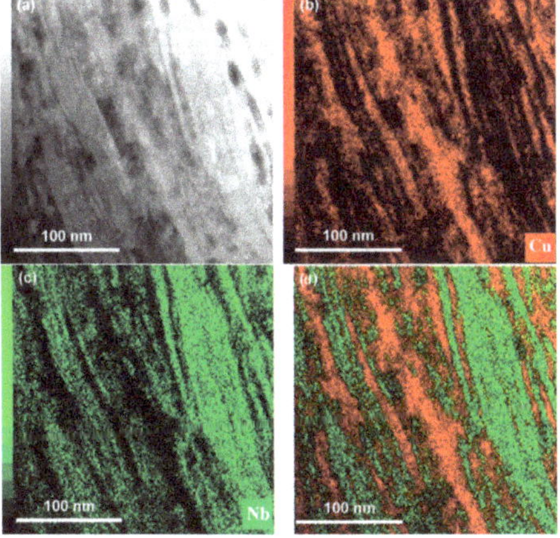

Figure 14. BF-STEM image (**a**) of the sample deformed by HPT to $N = 4$ and X-ray element distribution maps: copper (**b**), niobium (**c**), superposition of Cu and Nb (**d**).

The sequence of structural-phase transformations in the Cu-Nb nanolaminate after HPT was analyzed by XRD, along with TEM/STEM methods. The XRD spectra from the planes and the end faces of the samples before and after HPT in the Bridgman chamber are presented in Figure 16. The XDR intensities in Figure 16 are scaled linearly. With the aid of standard indexing procedures, we established that the experimental set of X-ray lines corresponds to copper (ICDD card No. 00-004-0836, Fm-3m, $a = 0.3615$ nm) and niobium (ICDD card No. 00-034-0370, Im-3m, $a = 0.3303$ nm).

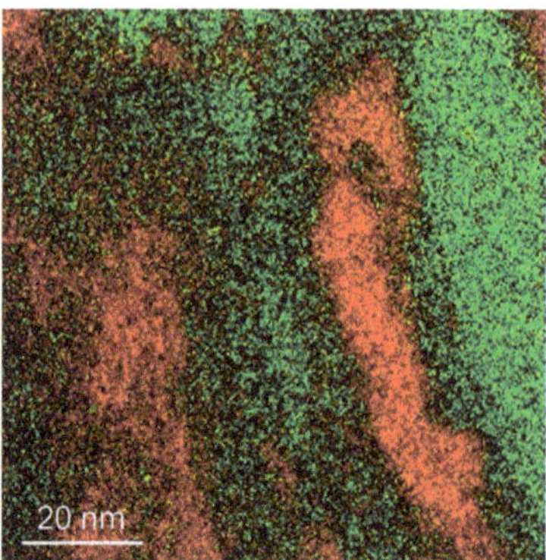

Figure 15. Fragment of the map of element distribution in the sample after HPT to $N = 6$: Cu and Nb are displayed in red and green colors, respectively.

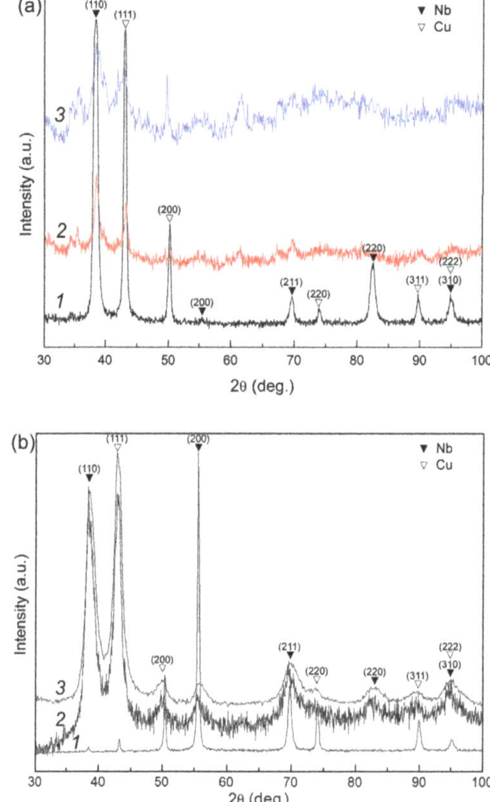

Figure 16. XRD spectra of the initial Cu-Nb nanolaminate *1–N* = 0 and after HPT to *2–N* = 2 and *3–N* = 4: (**a**) plane of sample, (**b**) end face of sample.

Comparing the profiles of the spectra, we see that the intensity of the X-ray maxima after deformation decreases. This fact can be associated with the intense fragmentation of grains and the increase of the defect density. In addition, some maxima are "blurred". This is a consequence of strain-induced amorphization processes beginning at $N = 4$, which are also revealed by the TEM studies (Figure 13a,b). Thus, the accumulation of deformation is accompanied by the degradation of the nanolaminite structure of the samples: the regularity of the layer alternation is disrupted, the layers become fragmented and curved, and the Cu/Nb system partially undergoes the transition from crystalline to amorphous state. As a prerequisite for the formation of a supersaturated solid solution during the HPT process, conditions can arise that are characterized by an inhomogeneous distribution of the components in the bulk of the material. Based on the fact that in the laminate there are areas with different concentrations of one of the components, this should partially change the lattice parameter and, accordingly, the position of the lines on the XRD spectra. The line will shift towards smaller angles if the volume is enriched in the second component. At the same time, the depletion of the remaining volume by this component causes a decrease in the parameter, and, therefore, should shift the diffraction line toward large angles. In a word, such heterogeneity is capable of causing some broadening of the lines of the XRD spectra. In Figure 16, the Cu (200) line shifts toward smaller angles by ~0.5 degrees 2θ with increasing strain during HPT. In the process of migration of the Cu/Nb interphase boundary is possible to incorporate of nanolamellas of one metal into another with the subsequent formation of isolated monatomic clusters. The crystal structure was distorted in cluster and lamella regions, which could be perceived as amorphization of the interphase boundary in studies by the methods of high-resolution microscopy [48].

The peaks at 35–38 and 63–69 degrees 2θ in Figure 16a may belong to tungsten carbide. Their appearance on the XRD spectra is due to the fact that Bridgman anvils heads are made of it. Accordingly, at large degrees of deformation, traces of WC can be observed on the surface of the deformed samples. When comparing the results, Figure 16b shows that when shooting at the end faces of the sample (at 35–38 and 63–69 degrees 2θ), no lines from the tungsten carbide phases are observed. A similar situation with the appearance of additional peaks in the Cu-Nb XDR spectra was observed in [52].

The appearance of an amorphous phase in the near-interface areas was first experimentally established in our investigation by the methods of direct-resolution TEM (Figure 13a,b) and XRD (Figure 16). This confirms the hypothesis [48–50] on the possibility of the clustered structure of the regions containing simultaneously copper and niobium atoms. In a rather detailed scientific work [47], no amorphous phases have been detected in the Cu-Nb nanolaminates after SPD, but nonequilibrium supersaturated solid solutions of Cu-Nb (up to 1.5 at. % Nb) and Nb-Cu (up to 10 at. % Cu) have been observed. The discrepancy between our results and the results of [47] could be caused by the fact that, because of somewhat different conditions of SPD in the Bridgman chamber, the supersaturated solid solutions in our case are less thermodynamically stable and undergo phase transition from crystalline (solid solution of Nb-Cu and/or Cu-Nb) to amorphous phase (Nb-Cu and/or Cu-Nb). It seems that, in our case, the following relation is satisfied:

$$U_{ss} > U_{ap}, \tag{3}$$

where U_{ss} and U_{ap} are the free energies of nonequilibrium crystalline solid solutions and Cu-Nb and Nb-Cu amorphous phases, respectively. Relation (1) is fulfilled at concentrations of solid solutions above critical. Apparently, such supercritical concentrations of the dissolved component were achieved in our experiments, but were not achieved by the authors of [47].

We established the pattern of the Vickers hardness variations in the Cu-Nb nanolaminate as a function of the degree of deformation (Figure 17).

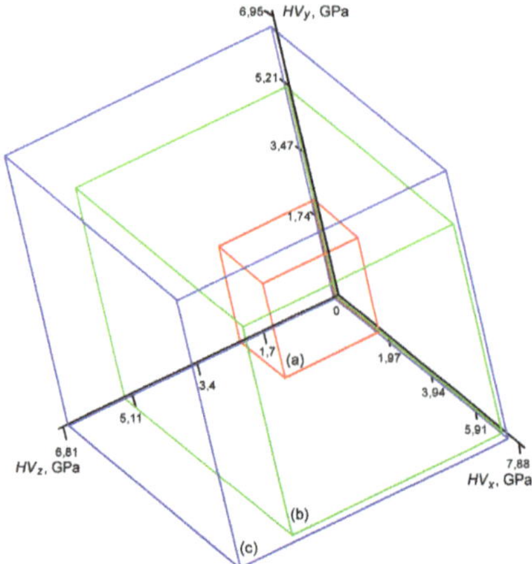

Figure 17. Behavior of microhardness *HV* of the Cu-Nb nanocomposites in three mutually orthogonal directions as a function of the degree of deformation: before HPT at $N = 0$ (**a**); after HPT to $N = 2$ (**b**) and $N = 4$ (**c**), respectively.

The anisotropy of *HV* is observed in the initial samples before HPT. The indentation results show that the microhardness of two mutually perpendicular ends is somewhat higher (see the *Y* and *Z* axes in Figure 17) than *HV* of the planar surface of the disk samples (see the *X* axis). Figure 17a shows that the *HV* edges of red rectangular parallelepiped are unequal. As the degree of deformation increases, a significant growth in microhardness is observed in all directions (see green rectangular parallelepiped in Figure 17b). The maximum difference in the microhardness *HV* on the surface (along the *X* axis) and on the ends (along *Y* and *Z* axes) is observed at $N = 2$. In other words, a strengthening surface effect of the nanolaminates was found after HPT to $N = 2$. The transition from anisotropy to isotropy of microhardness (blue cube in Figure 17c with equal *HV* edges) is observed after HPT to $N = 4$.

Thus, we observe two stages of the strengthening of nanolaminates upon HPT. First, there is a noticeable increase in microhardness (strength) under conditions of remaining structural anisotropy caused by the initial plate-like shape of the structure constituents of the composite. Then, at the second stage, there is a transition to isotropic growth of *HV*, due to much more uniform volume distribution of Cu and Nb nanoparticles, which underwent significant transformations of size and shape (degradation of the nanolaminated structure) upon HPT. The jump-like increase in *HV* is caused by complex changes in the Cu-Nb nanolaminated structure as the degree of SPD increases, namely, the thinning of nanolayers, the appearance of regions of the amorphous phase and the amorphization-induced modification of the interphase interfaces of Cu and Nb nanocrystals.

4. Conclusions

(1) For the first time, deformation-induced composites from alternating Fe-Ni-B and Co-Fe-Cr-Si-B AA layers were obtained by their consolidation upon HPT. It was found that the average microhardness of the composite obtained by HPT to $N > 4$ exceeds *HV* of its individual AA components, from which it is formed.

(2) The degree of the structure refinement of the amorphous nanocrystalline composites obtained by SPD depends on the processing regime and on the nature of the starting materials. The HPT method allows one to obtain nanocrystals of up to 5–20 nm in size in amorphous matrix in the AA composites. In the Cu-Nb nanolaminates, the gradual refinement of the nanocrystalline structure can be accompanied by local amorphization and the retention of the general orientation of the

structure constituents. The fact that an amorphous state is formed in Cu-Nb laminates during HPT requires further and more thorough investigation.

(3) The phenomenon of increase in ductility (K_{1c} increases by ~6%) of amorphous-nanocrystalline composites obtained by HPT of the $Fe_{53.3}Ni_{26.5}B_{20.2}$ and $Co_{28.2}Fe_{38.9}Cr_{15.4}Si_{0.3}B_{17.2}$ AA ribbons is established. SPD allows one to obtain strain-induced composites with satisfactory fracture toughness in combination with high hardness and high soft magnetic characteristics.

(4) For the first time, the formation of regions with an amorphous structure directly related to interphase interfaces was experimentally found in Cu-Nb nanolaminates by TEM and XRD methods after a high degree of SPD ($N = 4$).

(5) An increase in the degree of SPD by HPT increases strength of the Cu-Nb nanolaminates by a factor of three. A two-stage transition from the anisotropy of microhardness in the initial state to the isotropy of *HV* was detected at $N = 4$. Surface strengthening of the Cu-Nb nanolaminates is observed at $N = 2$; the difference between *HV* on the surface and in the volume is 29%.

Author Contributions: Conceptualization and Methodology, A.G. and I.P.; software, I.P.; validation, A.G. and I.P.; formal analysis, I.P.; investigation, A.G. and I.P.; resources, A.G.; data curation, I.P.; writing—original draft preparation, I.P.; writing—review and editing, A.G. and I.P.; visualization, I.P.; supervision and project administration, A.G.; funding acquisition, I.P. All authors have read and agreed to the published version of the manuscript.

Funding: The reported study was funded by RFBR, project numbers 17-02-00402 and 20-08-00341.

Acknowledgments: The authors are grateful to their colleagues, Michael Karpov, Victor Vnukov, Dmitry Shtansky, Igor Shchetinin, Michael Gorshenkov, Elena Savchenko and Elena Blinova for long-term fruitful cooperation.

Conflicts of Interest: The authors declare no conflict of interest.

References

1. Matthews, F.L.; Rawlings, R.D. *Composite Materials: Engineering and Science*; CRS Press: Boca Raton, USA, 1999; ISBN 0-8493-0621-3.

2. Gibson, R.F. *Principles of Composite Material Mechanics*, 4th ed.; CRC Press: Boca Raton, USA, 2016; ISBN 978-1-4987-2072-4.

3. Strong, A.B. *Fundamentals of Composites Manufacturing Materials, Methods and Applications*, 2nd ed.; Society of Manufacturing Engineers: Dearborn, MI, USA, 2008; ISBN 978-0872638549.

4. Barnett, S.A. Deposition and Mechanical Properties of Superlattice Thin Films. In *Physics of Thin Films. Mechanic and Dielectric Properties. Advances in Research and Development*; Francombe, M.H., Vossen, J.A., Eds.; Academic Press: New York, NY, USA, 1993; Volume 17, pp. 1–77. ISBN 978-0-1253-3017-6.

5. Vepřek, S. The search for novel, superhard materials. *J. Vac. Sci. Technol. A* **1999**, *17*, 2401–2420. [CrossRef]

6. Hovsepian, P.E.; Lewis, D.B.; Münz, W.-D. Recent progress in large scale manufacturing of multilayer/superlattice hard coatings. *Surf. Coat. Technol.* **2000**, *133*, 166–175. [CrossRef]

7. Barnett, S.A.; Madan, A.; Kim, I.; Martin, K. Stability of nanometer-thick layers in hard coatings. *MRS Bull.* **2003**, *28*, 169–172. [CrossRef]

8. Chung, Y. W.; Sproul, W.D. Superhard coating materials. *MRS Bull.* **2003**, *28*, 164–168. [CrossRef]

9. Münz, W.-D. Large-scale manufacturing of nanoscale multilayered hard coatings deposited by cathodic arc/unbalanced magnetron sputtering. *MRS Bull.* **2003**, *28*, 173–179. [CrossRef]

10. Abrosimova, G.E. Evolution of the structure of amorphous alloys. *Phys.-Usp.* **2011**, *54*, 1227–1242. [CrossRef]

11. Perrière, L.; Champion, Y. Phases distribution dependent strength in metallic glass-aluminium composites prepared by spark plasma sintering. *Mater. Sci. Eng. A* **2012**, *548*, 112–117. [CrossRef]

12. Wang, Y.; Li, J.; Hamza, A.V.; Barbee, T.W. Ductile crystalline-amorphous nanolaminates. *Proc. Natl. Acad. Sci. USA* **2007**, *104*, 11155–11160. [CrossRef]

13. Donohue, A.; Spaepen, F.; Hoagland, R.G.; Misra, A. Suppression of the shear band instability during plastic flow of nanometer-scale confined metallic glasses. *Appl. Phys. Lett.* **2007**, *91*, 241905. [CrossRef]

14. Molokanov, V.V.; Chueva, T.R.; Umnov, P.P.; Simakov, S.V.; Shalygina, E.E. "Thick" amorphous wires in the $Fe_{75}Si_{10}B_{15}$-$Co_{75}Si_{10}B_{15}$-$Ni_{75}Si_{10}B_{15}$ system: Fabrication, structure, properties. *Inorg. Mater. Appl. Res.* **2016**, *7*, 643–647. [CrossRef]

15. Shalygina, E.E.; Umnova, N.V.; Umnov, P.P.; Molokanov, V.V.; Samsonova, V.V.; Shalygin, A.N.; Rozhnovskaya, A.A. Specific features of magnetic properties of "thick" microwires produced by the Ulitovsky-Taylor method. *Phys. Solid State* **2012**, *54*, 287–292. [CrossRef]

16. Shelyakov, A.; Sitnikov, N.; Saakyan, S.; Menushenkov, A.; Korneev, A. Study of two-way shape memory behavior of amorphous-crystalline TiNiCu melt-spun ribbon. *Mater. Sci. Forum* **2013**, *738–739*, 352–356. [CrossRef]

17. Glezer, A.M.; Manaenkov, S.E.; Permyakova, I.E.; Shurygina, N.A. Effect of nanocrystallization on the mechanical behavior of Fe-Ni-based amorphous alloys. *Russ. Metal.* **2011**, *2011*, 947–955. [CrossRef]

18. Glezer, A.M.; Permyakova, I.E.; Shurygina, N.A.; Rassadina, T.V. Structural features of crystallization and hardening of amorphous alloy in the Fe-Cr-B system. *Inorg. Mater. Appl. Res.* **2012**, *3*, 23–27. [CrossRef]

19. Shurygina, N.A.; Glezer, A.M.; Permyakova, I.E.; Blinova, E.N. Effect of nanocrystallization on the mechanical and magnetic properties of Finemet-type alloy ($Fe_{78.5}Si_{13.5}B_9Nb_3Cu_1$). *Bull. Russ. Acad. Sci. Phys.* **2012**, *76*, 44–50. [CrossRef]

20. Permyakova, I.E.; Glezer, A.M.; Ivanov, A.A.; Shelyakov, A.V. Application of laser design of amorphous FeCo-based alloys for the formation of amorphous-crystalline composites. *Russ. Phys. J.* **2016**, *58*, 1331–1338. [CrossRef]

21. Mudry, S.I.; Nykyruy, Y.S.; Kulyk, Y.O.; Stotsko, Z.A. Influence of pulse laser irradiation on structure and mechanical properties of amorphous $Fe_{73.1}Nb_3Cu_{1.0}Si_{15.5}B_{7.4}$ alloy. *J. Achiev. Mater. Manufact. Eng.* **2013**, *61*, 7–11.

22. Sitnikov, N.N.; Shelyakov, A.V.; Khabibullina, I.A.; Borodako, K.A. Two-way shape memory effect in rapidly quenched highly doped alloys of TiNi-TiCu system upon laser treatment. *Bull. Russ. Acad. Sci. Phys.* **2018**, *82*, 1136–1142. [CrossRef]

23. Glezer, A.M.; Permyakova, I.E. *Melt-Quenched Nanocrystals*; CRC Press: Boca Raton, FL, USA, 2013; ISBN 978-1-4665-9414-2.

24. Glezer, A.M.; Shurygina, N.A. *Amorphous-Nanocrystalline Alloys*; CRC Press: Boca Raton, FL, USA, 2017; ISBN 978-1-1385-0237-6.

25. Inoue, A.; Louzguine, D.V. Bulk Nanocrystalline and Nanocomposite Alloys Produced from Amorphous Phase. In *Nanostructured Metals and Alloys. Processing, Microstructure, Mechanical Properties and Applications*; Whang, S.H., Ed.; Woodhead Publishing Ltd.: Cambridge, UK, 2011; pp. 152–177. ISBN 978-1-84569-670-2.

26. Wilde, G. Bulk Nanostructured Materials from Amorphous Solids. In *Bulk Nanostructured Materials*; Zehetbauer, M.J., Zhu, Y.T., Eds.; Wiley-VCH. Verlag GmbH & Co: Weinheim, Germany, 2009; pp. 293–310. ISBN 978-3-5273-1524-6.

27. Glezer, A.M.; Kozlov, E.V.; Koneva, N.A.; Popova, N.A.; Kurzina, I.A. *Plastic Deformation of Nanostructured Materials*; CRC Press: Boca Raton, FL, USA, 2017; ISBN 978-1-1380-7789-8.

28. Valiev, R.Z.; Zhilyaev, A.P.; Langdon, T.G. *Bulk Nanostructured Materials: Fundamentals and Applications*; John Wiley & Sons: Hoboken, NJ, USA, 2013; ISBN 978-1-118-09540-9.

29. Glezer, A.M. Creation principles of new-generation multifunctional structural materials. *Phys.-Usp.* **2012**, *55*, 522–529. [CrossRef]

30. Permyakova, I.E.; Blinova, E.N.; Shchetinin, I.V.; Savchenko, E.S. Amorphous-alloy-based composites prepared by high-pressure torsion. *Russ. Metal.* **2019**, *2019*, 994–1001. [CrossRef]

31. Permyakova, I.E.; Glezer, A.M.; Karpov, M.I.; Vnukov, V.I.; Shtansky, D.V.; Gorshenkov, M.V.; Shchetinin, I.V. Structural amorphization and mechanical properties of nanolaminates of the cooper-niobium system during high-pressure torsion. *Russ. Phys. J.* **2018**, *61*, 428–438. [CrossRef]

32. Glezer, A.M.; Permyakova, I.E.; Fedorov, V.A. Crack resistance and plasticity of amorphous alloys under microindentation. *Bull. Russ. Acad. Sci. Phys.* **2006**, *70*, 1599–1603.

33. Glezer, A.M.; Permyakova, I.E.; Manaenkov, S.E. Plasticizing effect in the transition from an amorphous state to a nanocrystalline state. *Dokl. Phys.* **2008**, *53*, 8–10. [CrossRef]

34. Karpov, M.I.; Gnessin, B.A.; Vnukov, V.I.; Medved, N.V.; Volkov, K.G. Texture and Mechanical Properties of the Bulk Multilayered Nb-Cu Composite. In *Proceedings of the Intrernational Conference "Advanced Metallic Materials", Smolenice Castle, Slovakia, 5–7 November 2003*; Slovak Academy of Sciences: Bratislava, Slovakia, 2003; pp. 141–143.

35. Karpov, M.I.; Vnukov, V.I.; Medved, N.V.; Volkov, K.G.; Khodos, I.I. Nanolaminate-Bulk Multilayered Nb-Cu Composite: Technology, Structure, Properties. In Proceedings of the 15-th International Plansee-Seminar, Reutte, Austria, 28 May–1 June 2001; Volume 4, pp. 97–107.

36. Yasuna, K.; Tarauchi, M.; Otsuki, A.; Ishihara, K.N.; Shingu, P.H. Bulk metallic multilayers produced by repeated press-rolling and their perpendicular magnetoresistance. *J. App. Phys.* **1997**, *82*, 2435–2438. [CrossRef]

37. Huang, B.; Ishihara, K.N.; Shingu, P.H. Bulk nano-scale Fe/Cu multilayers produced by repeated pressing-rolling and their magnetoresistance. *J. Mater. Sci. Lett.* **2000**, *19*, 1763–1765. [CrossRef]

38. Shtansky, D.V.; Kaneko, K.; Ikuhara, Y.; Levashov, E.A. Characterization of nanostructured multiphase Ti-Al-B-N thin films with extremely small grain size. *Surf. Coat. Technol.* **2001**, *148*, 206–215. [CrossRef]

39. Glezer, A.M.; Manaenkov, S.E.; Permyakova, I.E. Structural mechanisms of plastic deformation of amorphous alloys containing crystalline nanoparticles. *Bull. Russ. Acad. Sci. Phys.* **2007**, *71*, 1702–1707. [CrossRef]

40. Glezer, A.M.; Shurygina, N.A.; Zaichenko, S.G.; Permyakova, I.E. Interaction of deformation shear bands with nanoparticles in amorphous-nanocrystalline alloys. *Russ. Metall.* **2013**, *2013*, 235–244. [CrossRef]

41. Permyakova, I.E.; Glezer, A.M.; Grigorovich, K.V. Deformation behavior of amorphous Co-Fe-Cr-Si-B alloys in the initial stages of severe plastic deformation. *Bull. Russ. Acad. Sci. Phys.* **2014**, *78*, 996–1000. [CrossRef]

42. Gogotsi, G.A. Fracture resistance of ceramics: Base diagram and R-line. *Strength Mater.* **2006**, *38*, 261–270. [CrossRef]

43. Schuller, I.K. New class of layered materials. *Phys. Rev. Lett.* **1980**, *44*, 1597–1600. [CrossRef]

44. Beyerlein, I.J.; Mara, N.A.; Carpenter, J.S.; Nizolek, T.; Mook, W.M.; Wynn, T.A.; McCabe, R.J.; Mayeur, J.R.; Kang, K.; Zheng, S.; et al. Interface-driven microstructure development and ultra-high strength of bulk nanostructured Cu-Nb multilayers fabricated by severe plastic deformation. *J. Mater. Res.* **2013**, *28*, 1799–1812. [CrossRef]

45. Carpenter, J.S.; Vogel, S.C.; LeDonne, J.E.; Hammon, D.L.; Beyerlein, I.J.; Mara, N.A. Bulk texture evolution of Cu-Nb nanolamellar composites during accumulative roll bonding. *Acta Mater.* **2012**, *60*, 1576–1586. [CrossRef]

46. Betekhtin, V.I.; Kolobov, Y.R.; Kardashev, B.K.; Golosov, E.V.; Narykova, M.V.; Kadomtsev, A.G.; Klimenko, D.N.; Karpov, M.I. Elasto-plastic properties of Cu-Nb nanolaminate. *Tech. Phys. Lett.* **2012**, *38*, 144–146. [CrossRef]

47. Ekiz, E.H.; Lach, T.G.; Averback, R.S.; Mara, N.A.; Beyerlein, I.J.; Pouryazdan, M.; Hahn, H.; Bellon, P. Microstructural evolution of nanolayered Cu-Nb composites subjected to high-pressure torsion. *Acta Mater.* **2014**, *72*, 178–191. [CrossRef]

48. Lipnitskii, A.G.; Nelasov, I.V.; Golosov, E.V.; Kolobov, Y.R.; Maradudin, D.N. A Molecular-dynamics simulation of grain-boundary diffusion of niobium and experimental investigation of its recrystallization in a niobium-copper system. *Russ. Phys. J.* **2013**, *56*, 330–337. [CrossRef]

49. Lipnitskii, A.G.; Nelasov, I.V.; Klimenko, D.N.; Mapadudin, D.N.; Kolobov, Y.P. Moleculardynamical simulation of multilayered Cu/Nb composite. *Materialovedenie* **2009**, *6*, 7–10.

50. Nelasov, I.V.; Lipnitskii, A.G.; Kolobov, Y.R. Study of the evolution of the Cu/Nb interphase boundary by the molecular dynamics method. *Russ. Phys. J.* **2009**, *52*, 1193–1198. [CrossRef]

51. Wei, S.; Oyanagi, H.; Wen, C.; Yang, Y.; Liu, W. Metastable structures of immiscible Fe_xCu_{100-x} system induced by mechanical alloying. *J. Phys. Condens. Matter.* **1997**, *9*, 11077–11083. [CrossRef]

52. Advani, A.H.; Thadhani, N.N. Shock-induced reaction synthesis of isomorphous (Cu-Ni) and immiscible (Cu-Nb) compounds. *Metall. Mater. Trans. A* **1999**, *30*, 1367–1379. [CrossRef]

Article

Shear Bands Topology in the Deformed Bulk Metallic Glasses

Mikhail Seleznev [1,*] and Alexei Vinogradov [2]

[1] Institute of Materials Engineering, Technische Universität Bergakademie Freiberg, 09599 Freiberg, Germany
[2] Department of Department of Mechanical and Industrial Engineering, Norwegian University of Science and Technology - NTNU, 7491 Trondheim, Norway; alexei.vinogradov@ntnu.no
* Correspondence: Mikhail.Seleznev@iwt.tu-freiberg.de; Tel.: +49-3731-394034

Received: 18 February 2020; Accepted: 12 March 2020; Published: 14 March 2020

Abstract: Recent experimental studies revealed the presence of Volterra dislocation-type long-range elastic strain/stress field around a shear band (SB) terminated in a bulk metallic glass (BMG). The corollary from this finding is that shear bands can interact with these stress fields. In other words, the mutual behaviour of SBs should be affected by their stress fields superimposed with the external stresses. In order to verify this suggestion, the topography of the regions surrounding SBs terminated in the BMGs was carefully analysed. The surfaces of several BMGs, deformed by compression and indentation, were investigated with a high spatial resolution by means of scanning white-light interferometry (SWLI). Along with the evidence for the interaction between SBs, different scenarios of the SB propagation have been observed. Specifically, the SB path deviation, mutual blocking, and deflection of SBs were revealed along with the significant differences between the topologies of the mode II (in-plane) and mode III (out of plane) SBs. While the type II shear manifests a linear propagation path and a monotonically increasing shear offset, the type III shear is associated with a curved, segmented path and a non-monotonically varying shear offset. The systematic application of the "classic" elastic Volterra's theory of dislocations to the behaviour of SBs in BMGs provides new insight into the widely reported experimental phenomena concerning the SB morphology, which is further detailed in the present work.

Keywords: bulk metallic glass; shear band; dislocation theory; scanning white light interferometry

---

1. Introduction

Bulk metallic glasses (BMGs) possess a unique set of properties such as an unrivalled combination of toughness and strength [1], very high elastic limit and elastic energy capacity [2], biocompatibility [3], corrosion resistance, etc. [4]. One of the main disadvantages is the localisation of the BMG plastic deformation at room temperature in the shear bands (SB) [5]. SB is a planar defect with a thickness of about 10–20 nm [6] accumulating the excess-free volume (EFV) during deformation [7]. Such a "decompaction" of the SB structure induces strain softening [8], with the formation of voids and microcracks merging into the main crack as load increases [9,10]. However, significant experimental evidence has been put forward on strain hardening and material densification occurring in BMGs in association with shear bands during mechanical compression [11], tension [12], and rolling [13]. Moreover, the presence of EFV trapped in the glassy state does not necessarily lead to strain softening. The finite element modelling shows that the plasticity of metallic glasses can be enhanced with the EFV-induced heterogeneity, and even apparent hardening can be achieved [14]. The cyclic loading of BMGs in the range of apparently elastic strains leads to the formation of the elastic nano-scale heterogeneity in the bulk [15]. The nano-scale heterogeneity, which is inherent to SB, has been widely approved not only by modelling [16], but also experimentally by high-resolution transmission electron microscopy (HRTEM) and digital image correlation (DIC). In particular, it was convincingly demonstrated that the SB structure consists of alternating ~0.1–0.4 μm-long segments representing compressed and stretched areas with the density, which is either larger or smaller than that in the matrix, respectively [17,18]. Thus, the material within the SB can experience both local softening due to decompaction, and hardening due to compaction of atoms.

The successful observations of the deformation hardening in BMGs commonly refer to the high density of SBs. The greater the number of SBs and the total sheared area, the larger the deformation capacity and the apparent ductility [1,11]. The emergence of the primary SB always leads to softening and failure due to the concentration of all shear deformation in one band [19]. The intermittent hardening of BMG is witnessed by multiple observations of intersecting SBs blocking each other similarly to slip bands in crystals [11,13]. Thus, the hardening/softening behaviour is governed by the development of SBs and their interactions with each other as well as with the applied stress.

The underlying mechanism of non-uniform deformation of BMGs is still under debates due to a great complexity of direct observations. Currently, the most widespread concept is based on the hypothetical elementary deformation units—the shear-transformation zones (STZ) [20]. Microscopic STZs are supposed to interact elastically by triggering slip avalanches and forming macroscopic SBs. Ironically, the critical interaction parameter—the elastic field of the STZ—is a priori undefinable in the static state [5]. One can approximate the STZ-induced elastic fields by the infinite-range mean-field [21], which is in good numerical agreement with experiments even though it is still largely speculative. Along with the family of STZ models, the approaches based on atomistic simulations are actively discussed and used for the interpretation of the SB behaviour in metallic glasses [22]. Among these approaches, the dislocation-based model [23], the free volume model [24], and the percolation approach [25] are particularly noteworthy.

The alternative to simulations is the use of empirical studies of elastic displacements created by SB. The recently disclosed long-range elastic fields around the SB tip [26] allow assuming reasonably that the interaction between SBs occurs across the scales not only on the atomic level but also on the macroscopic scale, which resembles the behaviour of dislocations in crystals, despite fundamental microstructural differences between crystals and amorphous solids.

The SB activity manifests itself as shear steps on a polished surface of a deformed sample and can be revealed with the surface profile analysis. The surface morphology of sheared BMGs has been frequently observed in the 2D-mode by means of scanning electron microscopy (SEM) [27–29], or in the 3D-mode by atomic force microscopy (AFM) [30–32] or confocal laser scanning microscopy (CLSM) [33]. In comparison with CLSM and AFM, scanning white light interferometry (SWLI) is more suitable for the 3D mapping of the SB topology due to a combination of the unbeatable vertical resolution up to 0.1 nm and rapid scanning of the area up to 1 cm^2 [34].

Both AFM and SWLI methods have been used to investigate the influence of the applied stress and pre-strain history on the SB morphology during BMG indentation [32,34]. The results clearly suggest that the interaction exists between the developing SBs, previously formed SBs and residual stresses, which results in SB reactivation, blockage, intersection, and termination. However, all these effects were described without taking into account that the SB is the imperfection with its own self-induced elastic stress field. The present work reports on the results of detailed investigations of the macro-scale 3D surface topology of the deformed BMG aiming at shedding light on the formation of SB's features and mutual SBs interactions on account of their long-ranged elastic stresses.

The paper is organised as follows. In the following section, the material properties and testing methods are briefly described. For the sake of consistency, the results and discussion are provided together within each sub-section of Section 3. Section 3.1 highlights the experimental evidence to date supporting the presence of long-range elastic fields. SBs morphologically classified as shear and tear mode types in Section 3.1 are further analysed by means of the 3D surface mapping in Sections 3.2 and 3.3, respectively. Section 3.4 is focused on the local offset and path deviations within SBs. Whereas Sections 3.1–3.4 describe the results of SB investigations after mechanical compression,

Section 3.5 reveals new facts of the interaction between SBs formed during micro-indentation of BMGs. Lastly, all findings are summarised in Section 4.

2. Materials and Methods

Pd-based and Zr-based BMG ingots were produced by a suction casting process in a copper mould under a purified argon atmosphere. Processing details are given in Reference [35] for Pd40Ni40P20, Reference [36] for Zr48Cu45Al7, and Reference [37] for Pd40Cu30Ni10P20, respectively. All specimens for mechanical testing were shaped by spark erosion. The samples with $2 \times 2 \times 4$ mm^3 dimensions were prepared from the Pd40Ni40P20 (PdNiP) glassy ingot. The surfaces were polished to a mirror finish for shear band observations at different scales. The samples of Pd40Cu30Ni10P20 (PdCuNiP) were shaped to $2.66 \times 2.66 \times 5.5$ mm^3 dimensions. U-shaped and square notches were introduced in these samples to localise the deformation zone and provoke the SB nucleation in the field of view of the camera used for in-situ imaging of the deformation surface and the digital image correlation (DIC) analysis. One batch of the samples was polished to the mirror finish for SB observations and the SWLI analysis of the shear band profiles. The other batch was kept in the as-cast state for the measurement of displacement fields around SB tips with the help of DIC analysis.

Compression experiments were carried out using the rigid mechanical testing system comprising of a screw-driven tension/compression module Kammrath and Weiss (Kammrath and Weiss, Dusseldorf, Germany), high-speed video camera Photron SA3 (Photron Ltd, Tokio, Japan) and the acoustic emission triggering circuit designed to interrupt the experiment when a new SB emerges. Details of the experimental setup have been documented in Reference [38]. Samples were loaded in compression between two parallel lubricated tungsten carbide plates with a speed of 1 µm/s. When the new SB appeared at the free surface, the testing system was stopped automatically, the samples were unloaded, and the surface topology was investigated by SWLI (Zygo NewView 7100, Zygo Corporation, Middlefield, CT, USA).

A piece of the as-cast Zr48Cu45Al7 (ZrCuAl) alloy was polished to a mirror finish for micro-indentation experiments performed using the Nanovea Scratch tester (Nanovea, Irvine, CA, USA) at 20 N force applied to the diamond Berkovich indenter (three-sided pyramid) with the loading/unloading speed of 1 N/s and the load holding time of 10 s. The shear bands emerging around the indenter were observed using the SWLI technique. To reveal the possible interactions between the SBs, the indentations were performed close to each other, of 50 µm apart, so that the indenter affected deformation zones overlapped. After indentation, the sample surface was scanned by SWLI.

3. Results and Discussion

3.1. Measurement of the Long-Range Stress Fields Produced by Shear Bands

The first brief report on the direct observations of the long-range elastic strain and stress fields created by a shear band terminated inside the BMG was made by the same authors in Reference [26]. For consistency with what follows, we repeat the same experiment and provide new additional insight into this topic. The specially designed sample geometry promotes obtaining SBs with either the dominating shear mode (mode II, by analogy with that of the shear crack), c.f., the inset in Figure 1g, or the tearing mode (mode III out-of-plane shear), c.f., the inset in Figure 1c, during compression. After in-situ imaging of emerging SBs in PdCuNiP specimens, the displacement fields around the SB tips were calculated using the DIC algorithm (Figure 1a,d).

Notice that the presented in Figure 1a,d displacements U_y and U_x are "raw", i.e., they are calculated using only the correlation coefficient method without any smoothing or filtering. The displacement of the mode II shear was calculated directly from the XY-plane. The mode III shear offset was inclined to the observed surface at 45°, which projects itself on the XY-plane. This made it possible to quantify the out-of-plane shear offset via its XY-projection. Using the well-known expressions for the strain and stress fields created by perfect screw and shear dislocations in elastic continuum [39] with account of the actual shear geometry in the present experiments, the expected displacement fields have been calculated [26] and plotted in Figure 1b,f for comparison with experimental data, (Figure 1a,e) respectively. Even a simple juxtaposition of these respective figures suggests good qualitative agreement between the experimental findings and the predictions from the dislocation theory both for mode III (Figure 1a,b) and mode II shear (Figure 1e,f). The quantitative comparison is represented in Figure 1d,h using a circular contour around the shear band tip (with the tip in the center and the arbitrarily chosen radius). The dislocation-based elastic model predictions exhibit impressive numerical agreement with the experimental data within the error of less than 10%, which is a remarkable result.

Both experimental data and theoretical calculations show that elastic fields around the SB associated with the Volterra-type macro-dislocation [40] are large enough to increase the local stresses significantly, i.e., of 10–100 MPa stress hundreds of micrometres apart from the shear tip (Figure 1c,g).

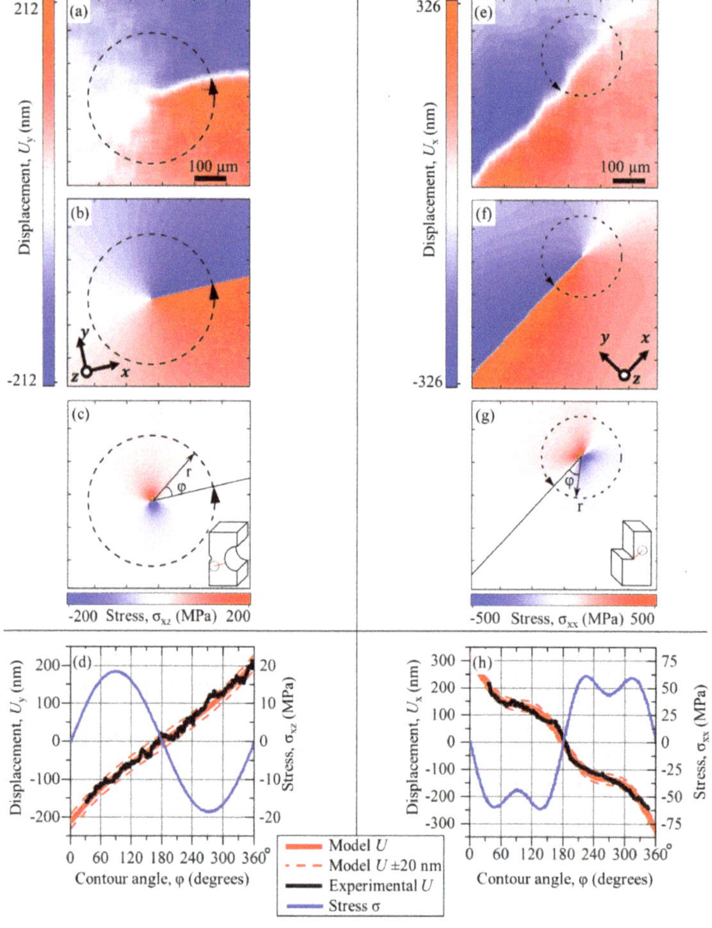

Figure 1. Examples of type III (Left) and type II (Right) shear bands (SBs) in the amorphous Pd40Cu30Ni10P20 deformed in compression. The displacement fields U_y (**a,b**) and U_x (**e,f**) were measured experimentally (**a,e**) and modelled as Volterra's screw (**b**) and edge (**f**) dislocations [26]. The σ_{xz} and σ_{xx} components were calculated for the modelled screw (**c**) and edge (**g**) dislocations, respectively. The comparison [26] of experimental (rounds) and model (red lines) displacements as well as

stresses (blue lines) is presented in (**d,h**) for the depicted circular paths around the SB tips. Excellent agreement can be seen between the experimental data and model predictions within a ± 20 nm band (red dash lines). The insets on (**c,g**) represent the schemes of used samples geometry: the red line indicates the SB with a tip surrounded by an investigated circle cross-section.

The experimental evidence for the existing Volterra dislocation-type long-range elastic fields around the SB front offers a new insight into the interpretation of previously reported experimental results regarding the SB-related properties of metallic glasses, such as apparent work hardening [11–13], internal friction relaxation peak [41–43], reactivation and suppression of the SB activity [34], etc. While there are multiple similarities in the shear behaviour in crystalline and amorphous solids, there are also major differences. Due to the absence of long-range order within the atomic structure of a glass, BMGs do not have crystallographically defined slip planes and directions. Therefore, (i) the shear slip in the amorphous structure requires dilatation [44], which is notably more pronounced than that in the core of the crystal dislocations [45], and (ii) the SB's offset and propagation direction are not constrained by a crystal structure. These factors collectively lead to considerable topological differences between the mode II and mode III SBs, which is further investigated in Sections 3.2 and 3.3, respectively.

3.2. Mode II Shear Morphology

The offset of the mode II SB aligns with the shear propagation direction. This promotes the straight path of the shear front through the specimen. Therefore, under the uniform far-field stress (e.g., during compression of an unnotched specimen), the SB appears commonly as a straight line. Since such an SB does not produce a step at the lateral surface, it is challenging to observe it by any topology-sensitive microscopic imaging technique. The offsets of plastic shear along the SB can be, however, easily noted on the lightly scratched surface, as exemplified by Figure 2d. This figure shows the topology map of the surface fragment represented in Figure 2a (the red line indicates the scratch intersecting the SB and the red arrows indicate 1 μm plastic shear offsets associated with the bands).

The pure mode II shear is rarely observed. Many, if not most, SBs of this mode contain some contribution from the mode III component, which manifests itself as a surface step (Figure 2b). This step is nicely resolvable with aid from SWLI (Figure 2) where three shear bands marked #1, #2, and #3, respectively, are shown (SBs #1 and #2 are parallel to each other while SB #3 intersects SB #1 at the right angle). It has been shown that a mode II shear offset changes linearly along the SB from the maximum at the surface step to zero at the shear tip [27]. However, the mode III offset component within the primary mode II SB (Figure 2a) exhibits different types of behaviour, which can be seen as

(1) the monotonic decrease in a shear step height to a zero (SB #1),

(2) the non-monotonic decrease in the shear step height alongside the shear band (SB #2, notice the altering step direction in the middle of the figure),

(3) propagation blocking due to the SB-SB interaction (SB #3).

The offset of the off-plane component of the predominantly mode II SB can linearly decrease from maximum (step 7 ≈ 250 nm, Figure 2c) to zero at the tip of SB #1, similarly to the mode II offset reported in Reference [27]. The step direction can change the sign and becomes negative, c.f., SB #2, step 6 ≈ −10 nm (Figure 2c). Figure 2 illustrates the known fact [46] that one SB can block the propagation of another, c.f., SB #3 is terminated at SB #1 (Figure 2). One can see that SB #3 exhibits the step height 150 nm (step 1) at the initiation site at the specimen edge, and it ends sharply when meeting SB #1 with the step height of 42 nm (step 2). The site is featured by a local plastic pile-up of 90 nm (step 4) visible in the inset of Figure 2a between steps 3 and 5, being of 50–60 nm each. On the other hand, SB #1 propagates straight by 0.5 mm into the specimen with the gradually vanishing shear offset.

Considering the macroscopic homogeneity of the glassy microstructure, the fact that one SB can block the other, which has been frequently observed in abundant literature, is yet to be understood. One of the plausible explanations to this blockage effect can be given based on the possible SB-SB interaction via their elastic stress fields. As mentioned in the introduction, the experimentally revealed nano-scale SBs heterogeneities create alternating tension/compression regions along the shear path [17,18]. Thus, the elastic field of the tip of SB #3 can interact with SB #1, which results in the effective blockage of shear bands.

If the SB-induced elastic stresses extend across the scales from nano-range to macro-range, one should be able to observe the following related effects.

(1) The long-range elastic stress field around the SB-tip (c.f., Figure 1) should influence the SB step morphology if the shear offset is out-of-plane (mode III).

(2) The local deviations should manifest themselves not only along the SB path but also in the shear offset.

(3) The SB propagation should be sensitive to the superposition of the local stresses and the self-stresses arising from the SB tip.

These effects are overviewed in Sections 3.3–3.5, respectively.

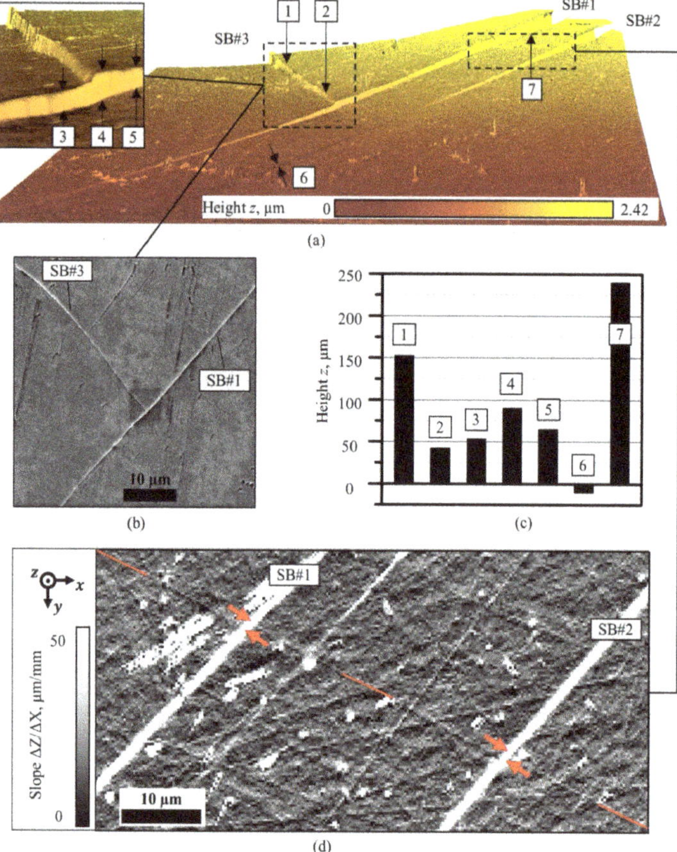

Figure 2. Typical behaviour of primarily mode II shear bands (SBs) in the compressed Pd-based bulk metallic glass (BMG). The 3D surface map (330×300 µm^2) obtained with the scanning white-light interferometry (SWLI) technique (**a**) shows three SBs (#1, #2, #3) and their topology features. The site of the collision of SB #1 and SB #3 is shown on the inset (**a**) and obtained with scanning electron microscopy (SEM) (**b**) for comparison. Shear steps No. 1 to 7 heights, measured by SWLI and pointed on (**a**) are represented on the bar chart (**c**). The slope map reveals the breaks in the fine scratch (marked with red arrows) by witnessing the mode II shear (**d**).

3.3. Mode III Shear Morphology

Unlike the mode II shear, the mode III shear deformation occurs in the direction perpendicular to the shear propagation plane. This shear mode is less constrained and has some "degree of freedom" in the propagation direction, which can result not only in the appearance of local deviations of the SB path but also in its segmentation. A typical example of such an SB is represented in Figure 3. Each segment of the mode III SB is labelled numerically from #1 to #7 and consists of two parts: (i) the arched step with a constant shear offset ranging from 500 to 50 nm and (ii) the tip with a monotonic decrease in the offset height from 50 nm to zero (Figure 3a).

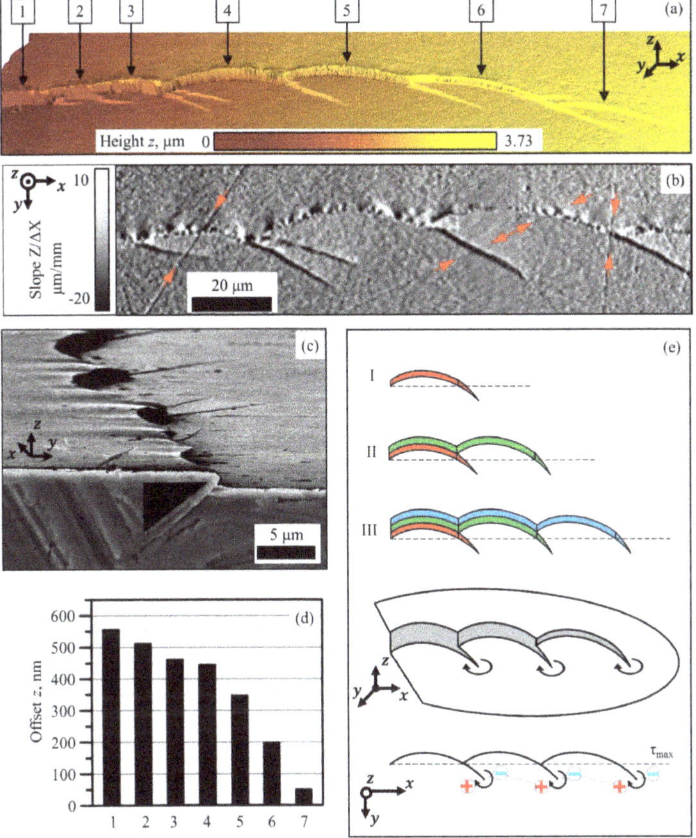

Figure 3. The typical behaviour of the mode III shear band (SB) in the compressed Pd-based bulk metallic glass (BMG) specimen. The 3D surface map obtained with the scanning white-light interferometry (SWLI) technique (**a**) shows a segmented nature of a mode III shear (275 μm length). The slope map reveals fine scratches (marked by red arrows) extending through SBs and showing the absence of a mode II shear component (**b**). The scanning electron microscope (SEM) image of the mode III SB shows 40° inclination to the surface (**c**). Shear offsets 1 to 7 measured by SWLI are represented on the bar chart (**d**). The schematic (**e**) illustrates a dislocation-based interpretation of the shear morphology formed in three consecutive steps I, II, and III, and the corresponding isometric and top views of the resulting shear. In this case, the circled arrows indicate the positions of screw dislocation lines, and the "plus"/"minus" signs indicate the tensile and compressive stress components of a macro-dislocation, respectively.

One should keep in mind that the shear plane of the observed SB is inclined with respect to the scanned specimen surface (Figure 3c). Thus, the shear offset values (Figure 3d) are larger than the vertical (Z-axis) step height values by a factor of around 1.4. This, however, does not change the mode III shear behaviour qualitatively. Such a shear band propagates in a progressive manner (at sharp contrast with simultaneous sliding), which has been shown in Reference [27]. Unlike the straight single-line mode II shear, the mode III shear path is segmented into short consequent fragments, so that each propagation event occurs one after another.

Considering that the SB tip induces a long-range stress field, one can easily rationalize the above observations. The mode III shear initiates from a stress concentrator, located at the edge of the specimen (Figure 3c). The plane of the maximum shear stresses is inclined at approximately 45° to the loading axis (40° in the case of compression). The SB tends to align itself with this plane, but the resulting trajectory is not straightforward. This is due to the fact that a path of the mode III shear in a BMG is the result of the superposition of external far-field stress and the stress field of the mode III shear itself (Figure 1c). A sum of two perpendicular acting forces—one from the external stress and one from the dislocation-induced peak stress—results in the tortuous path, which one can observe clearly for each segment.

Even in polycrystalline materials, the planar discontinuities, such as cracks, may propagate along a smoothly curved non-crystallographic path due to a specific superposition of triaxial stresses. Compare, for example, the S-shaped cracks, which are frequently observed in hydrogen embrittled steels [47].

The curved morphology of the SB indicates that the Volterra-type macro-dislocation associated with the shear band deviates from the preferred maximum macroscopic shear stress plane. The more the SB deviates from this plane, the lesser is the driving force for its propagation, which vanishes at some point, and the shear front stops (Figure 3e(I)). The next shear segment starts from the stress concentrator located at the intersection of the maximum shear plane and the previous shear path (Figure 3e(II)). In this scenario, each new SB segment increases the total offset at the surface, extends the overall SB length, deviates from the general mode III plane, and stops (Figure 3e(III)).

Another exciting feature of the mode III SB step morphology is that the individual segments tend to curve in the same direction. This observation can be logically explained by the stress fields of SB's tips. Each tip of the terminated SB can be described as a Volterra-type screw macro-dislocation, which is shown in Figure 3e—isometric view. The screw shear tips form an energetically favourable dislocation-like ladder (just like that in crystals) with alternating negative and positive stresses (see Figure 3e, top view). This happens due to the attraction of the opposite sign stress components—an example of dislocations interaction in a BMG.

3.4. Shear Offset Fluctuations

Considering the shear propagation mechanism proposed in Section 3.3, the nano-scale heterogeneities of the SB [17,18] should somehow accumulate with each segmental length increment. The regions of low density along the SB evolve to from nanovoids and microcracks, which leads to fracture [9,10,19,48]. This density fluctuation growth should then influence the observed SB step morphology. The SWLI technique allows us to detect such an influence, which manifests itself through the wavy fluctuation of the SB step height (Figure 4). The "old" SBs (i.e., bands formed by several shear events) exhibit these fluctuations, which are clearly visible on the tilted map (Figure 4a). Isometric view of the SB part zooms in the morphology and shows an inhomogeneous nature of BMG shear. The observed SB fluctuates both along its path up to ± 1.5 μm (Figure 4c) and the step height around ± 200 μm (Figure 4d).

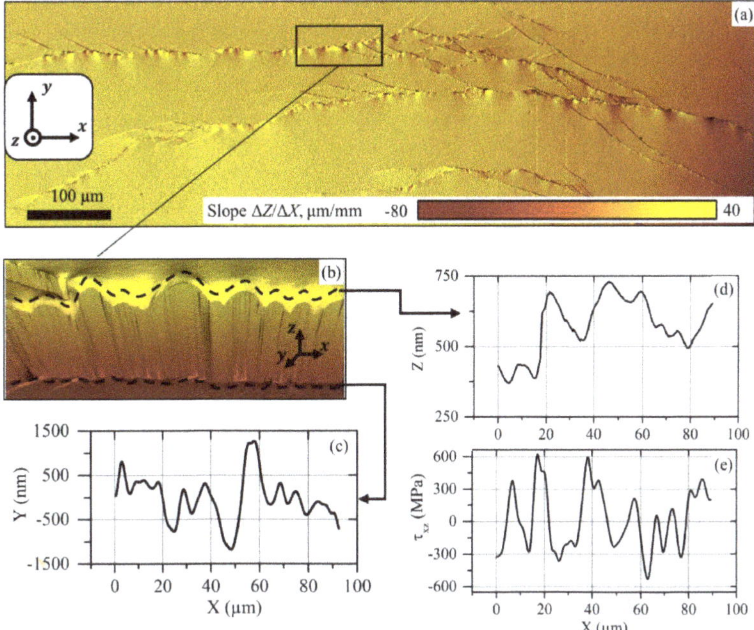

Figure 4. Fluctuations of the shear band (SB) steps on the polished surface of the Pd-based bulk metallic glass (BMG) specimen tested in compression. The 3D surface map obtained with the scanning white-light interferometry (SWLI) technique is represented as a slope map, clearly showing a wavy character of SB steps (**a**). A magnified view of the area in a black rectangular (**a**) is shown in the isometric view below (**b**), where colour scale represents height Z in the range of 750 nm. Profiles of the SB step in XY and XZ planes are shown qualitatively with the dashed lines (**b**) and quantitatively on graphs (**c,d**), respectively, including the shear strain introduced by shear step deviations (**e**).

As suggested by Gilman, the observed path deviations up to 30% would produce the elastic energy increase of less than 10% [49]. The shear stress τ_{xz} introduced by the offset fluctuation can be estimated via the Cauchy's infinitesimal strain γ_{xz} tensor as:

$$\tau_{XZ} = G{\cdot}\gamma_{XZ} = G\frac{1}{2}\left(\frac{dx}{dz} + \frac{dz}{dx}\right) = \frac{G}{2}\frac{dz}{dx}, \tag{1}$$

where G = 35.5 GPa is the shear modulus of the $Pd_{40}Cu_{30}Ni_{10}P_{20}$ alloy [2]. The calculated local shear stress value varies in the range of ± 600 MPa (Figure 4e). It is worth mentioning that a real stress range is likely even larger, considering the of 45° angle between a shear plane and the scanned surface. Such large stress values up to 1 GPa result from the accumulation of several (up to 10 in observed "old" SB) successive shear events on one plane. This means that the stress oscillations along the SB, introduced by one mode III shear event, should be of an order of magnitude smaller, i.e., around 100 MPa. However, the calculation of strains performed by Binkowski et al. [17] shows that the strain variations occur in the range from +0.30 to −0.10 alongside an individual mode II shear band, which yields the local shear stress values up to several GPa.

Nevertheless, the above described quantitative estimations strongly suggest the inherently fluctuating nature of the shearing processes occurring in BMGs. Local stress values along the SB are comparable with those at the SB tip, as shown in Section 3.1. This means that the SB stress interaction can occur not only around the SB tip but also along a whole SB length.

3.5. Shear Bands' Behaviour during Indentation

Several indentation experiments were conducted on the surface of Zr-based BMG, as described in Section 2. Examples of morphology of indented surfaces are presented in Figure 5. A typical triangular pyramidal indentation footprint is surrounded by SBs forming stepwise pile-ups with the height variations observable on 3D maps (Figure 5d,e). The 20–30 µm wide pile-ups form near each side of the indent. To reveal the influence of shear deformation prehistory on the formation of SBs, the distance between the indentation centers was chosen to be close enough to ensure that the shear zones created by two neighbouring indents overlap. The first indentation in the non-deformed material is labelled #1 and the subsequent indentation is marked #2. Three pairs of indents were made so that the edges of their footprints were aligned parallel to each other at 43 µm, 35 µm, and 28 µm apart, respectively (Figure 5a–c). The second indentation was also performed with the indenter tip perpendicular to the edge of the first imprint (Figure 5f).

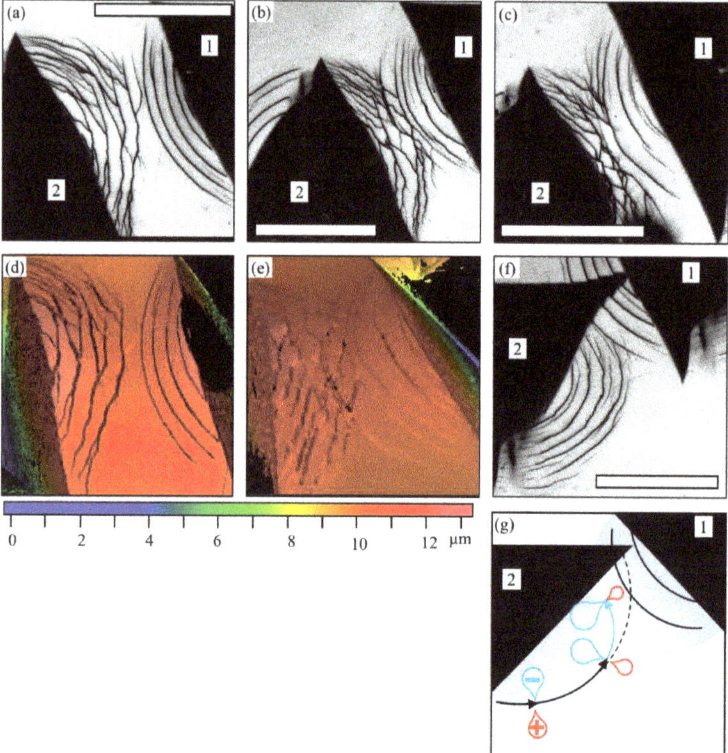

Figure 5. The topology of the surface in the vicinity of the indenter footprint reflecting the different behaviour of shear bands (SBs) formed by the indentation of the polished surface of the Zr-based bulk metallic glass (BMG). The optical images (**a–d**) and 3D surface maps (**d,e**) were obtained with the scanning white-light interferometry (SWLI) technique. The shear bands form regular arc-shaped steps when indentation is performed on a "fresh" non-deformed surface (marked as "1" on (**a–c,f**)). Indenting the area near the previous indents forms the shear bands deviated from arc-shape (marked as "2" on (**a–f**)). The scale bar refers to 50 µm on (**a–c,f**). 3D surfaces of images on (**a**) and (**b**) are coloured according to the height magnitude (**d,e**). The schematics of SB deflection due to the stress field gradient observed in (**f**) is shown in the subfigure (**g**).

Indentation of the non-deformed BMG forms regular concentric arcs of shear offsets around the pyramidal imprint, indentations #1 in Figure 5a–c. The monotonic stress field gradient form accurately curved shear traces, which can be considered as iso-stress lines (Figure 5a–f, indentation #1). The picture changes significantly when the plastic shearing is enforced in the pre-deformed area. In this case, the formation SBs caused by the second indentation is driven by the superposition of their stress fields with the external stress field created by the indenting pyramid tip and the stress field existing in the specimen from the previously formed SBs around indent #1. The most clear effect is seen as shear branching and the dramatic change in the SBs morphology form a regular circle line to a set of segmented, significantly distorted lines (Figure 5a–f). Such a complex branched morphology cannot be obtained as a geometrical sum of slip events occurring in two-slip systems [50]. A new SB can swap its arc curvature to the opposite one from the previous indent (Figure 5a). This effect can be related to the reactivation of the "old" preexisting SBs, as it was observed in Reference [34]. The significant residual compressive stresses from the first indentation can even suppress the formation of new SBs when the edges of the indents are aligned to each other (Figure 5c). In the case of the normal orientation of imprints (Figure 5f), one can notice the deflection of the second set of SBs from the ideal circumferential shape in the vicinity of the previously formed SBs. The possible mechanism of this effect is sketched schematically in Figure 5g. Although the exact initiation point of the SB is not known, it is likely to be located on the plane perpendicular to the centre of the indent side as the point of the maximum stress. The propagating SB reaches the pre-deformed area where the stresses of both indents superimpose (Figure 5g, blue), and the resulting stress field at the SB tip shifts significantly toward the compressive direction. This forces the SB to turn in the direction of the compressive stress maximum.

4. Summary and Conclusions

The results presented in this study testify to the existence and significance of Volterra dislocation-type long-range elastic strain/stress fields around a shear band tip in deformed BMGs. Such a stress field affects the behaviour of the SB under external load and can be revealed by examining the surrounding surface topography. In order to study the SB's topological features, 3D surface scanning of the deformed BMGs was conducted with aid from the SWLI technique. Different stress conditions were achieved in BMGs under mechanical compression and indentation testing to stimulate different shear modes and SB's interactions. The results of the 3D surface mapping of the deformation-induced shear zones were analysed by taking into account the existence of long-range strain/stress field around each SB tip. The results can be summarised as follows.

(1) The elastic field around a tip of the SB in a BMG was experimentally approved to be similar to the elastic field of the Volterra-type macro-dislocation, revealing edge-like (mode II shear) and screw-like (mode III shear) shear components. Thus, SBs in BMGs can be described in terms of the Volterra's macro-dislocations and their behaviour under load can be rationalised accordingly.

(2) The long-range stress fields produced by macro-dislocations are supposed to be responsible for the frequently reported SB-SB interactions causing branching, deflecting, mutual blocking, and local strain hardening.

(3) The offset of a mode II shear under uniform compression coincides with the shear propagation direction. This determines the SB's path and offsets deviations, which gives rise to the mode II shear features including the progression of a solitary shear band along the straight path without significant deviations, and a monotonic offset change from the tip to the specimen edge without significant offset deviations.

(4) The offset of the mode III shear under uniform compression is perpendicular to the shear propagation direction. Having no strong constraints on the path or offset, the mode III shear tend to curve exhibiting significant deviations of the path and in the offset. This mode of SBs shows a segmented morphology, where each segment represents a separate mode III shear, which bends away from the primary shear plane toward the side by forming a sort of a "dislocation-like" wall.

(5) The SWLI technique applied to compressed BMG specimens reveals local deviations both in the shear path and shear step height in qualitative agreement with suggestions by Gilman [49] and Binkowksi et al. [17]. The shear offset deviations were found to accumulate with every following shear event. Shear deviation accumulation is believed to be the main reason for the crack initiation and failure of BMGs.

(6) The behaviour of SBs in deformed BMGs is strongly affected by their self-induced local strain/stress fields leading to frequently observed and, reported in the literature, SB interactions revealed as branching, rejuvenation, mutual blocking, and deflection of propagating SBs.

Author Contributions: Conceptualization, methodology, software, validation, analysis, investigation, visualization, writing, original draft preparation—M.S. Writing—review and editing, supervision, project administration—A.V. Both authors have read and agreed to the published version of the manuscript.

Funding: This research received no external funding.

Acknowledgments: The authors gratefully acknowledge the Institute of Advanced Technologies of Togliatti State University, Russia, and, particularly, Dmitriy L. Merson for granting access to the experimental equipment. The authors also would like to thank the Institute of materials physics of the University of Münster, particularly, the group of Gerhard Wilde and Sergiy V. Divinski for the provision of materials and manufacturing facility for the fabrication of Pd-based metallic glasses.

Conflicts of Interest: The authors declare no conflict of interest.

References

1. Demetriou, M.D.; Launey, M.E.; Garrett, G.; Schramm, J.P.; Hofmann, D.C.; Johnson, W.L.; Ritchie, R.O. A damage-tolerant glass. *Nat. Mater.* **2011**, *10*, 123–129. [CrossRef] [PubMed]

2. Liu, Z.Q.; Zhang, Z.F. Strengthening and toughening metallic glasses: The elastic perspectives and opportunities. *J. Appl. Phys.* **2014**, *115*, 163505. [CrossRef]

3. Schroers, J.; Kumar, G.; Hodges, T.M.; Chan, S.; Kyriakides, T.R. Bulk metallic glasses for biomedical applications. *JOM* **2009**, *61*, 21–29. [CrossRef]

4. Trexler, M.M.; Thadhani, N.N. Mechanical properties of bulk metallic glasses. *Prog. Mater. Sci.* **2010**, *55*, 759–839. [CrossRef]

5. Greer, A.L.; Cheng, Y.Q.; Ma, E. Shear bands in metallic glasses. *Mater. Sci. Eng. R* **2013**, *74*, 71–132. [CrossRef]

6. Zhang, Y.; Greer, A.L. Thickness of shear bands in metallic glasses. *Appl. Phys. Lett.* **2006**, *89*, 071907. [CrossRef]

7. Park, E.S. Understanding of the Shear Bands in Amorphous Metals. *Appl. Microsc.* **2015**, *45*, 63–73. [CrossRef]

8. Maaß, R.; Birckigt, P.; Borchers, C.; Samwer, K.; Volkert, C.A. Long range stress fields and cavitation along a shear band in a metallic glass: The local origin of fracture. *Acta Mater.* **2015**, *98*, 94–102. [CrossRef]

9. Qu, R.T.; Wu, F.; Zhang, Z.-F.; Eckert, J. Direct observations on the evolution of shear bands into cracks in metallic glass. *J. Mater. Res.* **2009**, *24*, 3130–3135. [CrossRef]

10. Qu, R.T.; Wang, S.G.; Wang, X.D.; Liu, Z.Q.; Zhang, Z.F. Revealing the shear band cracking mechanism in metallic glass by X-ray tomography. *Scr. Mater.* **2017**, *133*, 24–28. [CrossRef]

11. Das, J.; Tang, M.B.; Kim, K.B.; Theissmann, R.; Baier, F.; Wang, W.; Eckert, J. "Work-Hardenable" ductile bulk metallic glass. *Phys. Rev. Lett.* **2005**, *94*, 205501. [CrossRef] [PubMed]

12. Wang, Z.T.; Pan, J.; Li, Y.; Schuh, C.A. Densification and strain hardening of a metallic glass under tension at room temperature. *Phys. Rev. Lett.* **2013**, *111*, 1–5. [CrossRef] [PubMed]

13. Takayama, S. Work-hardening and susceptibility to plastic flow in metallic glasses (rolling deformation). *J. Mater. Sci.* **1981**, *16*, 2411–2418. [CrossRef]

14. Wang, Y.; Li, M.; Xu, J. Toughen and harden metallic glass through designing statistical heterogeneity. *Scr. Mater.* **2016**, *113*, 10–13. [CrossRef]

15. Ross, P.; Küchemann, S.; Derlet, P.M.; Yu, H.B.; Arnold, W.; Liaw, P.; Samwer, K.; Maaß, R. Linking macroscopic rejuvenation to nano-elastic fluctuations in a metallic glass. *Acta Mater.* **2017**, *138*, 111–118. [CrossRef]

16. Hassani, M.; Lagogianni, A.E.; Varnik, F. Probing the degree of heterogeneity within a shear band of a model glass. *Phys. Rev. Lett.* **2019**, *123*, 195502. [CrossRef]

17. Binkowski, I.; Schlottbom, S.; Leuthold, J.; Ostendorp, S.; Divinski, S.V.; Wilde, G. Sub-micron strain analysis of local stick-slip motion of individual shear bands in a bulk metallic glass. *Appl. Phys. Lett.* **2015**, *107*, 221909. [CrossRef]

18. Rösner, H.; Peterlechner, M.; Kübel, C.; Schmidt, V.; Wilde, G. Density changes in shear bands of a metallic glass determined by correlative analytical transmission electron microscopy. *Ultramicroscopy* **2014**, *142*, 1–9. [CrossRef]

19. Zhao, Y.; Zhang, G.; Estévez, D.; Chang, C.; Wang, X. Evolution of shear bands into cracks in metallic glasses. *J. Alloys Compd.* **2015**, *621*, 238–243. [CrossRef]

20. Hufnagel, T.C.; Schuh, C.A.; Falk, M.L. Deformation of metallic glasses: Recent developments in theory, simulations, and experiments. *Acta Mater.* **2016**, *109*, 375–393. [CrossRef]

21. Dahmen, K.A.; Ben-Zion, Y.; Uhl, J.T. A simple analytic theory for the statistics of avalanches in sheared granular materials. *Nat. Phys.* **2011**, *7*, 554–557. [CrossRef]

22. Takeuchi, S.; Edagawa, K. Atomistic simulation and modeling of localized shear deformation in metallic glasses. *Prog. Mater. Sci.* **2011**, *56*, 785–816. [CrossRef]

23. Shi, L.T. Dislocation-like defects in an amorphous Lennard-Jones solid. *Mater. Sci. Eng.* **1986**, *81*, 509–514. [CrossRef]

24. Spaepen, F. A microscopic mechanism for steady state inhomogeneous flow in metallic glasses. *Acta Metall.* **1976**, *25*, 407–415. [CrossRef]

25. Shrivastav, G.P.; Chaudhuri, P.; Horbach, J. Yielding of glass under shear: A directed percolation transition precedes shear-band formation. *Phys. Rev. E* **2016**, *94*, 1–10. [CrossRef]

26. Vinogradov, A.; Seleznev, M.; Yasnikov, I.S. Dislocation characteristics of shear bands in metallic glasses. *Scr. Mater.* **2017**, *130*, 138–142. [CrossRef]

27. Qu, R.T.; Liu, Z.Q.; Wang, G.; Zhang, Z.F. Progressive shear band propagation in metallic glasses under compression. *Acta Mater.* **2015**, *91*, 19–33. [CrossRef]

28. Xu, Y.; Shi, B.; Ma, Z.; Li, J. Evolution of shear bands, free volume, and structure in room temperature rolled Pd40Ni40P20 bulk metallic glass. *Mater. Sci. Eng. A* **2015**, *623*, 145–152. [CrossRef]

29. Huo, L.S.; Wang, J.Q.; Huo, J.T.; Zhao, Y.Y.; Men, H.; Chang, C.T.; Wang, X.M.; Li, R.W. Interactions of Shear Bands in a Ductile Metallic Glass. *J. Iron Steel Res. Int.* **2016**, *23*, 48–52. [CrossRef]

30. Kovács, Z.; Schafler, E.; Szommer, P.; Révész, Á. Localization of plastic deformation along shear bands in Vitreloy bulk metallic glass during high pressure torsion. *J. Alloys Compd.* **2014**, *593*, 207–212. [CrossRef]

31. Bouzakher, B.; Benameur, T.; Yavari, A.R.; Sidhom, H. In situ characterization of a crack trajectory and shear bands interaction in metallic glasses. *J. Alloys Compd.* **2007**, *434–435*, 52–55. [CrossRef]

32. Haag, F.; Beitelschmidt, D.; Eckert, J.; Durst, K. Influences of residual stresses on the serrated flow in bulk metallic glass under elastostatic four-point bending—A nanoindentation and atomic force microscopy study. *Acta Mater.* **2014**, *70*, 188–197. [CrossRef]

33. Merson, E.; Danilov, V.; Merson, D.; Vinogradov, A. Confocal laser scanning microscopy: The technique for quantitative fractographic analysis. *Eng. Fract. Mech.* **2017**, *183*, 147–158. [CrossRef]

34. Huang, H.; Yan, J. Investigating shear band interaction in metallic glasses by adjacent nanoindentation. *Mater. Sci. Eng. A* **2017**, *704*, 375–385. [CrossRef]

35. Nollmann, N.; Binkowski, I.; Schmidt, V.; Rösner, H.; Wilde, G. Impact of micro-alloying on the plasticity of Pd-based Bulk Metallic Glasses. *Scr. Mater.* **2015**, *111*, 119–122. [CrossRef]

36. Li, H.F.; Zheng, Y.F.; Xu, F.; Jiang, J.Z. In vitro investigation of novel Ni free Zr-based bulk metallic glasses as potential biomaterials. *Mater. Lett.* **2012**, *75*, 74–76. [CrossRef]

37. Vinogradov, A.; Danyuk, A.; Khonik, V.A. Localized and homogeneous plastic flow in bulk glassy Pd40Cu30Ni10P20: An acoustic emission study. *J. Appl. Phys.* **2013**, *113*, 153503. [CrossRef]

38. Seleznev, M.; Vinogradov, A. Note: High-speed optical imaging powered by acoustic emission triggering. *Rev. Sci. Instrum.* **2014**, *85*, 076103. [CrossRef]

39. Hirth, J.P.; Lothe, J. *Theory of Dislocations*, 2nd ed.; Wiley: New York, NY, USA, 1982.

40. Volterra, V. On the equilibrum of multiply-connected elastic bodies. *Ann. Sci. Éc. Norm. Supér..* **1907**, *24*, 401–517. [CrossRef]

41. Zolotukhin, I.V.; Belyavskii, V.I.; Khonik, V.A.; Ryabtseva, T.N. Internal friction in cold-rolled metallic glasses Cu50Ti50 and Ni78Si8B14. *Phys. Status Solidi Appl. Res.* **1989**, *116*, 255–265. [CrossRef]

42. Khonik, V.A.; Spivak, L.V. On the nature of low temperature internal friction peaks in metallic glasses. *Acta Mater.* **1996**, *44*, 367–381. [CrossRef]

43. Khonik, V.A. Dislocation-like relaxations in cold deformed metallic glasses. *J. Alloys Compd.* **1994**, *211–212*, 114–117. [CrossRef]

44. Liu, C.; Roddatis, V.; Kenesei, P.; Maaß, R. Shear-band thickness and shear-band cavities in a Zr-based metallic glass. *Acta Mater.* **2017**, *140*, 206–216. [CrossRef]

45. Clouet, E.; Ventelon, L.; Willaime, F. Dislocation core energies and core fields from first principles. *Phys. Rev. Lett.* **2009**, *102*, 055502. [CrossRef]

46. Liu, Y.H.; Wang, G.; Pan, M.X.; Yu, P.; Zhao, D.Q.; Wang, W.H. Deformation behaviors and mechanism of Ni-Co-Nb-Ta bulk metallic glasses with high strength and plasticity. *J. Mater. Res.* **2007**, *22*, 869–875. [CrossRef]

47. Laureys, A.; Depover, T.; Petrov, R.; Verbeken, K. Influence of sample geometry and microstructure on the hydrogen induced cracking characteristics under uniaxial load. *Mater. Sci. Eng. A* **2017**, *690*, 88–95. [CrossRef]

48. Wang, X.D.; Qu, R.T.; Liu, Z.Q.; Zhang, Z.F. Evolution of shear-band cracking in metallic glass under cyclic compression. *Mater. Sci. Eng. A* **2017**, *696*, 267–272. [CrossRef]

49. Gilman, J.J. Flow via dislocations in ideal glasses. *J. Appl. Phys.* **1973**, *44*, 675–679. [CrossRef]

50. Xie, S.; George, E.P. Hardness and shear band evolution in bulk metallic glasses after plastic deformation and annealing. *Acta Mater.* **2008**, *56*, 5202–5213. [CrossRef]

Article

Metallic Glasses: A New Approach to the Understanding of the Defect Structure and Physical Properties

Vitaly Khonik [1,*,†] and Nikolai Kobelev [2,†]

[1] Department of General Physics, State Pedagogical University, Lenin St. 86, 394043 Voronezh, Russia

[2] Institute for Solid State Physics, Russian Academy of Sciences, Chernogolovka, 142432 Moscow, Russia; kobelev@issp.ac.ru

* Correspondence: v.a.khonik@vspu.ac.ru

† These authors contributed equally to this work.

Received: 25 April 2019; Accepted: 20 May 2019; Published: 24 May 2019

Abstract: The work is devoted to a brief overview of the Interstitialcy Theory (IT) as applied to different relaxation phenomena occurring in metallic glasses upon structural relaxation and crystallization. The basic hypotheses of the IT and their experimental verification are shortly considered. The main focus is given on the interpretation of recent experiments on the heat effects, volume changes and their link with the shear modulus relaxation. The issues related to the development of the IT and its relationship with other models on defects in metallic glasses are discussed.

Keywords: metallic glasses; defects; structural relaxation; heat effects; shear elasticity; interstitialcy theory

1. Introduction

Metallic glasses (MGs) constitute an amazing example of a man-made non-crystalline state, which is not observed in nature. These materials are very promising in the sense of technological applications that results in the growing interest to the investigation of their physical properties. At first glance, since MGs do not have directional chemical bonding, one can expect that the understanding of their structure and properties should be a simpler task than for other types of glasses. However, any commonly accepted theory describing their formation and main structural features has been absent thus far, and any general theory of non-crystalline substances is still lacking as well. This largely constrains the development of new type MGs with the physical properties predicted in advance. Since MGs are prepared by melt quenching while the melt is obtained by fusion of the crystalline state, one should naturally expect a relationship between these states, including a connection between the properties of the glass and those of the maternal crystal (whose melt is used for the glass production). These interdependencies should be taken into account by a theoretical model of glass. On the other hand, since it is commonly accepted that glass is a frozen liquid, this model should also imply certain relation to the melting mechanism and formation of the liquid state. By that, it should include as a major ingredient the notions on structural defects, which are intrinsically related to the whole glass prehistory, i.e., maternal crystal→melt→glass. To date, quite a few theoretical models describing the structural features, defects and different properties of MGs have been suggested [1–12]. In our opinion, however, the above general requirements are satisfied by another approach—the Interstitialcy theory (IT) of condensed matter states suggested by Granato [13,14]. It was demonstrated in recent years that the IT provides a powerful tool for the understanding and predicting different relaxation phenomena in MGs and unambiguously shows a genetic relationship of the glass with the maternal crystal. Therefore, this brief overview is firstly devoted to an analysis of the major hypotheses of the IT and main experiments related to its verification. Since certain issues related to the IT have already been discussed [15,16], this work is largely devoted to the new experiments and their interpretations while earlier information is mentioned briefly. Finally, a relation of the IT to other models of the metallic glass structure and its defects is discussed.

2. The Interstitialcy Theory

The Interstitialcy theory is essentially based on the experimental investigations of the effect of low-temperature (4 K), low-dose soft neutron irradiation on the elastic moduli of single crystal copper carried out by Granato's group in the 1970s [17,18]. The irradiation leads to the formation of isolated Frenkel pairs (vacancy+interstitial) remaining mostly immobile in the liquid helium temperature range. A careful analysis of the experimental data led to the conclusion that the formation of interstitial defects results in a strong decrease of the shear modulus (diaelastic effect) C_{44} according to $\Delta C_{44}/C_{44} = -B_i c_i$, where c_i is the interstitial concentration and $B_i \approx 30$ is the shear sensitivity. For the vacancies, the shear sensitivity is by an order of magnitude smaller, $B_v \approx 2$. The effect of point defects on the bulk modulus is also small, comparable to the diaelastic effect produced by the vacancies. At the same time, similar results were obtained upon electron irradiation of single crystal aluminum [19]. It was argued that a strong diaelastic effect can be observed only if the interstitials do not make up an octahedral configuration, as considered before the 1970s (red circle in Figure 1a), but form a dumbbell (split) structure, which is characterized by the two atoms trying to occupy the same lattice cite (red circles in Figure 1b). To underline the difference between these defects, the latter defect is also called an interstitialcy [20]. It is of major importance that the octahedral interistitial is a spherically symmetrical defect (similarly to the vacancy) while the dumbbell interstitial is clearly anisotropic and, thus, constitutes an elastic dipole strongly interacting with the external shear stress [19,21]. It is this feature of the defect, which leads to a strong diaelastic effect. It is now generally accepted that dumbbell intertitials exist in all main lattices and constitute the defect state with the lowest formation enthalpy [22–26].

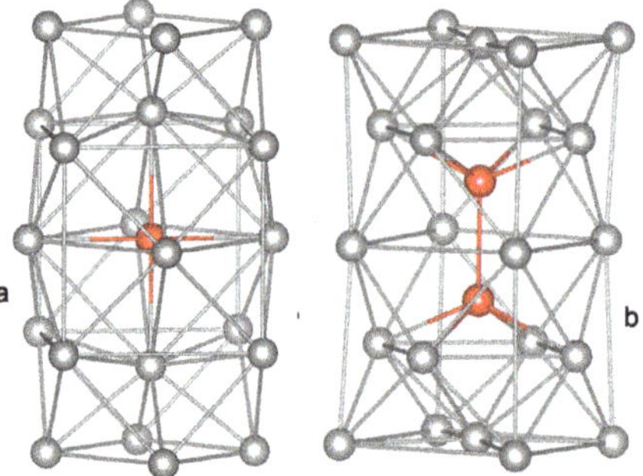

Figure 1. Octahedral (**a**) and dumbbell (split) (**b**) interstitial defects in a computer model of a face-centered cubic lattice [27]. All of the dumbbell atoms (marked by red circles) are characterized by $< 0,2,8,0 >$ Voronoi indexes. With permission from JETP Letters, 2019.

An important consequence of the interstitial dumbbell structure consists in the appearance of the low-frequencies in the vibrational spectrum, which are by a few times smaller than the Debye frequency [28]. This in turn leads to the high formation entropy of the defect (for copper, $S_f \approx (5-10)$ k_B according to the Granato's estimate [14]) and, respectively, to a decrease of the Gibbs formation free energy at high temperatures. The extrapolation of the elastic moduli of irradiated crystals towards high defect concentration in the experiments [17,18] showed that $C_{44} \to 0$ at $c_i \approx 0.02 \div 0.03$ providing a guess that such big defect concentration should lead to the crystal→liquid transition, because the liquid is characterized by a vanishing (or very small) shear modulus [29,30]. This allowed Granato to suggest that melting of metals should be related to the rapid thermoactivated generation of dumbbell interstitials. Another important point realized by Granato was the understanding that the defect formation enthalpy is proportional to the unrelaxed shear modulus G, in line with earlier investigations [31,32]. The above hypotheses led Granato to the formulation of the interstitialcy theory, which includes melting as an integral part [13,14] (see also a discussion given below). In fact, the mathematical formalism of the IT is based on the two equations,

$$\frac{\partial H_i^F}{\partial c_i} = \alpha G \Omega, \tag{1}$$

$$G = \mu \, exp(-B_i c_i), \tag{2}$$

where H_i^F is the interstitial formation enthalpy, α is a dimensionless parameter close to unity, μ is the shear modulus of the defectless crystal, Ω is the volume per atom and $B_i = \alpha\beta$ with β being the dimensionless shear susceptibility. This quantity was estimated by Granato as $\beta = -3C_{4444}/C_{44} \approx 40$, where C_{44} is the shear modulus of the crystal and C_{4444} is its fourth-rank (anharmonic) shear modulus. On the other hand, the shear susceptibility is related to the internal energy U of the crystal as $\beta = \dfrac{1}{16\mu}\dfrac{\partial^4 U}{\partial \varepsilon^4}$, where ε is the shear strain. The above equations show that the shear susceptibility constitutes a fundamental parameter of the material since it is proportional to the ratio of the fourth-rank shear modulus to the "usual" shear modulus.

The general approach of the IT to the crystal→melt→glass transformation and the relationship between these states consists in the following. Crystal melting is related to a rapid increase of the concentration of dumbbell interstitials, which remain identifiable structural units in the liquid state (as confirmed by later computer modeling [33]). Rapid melt quenching freezes the melt defect structure in the solid glass and different relaxation processes occurring in it upon heat treatment can be interpreted in terms of the changes of the defect concentration by using Equations (1) and (2). For the glassy state, the quantities G and μ in these equations correspond to the shear moduli of glass and maternal crystal, respectively. Despite the simplicity of these equations, it has been found that the IT, although originally derived for simple metals, provides explanations for many experiments on multicomponent MGs, as reviewed earlier [15,16] and discussed below.

3. Verification of the Main Starting Hypotheses of the Interstitialcy Theory

3.1. Shear and Dilatation Contributions into the Defect Elastic Energy

The IT is actually built on the hypothesis that the shear component of the elastic energy created by interstitial defects is predominant while the dilatation contribution can be neglected. This agrees with later calculations by Dyre [34] who showed that it is the shear strain component, which produces the main contribution into the elastic energy far from a point defect in a solid. He derived a simple relation for the ratio of the dilatation U_{dil} and shear U_{shear} components of this energy and concluded that the former is much smaller, i.e.,

$$\frac{U_{dil}}{U_{shear}} = \frac{2B/G}{\left[9\left(\dfrac{B}{G}\right)^2 + 8\dfrac{B}{G} + 4\right]} \leq 0.1, \tag{3}$$

where B is the bulk modulus. This equation, however, was derived within the linear elasticity approach for a spherically symmetric defect and does not account for the energy of the defect nucleus. For further verification of the above Granato's hypothesis, a molecular-static modeling of interstitial defects in four FCC metals was performed [27]. To compare the contributions U_{dil} and U_{shear} into the total elastic energy, the local relative change of the Voronoi polyhedra volume V_i for each atom with respect to the Voronoi polyhedra volume of an atom in the ideal lattice V_0 was accepted as a measure of the volume change upon defect formation, i.e., $U_{dil} = \frac{B}{2V_0} \int (\Delta V/V)^2 dV \approx \frac{B}{2} \Sigma_i \frac{(V_0 - V_i)^2}{V_0^2}$. The shear component of the elastic energy was taken as a difference $U_{shear} = H_f - U_{dil}$ with H_f being the interstitial formation enthalpy. The calculations showed that the ratio U_{dil}/U_{shear} for the interstitial dumbbell is nearly twice as that given by Equation (3) due to the accounting of the elastic energy of the defect nucleus. This correction allows using this equation for 63 elemental metals. The result shown in Figure 2 implies that this ratio does not exceed 0.15 for more than 90% of metals. Assuming that MGs contain similar interstitial-type defects in line with the IT, the same calculation procedure was applied to 189 metallic glass composition and nearly the same result was obtained (see Figure 2). Thus, in both cases, the contribution of the dilatation energy is indeed much smaller than that given by the shear energy, just as supposed by Granato [13,14].

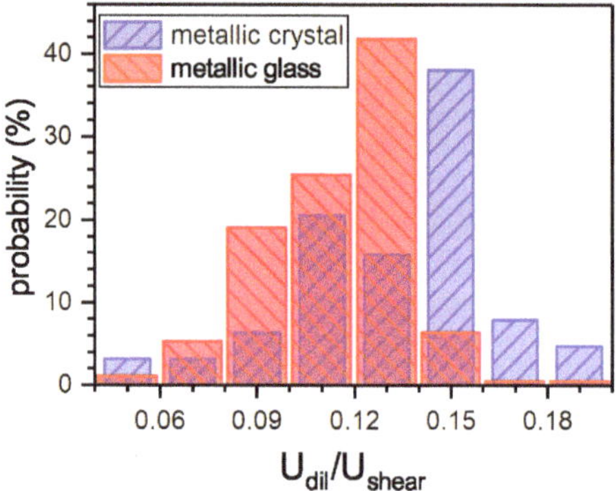

Figure 2. Histogram illustrating the distribution of the ratio of dilatation U_{bulk} and shear U_{shear} components of the elastic energy for dumbbell interstitials in 63 polycrystalline elemental metals. The same data for interstitial-type defects in 189 metallic glasses are also shown [27]. With permission from JETP Letters, 2019.

3.2. Increase of the Interstitial Concentration before Melting

As mentioned above, the IT argues that melting of metals is related to a rapid increase of the concentration of dumbbell interstitials. This is a crucial statement of the theory. Since dumbbell interstitials produce tenfold bigger shear softening as compared with vacancies, this effect can be detected experimentally provided that there are no contributions coming from other defects (e.g., dislocations) in the crystal. Such experiments were recently performed on single crystal aluminum and coarse-crystalline indium [35,36]. The main results of these measurements are shown in Figure 3. Despite the usual opinion that the equilibrium interstitial concentration c_i is negligible at any temperature [37], it is seen that c_i rapidly increases for both metals upon approaching the melting temperature T_m. In aluminum, this concentration remains smaller than the vacancy concentration c_v while for indium c_i becomes even bigger than c_v near T_m. Besides that, the data on Al clearly demonstrate an increasing tendency of c_i-growth at high temperatures $T \geq 926$ K (0.99 T_m) while the formation Gibbs free energy start to rapidly decrease in this region [35]. These features agree with the predictions of the IT. It is worth noting that increasing concentration of dumbbell interstitials can explain the non-linear growth of the heat capacity of simple metals near T_m, whose nature remains hitherto unclear [38]. It should also be mentioned that temperature dependence of the interstitial concentration obtained for aluminum [35] allowed an estimate of their formation entropy S_i^F, which was found to be $\approx 7 \, k_B$ [39], in full agreement with Granato's value [14].

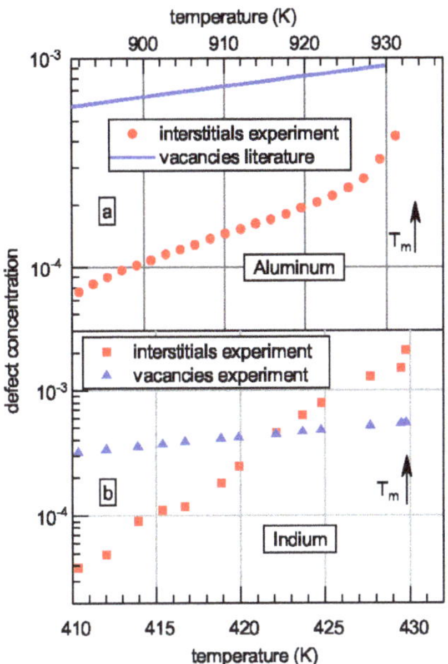

Figure 3. Estimates of interstitial and vacancy concentrations in crystalline aluminum (**a**) and indium (**b**) derived from the diaelastic effect measurements. The melting temperatures are indicated [35,36]. With permission from Pleiades Publishing, LTD, 2019.

3.3. Identification of Interstitial-Type Defects in the Glassy State

The topological pattern of dumbbell interstitials in crystals is very clear—two atoms, trying to occupy the same lattice cite (Figure 1b). However, any similar topological picture in the glassy state is absent. In this case, one can try to identify these defects by searching structural regions, which display the properties similar to those of dumbbell interstitials. Thus far, such attempts have been performed using computer models of glassy copper and aluminum [40,41]. It was found that certain nano-sized regions reveal large non-affine displacements and can be characterized by a strong sensitivity to the applied shear stress and distinctive local shear strain fields, which are described by the local shear susceptibility as well as by the diaelastic compliance and diaelastic polarizability tensors. Another feature consists in the characteristic low- and high-frequency modes (far below and above the Debye frequency, respectively) in the vibration spectra of atoms belonging to these regions. All these peculiarities are quite similar to those of dumbbell interstitials in crystals. Thus, interstitial-type defects indeed exist in non-crystalline simple metals. Two-component metallic structures should be analyzed in a similar way. Besides that, numerous simulations of MGs show the presence of the atoms characterized by $< 0, 2, 8, 0 >$ Voronoi indexes (or close to them) [7], which constitute a characteristic feature of dumbbell interstitials in crystals (Figure 1b). One the other hand, an interstitial defect is characterized by the two atoms with these indexes. We are unaware of any attempts for searching two adjacent atoms with $< 0, 2, 8, 0 >$ or close indexes in computer models of MGs.

3.4. Shear Susceptibility

The large magnitude of the shear susceptibility β (see Equation (2)), which determines the influence of defects on the shear modulus (diaelastic effect), constitutes a salient ingredient of the IT. This quantity is controlled by the non-linearity of the solid, specifically by the magnitude of the non-linear shear modulus C_{4444} or, in an isotropic approximation, by the quantity $\gamma_4 = \dfrac{1}{16} \dfrac{\partial^4 U_{el}}{\partial \varepsilon^4}$, where U_{el} is the elastic energy and ε is the shear strain. To determine the value of the shear susceptibility

β, the effect of elastic loading on the ultrasound velocity was studied on two (Zr- and Pd-based) MGs [42,43]. The shear susceptibility derived in these experiments is smaller than that originally estimated by Granato ($\beta \approx 40$) but has nonetheless the same order of magnitude. Moreover, this quantity is close to the estimates derived by other methods [44] (see Section 3.5 and Table 1).

3.5. Relation between the Shear Modulus and Heat Effects

A simple analysis of Equations (1) and (2) shows that the IT implies an intrinsic relation between shear modulus changes and heat effects in MGs. Indeed, integration of Equation (1) taking account of Equation (2) leads to an expression for the enthalpy increment ΔH upon insertion of interstitial-type defects, which can be expressed through the shear moduli of glass and maternal crystal, $\Delta H \approx (\mu - G)/\beta\rho$, where β is the shear susceptibility and ρ is the density. Then, one can arrive at an expression relating the heat flow with the change of the shear modulus,

$$W = \frac{\dot{T}}{\rho\beta}\left(\frac{G}{\mu}\frac{\partial\mu}{\partial T} - \frac{\partial G}{\partial T}\right), \tag{4}$$

where $\dot{T}$ is the rate of temperature change. This relationship was repeatedly tested and found to give an excellent description of the heat flow on the basis on shear modulus relaxation data not only upon structural relaxation below the glass transition temperature T_g and in the supercooled liquid state but upon crystallization as well [45,46]. The latter fact is really surprising and actually suggests that the whole excess internal energy ΔU (with respect to the crystalline maternal state) is mostly related to the interstitial-type defect system frozen-in upon glass production. An analysis gives a simple expression for this quantity [47],

$$\Delta U \approx \Delta H \approx \frac{\mu}{\beta\rho}\left(1 - \frac{G}{\mu}\right). \tag{5}$$

It is seen that the excess internal energy is simply controlled by the shear moduli of glass and maternal crystal. Upon crystallization, the defect system disappears and this energy is released as heat, i.e., $\Delta U \approx Q_{cryst}$, where Q_{cryst} is the heat of crystallization. A specially designed experiment confirmed this idea [48]. From a physical viewpoint, this result means that the whole heat content of glass (i.e., the excess enthalpy with respect to the crystalline maternal state) is mostly determined by the interstitial-type defect system frozen-in upon glass production.

On the other hand, considering the initial and relaxed states of MGs and calculating the corresponding differences for the heat flow and shear modulus, $\Delta W = W_{rel} - W$ and $\Delta G = G_{rel} - G$, Equation (4) after simplification can be rewritten as [49]

$$\Delta W = -\frac{\dot{T}}{\rho\beta}\frac{\partial\Delta G}{\partial T}. \tag{6}$$

This relationship shows that the quantities ΔW and $\partial\Delta G/\partial T$ should be proportional to each other. Figure 4 shows these quantities derived from calorimetric and shear modulus data for bulk glassy $Zr_{65}Al_{10}Ni_{10}Cu_{15}$. It is clearly seen that they are indeed proportional. This allows determination of proportionality constant and calculation of the shear susceptibility β. Table 1 shows β-values thus derived for a few MGs. It is seen that these values belong to a relatively narrow range $15 \leq \beta \leq 22$ that, in general, agrees with the IT. Other methods used for the determination of β give close results [49]. Since the shear susceptibility determines the shear softening effect (via Equation (2)), the heat effects (according to Equations (4) and (5)) and also related to the anharmonicity of the interatomic potential, it appears to be a major integral parameter of the glassy structure.

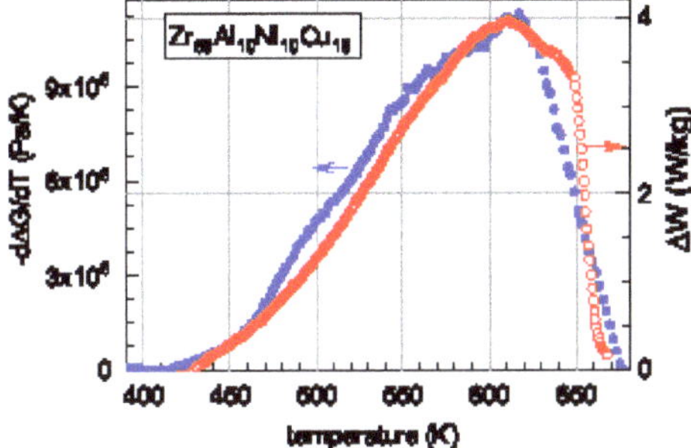

Figure 4. Temperature dependences of the quantities ΔW and $\partial \Delta G / \partial T$ entering Equation (6) derived from calorimetric and shear modulus measurements [49]. The data correspond to structural relaxation below the glass transition. With permission from Elsevier, 2019.

Table 1. Determination of the shear susceptiblity β on the basis of calorimetric and shear modulus data taken on MGs in the initial and relaxed state using Equation (6) [49]. With permission from Elsevier, 2019.

No	Metallic Glass	β
1	$La_{55}Al_{25}Co_{20}$	21
2	$Zr_{46}Cu_{46}Al_8$	19
3	$Zr_{46}Cu_{45}Al_7Ti_2$	21
4	$Zr_{56}Co_{28}Al_{16}$	17
5	$Zr_{65}Al_{10}Ni_{10}Cu_{15}$	22
6	$Pd_{40}Ni_{40}P_{20}$	15
7	$Pd_{41.25}Cu_{41.25}P_{17.5}$	21
8	$Pd_{40}Cu_{30}Ni_{10}P_{20}$	19

4. Refinement of the Parameters of the Interstitialcy Theory

As reviewed earlier [15,16] and discussed in the present work, the IT provides a good description of different aspects of MGs relaxation behavior. At the same time, some model parameters of the IT were introduced in a phenomenological way. It is therefore desirable to clarify their physical meaning and relationship with material parameters. In the initial model, an interstitial defect was considered by Granato as an elastic string. At the same time, a split interstitial can be treated as an elastic dipole [47,50]. The corresponding "dipole" approach is based on the expansion of the energy into a series in powers of the elastic strain created by the dipoles [50]. It was found that that the Granato and "dipole" approaches give practically identical expressions for the elastic energy and shear modulus [47]. A comparison of these approaches leads to an expression for the parameter α introduced in the original version of the IT (see Equations (1) and (2)) as $\alpha = \int (\varepsilon_{ij}\varepsilon_{ji} - \frac{1}{3}\varepsilon_{ii}^2)dV/\Omega$, where ε_{ij} is the elastic strain field created by the interstitial. Thus, the parameter α characterizes the "strength" of the defect. The shear susceptibility within the "dipole" approach was calculated as $\beta = -\gamma_4 \Omega_t / \mu$, where μ is the shear modulus of the maternal crystal and $\Omega_t = 1.38$ is a parameter characterizing the elastic anisotropy of the interstitial [50]. This estimate is about two times smaller than that given by Granato. It should be noted that the "dipole" approach was found very useful upon further IT development, especially in the part that accounts for the effect of the concentration of interstitials on their interaction (see Section 6).

5. Recent Experiments

5.1. Reconstruction of Temperature Dependence of the Shear Modulus Using Calorimetric Data

Equation (4) for the heat flow $W(T)$ can be used in the "opposite" way, i.e., for the calculation of the shear modulus relaxation using input calorimetric data . This leads to a relation [51]

$$G(T) = \frac{G_{rt}}{\mu_{rt}}\mu(T) - \frac{\rho\beta}{T}\int_{T_{rt}}^{T} W(T)dT. \tag{7}$$

where the subscript "rt" refers to the room temperature. Figure 5 gives experimental temperature dependences of the shear modulus of a Zr-based glass in the initial state, after relaxation obtained by heating into the supercooled liquid region and after full crystallization. The figure also shows temperature dependences of the shear modulus in the initial and relaxed states calculated with Equation (7) using experimental calorimetric heat flow $W(T)$, temperature dependence of the shear modulus in the crystalline state $\mu(T)$, experimental parameters and material constants entering this equation. It is seen that the calculation reproduces experimental $G(T)$-data quite well, including shear modulus growth due to structural relaxation below T_g and shear softening in the supercooled liquid region.

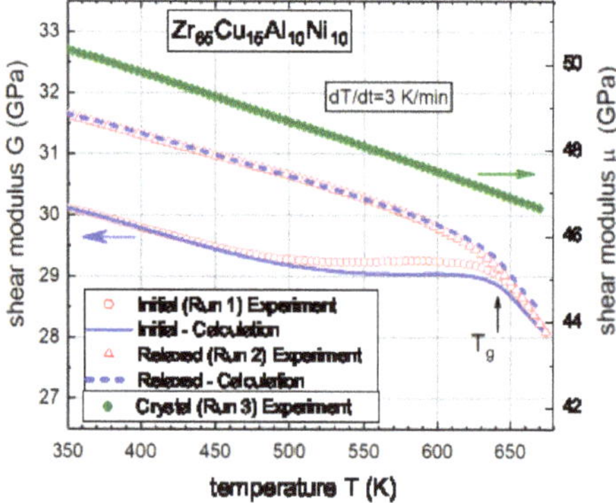

Figure 5. Experimental and calculated using Equation (7) temperature dependences of the shear modulus G of glassy Zr$_{65}$Cu$_{15}$Al$_{10}$Ni$_{10}$ in the initial and relaxed states. Temperature dependence of the shear modulus μ after full crystallization is also shown. Calorimetric T_g is indicated by the arrow [51]. With permission from Elsevier, 2019.

5.2. Heat Absorption Occurring upon Heating of Relaxed Glass

Within the IT framework, the heat absorbed upon heating from room temperature T_{rt} to a temperature T_{sql} in the supercooled liquid region can be calculated as [52]

$$Q \approx \frac{1}{\beta\rho}\left(G_{T_{rt}} - G_{T_{sql}} - \mu_{T_{rt}} + \mu_{T_{sql}}\right), \tag{8}$$

where $G_{T_{rt}}$, $G_{T_{sql}}$, $\mu_{T_{rt}}$ and $\mu_{T_{sql}}$ are the shear moduli of glass and maternal crystal at temperatures T_{rt} and T_{sql}, respectively, other quantities are the same as above. Since the shear moduli $G_{T_{sql}}$ and $\mu_{T_{sql}}$ in the supercooled liquid state do not depend on the thermal prehistory, the only quantity in Equation (8), which varies upon structural relaxation, is the room-temperature shear modulus. The moduli $G_{T_{sql}}$, $\mu_{T_{rt}}$ and $\mu_{T_{sql}}$ are only temperature dependent. The temperature T_{sql} can be accepted as a constant and the quantities $G_{T_{sql}}$, $\mu_{T_{rt}}$ and $\mu_{T_{sql}}$ are then constants as well. Thus, the heat Q is dependent on a single variable $G_{T_{rt}}$. The latter can be changed by preliminary heat treatment. Figure 6 shows the experimental data on the heat absorbed upon warming up of $Pd_{40}Ni_{40}P_{20}$ glass into the supercooled liquid as a function of the shear modulus measured at 330 K. The data points correspond to different preannealing treatments as indicated. It is seen that $Q(G_{T_{330}})$-data nicely fall onto a straight line, as implied by Equation (6). The slope of this line within a few percent error agrees with its theoretical value given by this equation as $\partial Q / \partial G_{330} = \dfrac{1}{\beta \rho}$ [52]. Similarly, one can describe [53] the widely known so-called "sub-T_g enthalpy relaxation" effect, which consists in the growth of the heat absorption near the glass transition temperature T_g in MGs subjected to prolonged preannealing well below T_g [54,55].

The data discussed above in Sections 3.5, 5.1 and 5.2 convincingly demonstrate a close relationship between shear modulus relaxation and heat effects in MGs, in full agreement with the IT predictions.

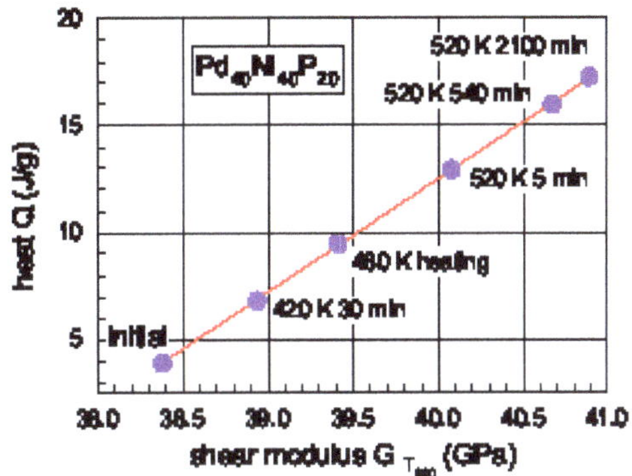

Figure 6. Dependence of the integral heat Q absorbed upon heating from 330 K to 610 K (supercooled liquid region) as a function of the shear modulus G_{330} measured at 330 K just after heating onset. The points correspond to different preannealing treatment applied for shear modulus measurements and DSC tests as indicated. The solid line gives the lest square fit [52]. With permission from Elsevier, 2019.

5.3. Density Changes upon Structural Relaxation and Crystallization

The interpretation of volume changes using the IT is based on the expected change of the volume ΔV upon creation of a dumbbell interstitial defect, which can be represented as $\Delta V / \Omega = -1 + r_i$, where r_i is the so-called relaxation volume reflecting the relaxation of the structure after defect creation, and Ω is the volume per atom [24,56]. Then, if a defect concentration c is created, the volume increases by ΔV and the relative volume change becomes $\Delta V / V = (r_i - 1)c$. Using the shear modulus given by Equation (2), one arrives at the relative change of the density upon isothermal structural relaxation at a particular temperature as

$$\left[\frac{\Delta \rho(t)}{\rho_0} \right]_{rel} = \frac{(r_i - 1)}{\alpha \beta} \ln \frac{G(t)}{G_0}, \tag{9}$$

where $\Delta \rho(t) = \rho(t) - \rho_0$, the densities $\rho(t)$ and ρ_0 correspond to the shear moduli $G(t)$ and $G_0 = G(t = 0)$, respectively, α and β are the same as in Equation (2). An example of relative density changes as a function of shear modulus change of glassy $Pd_{30}Cu_{30}Ni_{10}P_{20}$ measured at room temperature after isothermal annealing is given in Figure 7 [57]. It is seen that this dependence can be fitted by a straight line, in accordance with Equation (9). This equation implies the slope of this line equal to $\dfrac{(r_i - 1)}{\alpha \beta}$. With $\alpha = 1$ (as usually assumed), the shear susceptibility for this glass $\beta = 19$ (Table 1) and the relaxation volume $r_i = 1.6$ (as for FCC metals [24]), one arrives at the slope equal to 0.031, in a good agreement with the experimental slope of 0.037 (Figure 7).

Equation (9) can be then modified as [57]

$$\frac{d \ln G}{d \ln V} = \frac{\alpha \beta}{r_i - 1}. \tag{10}$$

With β and r_i given above, one arrives at $dlnG/dlnV = 32$, rather close to the experimental value $dlnG/dlnV = 25$ reported in Ref. [58] for the same glass.

For the density change upon crystallization, the IT gives [57]

$$\left(\frac{\Delta\rho}{\rho}\right)_{cryst} = \frac{(r_i - 1)}{\alpha\beta}\, ln\frac{\mu}{G}, \tag{11}$$

where $\Delta\rho = \rho_{cryst} - \rho$ with ρ and ρ_{cryst} being the densities of glass and maternal crystal, respectively. It is seen that, if the relaxation volume $r_i > 1$, the density change $\Delta\rho/\rho > 0$ upon crystallization. It was shown that Equation (11) then provides a reasonable explanation of crystallization-induced density changes of Zr-based MGs [57]. However, for loosely packed crystalline structures, one can expect that the relaxation volume r_i is less than unity. In this case, Equation (11) predicts a *decrease* of the density upon crystallization. Indeed, the literature gives a few examples of crystallization-induced density decrease of about 1% [59,60] or even more [61,62].

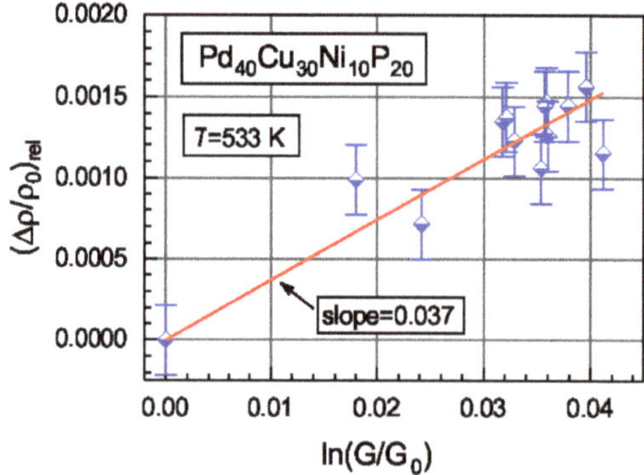

Figure 7. Dependence of the relative density change on the quantity $ln(G/G_0)$, where G_0 is the initial room-temperature shear modulus and G is the room-temperature shear modulus after annealing of bulk glassy $Pd_{40}Cu_{30}Ni_{10}P_{20}$ at $T = 533$ K [57]. With permission from Elsevier, 2019.

For warming up from room to the temperature of the full crystallization, the above reasoning leads to the relationship

$$\frac{\Delta\rho(T)}{\rho_{rt}} = \frac{r_i - 1}{\alpha\beta}\, ln\left[\frac{\mu_{rt}}{G_{rt}}\frac{G(T)}{\mu(T)}\right], \tag{12}$$

where $\Delta\rho(T)$ is the density change upon heating. It was shown that this equation provides a good description of $\Delta\rho/\rho$-changes occurring upon heating up to the temperature of the full crystallization [63]. It can be concluded that changes of the density are controlled by the shear moduli of glass and maternal crystal, which in turn reflect the evolution of the interstitial-type defect system.

5.4. Relation between the Enthalpies of Relaxation, Crystallization and Melting

According to the general IT approach discussed above (Section 2), the interstitial-type defects in glass are inherited from the melt. Provided that the melt quenching rate is big enough, one can expect that the defect concentrations in the glass and melt are nearly equal, $c_{glass} \approx c_{melt}$. The quantity c_{glass} determines the whole excess heat content (enthalpy) of the glass. Subsequent structural relaxation below T_g leads to a decrease of this concentration by Δc_{rel} that results in a growth of the shear modulus from G up to $G_{rel} = G\, exp(\alpha\beta\Delta c_{rel})$ and corresponding release of the enthalpy ΔH_{rel}. Upon crystallization, the remaining defect concentration $c_{melt} - \Delta c_{rel} \approx c_{cryst}$ drops down to zero (the defects disappear), the shear modulus increases up to its value μ in the crystalline state with simultaneous release of the enthalpy ΔH_{cryst}. The whole heat release of the initial glass after crystallization is then given by Equation (5), where $\Delta U \approx \Delta H_{cryst}$, as discussed above (Section 3.5). Since the defect concentration quenched-in from the melt is $c_{melt} \approx \Delta c_{rel} + c_{cryst}$, the heat absorbed upon melting should be approximately equal to the total heat release upon structural relaxation and crystallization , i.e., in terms of the corresponding enthalpy changes,

$$\Delta H_{melt} \approx - \left(\Delta H_{cryst} + \Delta H_{rel} \right) . \tag{13}$$

The result of a specially designed experiment [64] aimed at the verification of this relationship is reproduced in Figure 8, which shows that the data taken on eleven Zr-, Pd- and La-based MGs can be approximated by a straight line with the unity slope. Thus, the relationship in Equation (13) is indeed valid within the experimental error (about 10%) confirming the idea on a connection between the defects occurring upon melting of the maternal crystal and those disappearing upon structural relaxation and crystallization of the glass.

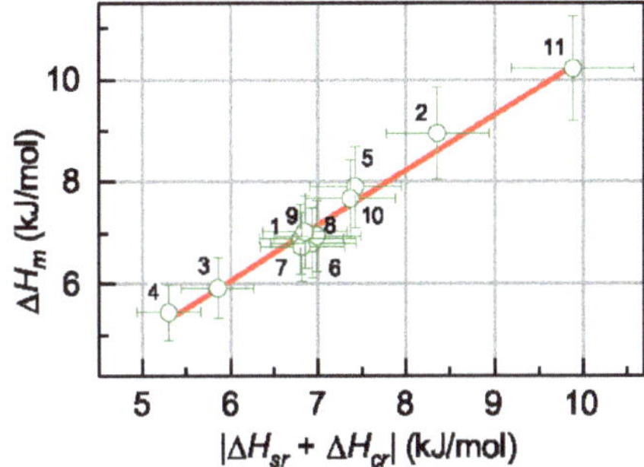

Figure 8. The melting enthalpy vs. the absolute value of sum of the enthalpies of structural relaxation and crystallization. The numbers correspond to different Zr-, Pd- and La-based MGs [64]. With permission from Elsevier, 2019.

5.5. Relation of the Boson Heat Capacity Peak to the Defect Structure

A peculiar universal feature of atomic dynamics in non-crystalline materials consists in the presence of excess (over the Debye contribution) low-frequency vibrational modes [65,66]. These low-frequency modes are detected as a peak in the low temperature (5–15 K) heat capacity C plotted as C/T^3 vs. T. These features are usually referred to as the boson peak, which is known for metallic glasses as well [67]. The nature of the boson peak constitutes a matter of intensive ongoing debates [68–71]. Granato argued that the boson peak originates from low-frequency resonance vibration modes of interstitial-type defects frozen-in upon glass production [72]. He showed that the boson peak height should be proportional to the defect concentration. A refined equation for the boson peak height has the form [73]

$$h_B = \frac{C^d}{T_B^3} = \frac{234\,R}{\Theta_D^3} \left[0.09 f \left(\frac{\omega_D}{\omega_r} \right)^3 + \frac{3}{2} \beta \right] c \equiv \Gamma c, \tag{14}$$

where T_B is the boson peak temperature, C^d is the heat capacity related to the interstitial-type defect system, Θ_D and ω_D are the Debye temperature and Debye frequency of the maternal crystal, respectively, ω_r is the characteristic frequency of interstitial resonance vibrations, f is the number of resonance modes per interstitial-type defect, R is the universal gas constant and other quantities are specified above. The defect concentration c can be monitored by measurements of the shear moduli of glass and maternal crystal as implied by Equation (2). Thus, the boson peak within the framework of the IT is considered to be a "fingerprint" of the defect glass structure.

An experiment aimed to check the prediction given by Equation (14) was carried out on glassy $Zr_{65}Cu_{15}Ni_{10}Al_{10}$ [73]. The main result of this experiment is shown in Figure 9, which gives the measured height of the boson peak as a function of the defect concentration c calculated with Equation (2) using room-temperature measurements of the shear modulus after different annealing treatments. The annealing protocol was designed to perform measurements on both fully amorphous and partially crystalline samples. It is seen that independent of the state of the samples (amorphous/partially crystalline), the boson peak height linearly increases with the defect concentration, in line with Granato's prediction [72]. The derivative dh_B/dc calculated from Figure 9 provides reasonable estimates for the resonant vibration frequencies of the defects assumed to be responsible for the boson peak [73].

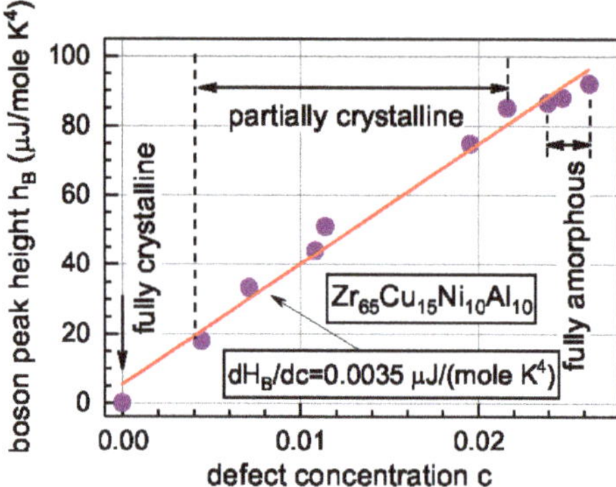

Figure 9. The height of the boson peak as a function of the defect concentration c calculated using Equation (2). The line gives the least square fit [73]. With permission from John Wiley and Sons, 2019.

On the other hand, since the interstitial-type defect structure determines the excess enthalpy ΔH of glass, as discussed above in Section 5.3, one can expect that the boson peak height h_B should also be related to ΔH. The calculation gives the relation between the boson peak height and excess enthalpy of glass ΔH as [74]

$$h_B = \frac{\Gamma}{\alpha\beta}ln\frac{\mu}{\mu - \rho\beta\Delta H},\tag{15}$$

where the quantity Γ is defined by Equation (14), ρ is the density, and μ, α and β are same as in Equation (2). The relationship in Equation (15) directly connects the boson peak height with the excess enthalpy of the glass. The latter may be considered as an independent variable, which can be changed by the annealing leading to either structural relaxation within the glassy state or partial crystallization. Using differential scanning calorimetry, the excess enthalpy can be determined as $\Delta H = \dot{T}^{-1}\int_{T_i}^{T_f} W(T)dT$, where W is the heat flow measured by the calorimeter, $\dot{T}$ is the heating rate, T_i can be accepted equal to room temperature and T_f is the temperature leading to the full crystallization. Figure 10 shows the dependence of the boson peak height h_B calculated using Equation (15) together with the experimental h_B-data as function of the variable $ln\left[\mu/(\mu - \rho\beta\Delta H)\right]$ [74]. It is seen, first, that the calculated and experimental h_B-points are quite close. The experimental dependence $h_B(\Delta H)$ nicely falls onto a straight line and the slope of this dependence equals to $(5.14 \pm 0.29) \times 10^{-4}$ J/(mol $\times$ K^4). This agrees with this slope given by Equation (15) as $\frac{\Gamma}{\alpha\beta} = 4.9 \times 10^{-4}$ J/(mol $\times$ K^4). Thus, the IT-based approach reproduces the boson peak height and provides a good description of its height on the excess enthalpy of the glass.

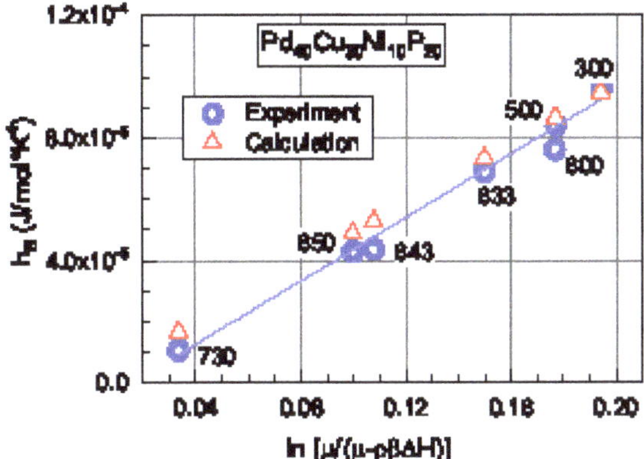

Figure 10. Experimental and calculated height of the boson peak h_B as a function of the excess enthalpy ΔH plotted according to Equation (15). The numbers near the data points indicate the corresponding preannealing temperatures in Kelvins [74]. With permission from John Wiley and Sons, 2019.

An interesting study of the boson peak was recently reported by Brink et al. [75]. They performed a molecular dynamic simulation of an equiatomic CuNiCoFe alloy in the crystalline and amorphous states alternatively with chemical disorder (high-entropy state), structural disorder and reduced density. They found that the density reduction and fluctuations of the elastic constants cannot be responsible for the boson peak. However, they revealed that the boson peak in the crystal increases with the concentration of dumbbell interstitials while other defects (e.g., dislocations) do not contribute to it. Interstitial atoms even at a small concentration lead to a boson peak, which is close to that in the glass of the same composition. At that, the vibrational modes of interstitial defects in the crystal resemble those of glass. Finally, the authors of [75] concluded that the softened regions provided by interstitials resemble the "soft spots" discussed in the literature on MGs [70,76] and the boson peak is due to quasi-localized defect-related modes.

5.6. Relation between the Properties of Glass and Maternal Crystal

In general, one can expect that the physical properties of MGs should be somehow related with the properties of their maternal crystalline states, which were used for the production of these glasses by melt quenching. Indeed, for instance, the properties of PdNiCuP glasses and their relaxation upon annealing should necessarily be related to the properties of intermetallic and/or metal-phosphide crystalline phases. To our knowledge, however, this issue was not raised in the literature.

Meanwhile, the IT is intrinsically based on the crystal→glass relationship. It starts from the main IT equation for the shear modulus of the glass G, which is scaled by the shear modulus of the maternal crystal μ (defectless state) as $G = \mu \, exp(-\alpha\beta c)$. That is why the shear moduli of glass and maternal crystal determine major thermodynamic parameters of the glass, including its excess internal energy (≈enthalpy) as given by Equation (5). As a result, the shear moduli G and μ explicitly enter all relations for the heat effects (Equations (4), (6), (8) and (13)), volume relaxations (Equations (9)–(12)) and low temperature excess heat capacity (boson peak) (Equations (14) and (15)). By that, in most cases, one must know not only G and μ at particular temperatures but also their exact temperature dependences, otherwise any quantitative agreement with the experiment cannot be achieved. Temperature dependences of the shear moduli of glass and maternal crystal reflect the interstitial-type defect structure of glass and the physical origin of crystal→glass relationship within the framework of the IT is intrinsically related to this defect structure, which controls the fundamental properties of the glass [77]. In a certain sense (not structural), one can accept Granato's statement that "… glasses and dense liquids are crystals containing a few percent of interstitials" [78].

6. Development of the Interstitialcy Theory

Despite the amazing matching of the IT predictions with a number of relaxation phenomena in MGs, there exist a few phenomena, which cannot be easily interpreted within the framework of the original Granato's theory. First, an increase of the apparent defect concentration above T_g due to the thermal activation, which follows from the original IT version (e.g., see Figure 10 in Ref. [16]), faces certain difficulties related to the high formation enthalpies necessary for this process. The same issue applies to the understanding of so-called rejuvenation of relaxation properties of MGs by quenching (or even relatively slow cooling) from the supercooled liquid region (i.e., from above T_g) [79,80]. However, this difficulty can be avoided by assuming that an interstitial-type defect in the glass has a few energy states and the transitions between them are accessible by thermal activation. In the original IT version, this possibility was not assumed. One can suggest that the occurrence of a spectrum of energy states is related to the high defect density in the glass and melt, which is estimated to be a few percent [13,16]. Such high density of the defects will result in their strong interaction and this must be taken into account. In Granato's original approach, the defect interaction was considered only qualitatively.

It is long known from the physics of crystals that this interaction can lead to the formation of interstitial clusters consisting of N individual interstitials, from $N = 2$ up to $N = 7$ and even more [23,28,81]. Clustering is energy profitable since the formation enthalpy per interstitial decreases with the number of interstitials [23,81]. The interstitial cluster consisting of $N = 7$ split interstitials in the FCC lattice represents a *perfect* icosahedron with $< 0, 0, 12, 0 >$ Voronoi indexes [81], as illustrated by Figure 11 for the FCC cell. Thus, if melting of metallic crystals is related to an increase of the concentration of split interstitials up to a few percent, as considered by the IT, then one can expect the formation of clusters consisting of $N = 2$ to $N = 7$ interstitial-type defects. It is to be emphasized in this relation the commonly accepted notion that icosahedral clusters constitute a major structural feature of metallic melts and their concentration increases upon supercooling defining thus the dynamic slowdown of the internal movements and eventual glass formation [7,82].

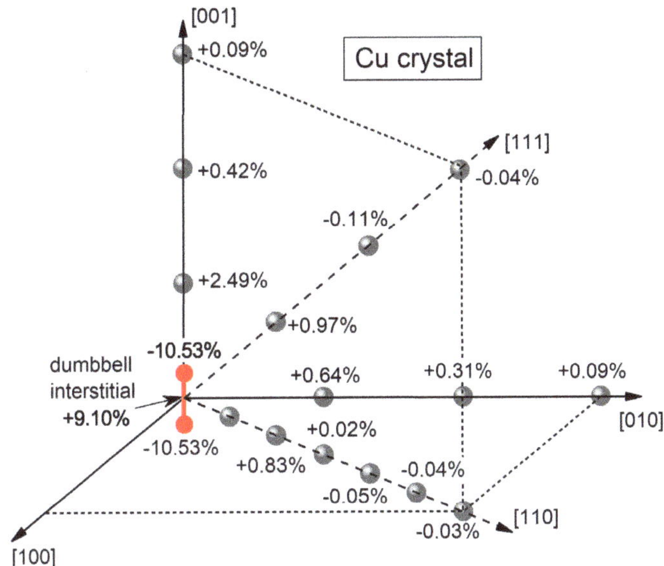

Figure 11. Formation of a perfect icosahedron by the creation of dumbbell interstitials on the opposite faces of the FCC cell: (**Left**) elementary FCC cell where the arrows show how two atoms are inserted instead of one atom; and (**Right**) perfect icosahedron with $< 0, 0, 12, 0 >$ Voronoi indexes formed by six dumbbell interstitials on the faces of the cell and one interstitial in the octahedral position (i.e., in center of the cell, see Figure 1a) [81].

One can expect, therefore, that the solid glass will contain, first, individual interstitial-type defects (adjacent atoms with $< 0, 2, 8, 0 >$ Voronoi indexes and/or close to them) and their small clusters ($N = 2, 3$), which correspond to the defect part of the structure. Larger clusters define the icosahedral-type structural backbone. Thus, the metallic glass constitutes a heterogeneous structure. The properties of the clusters should be evidently dependent on the number N. Specifically, the shear sensitivity B_i should decrease with N defining a reduction of the diaelastic effect produced by the clusters. By that, the vibrational entropy of interstitial-type defects in the clusters should decrease due to their interaction leading to the mutual damping of the low-frequency vibration modes, as was qualitatively noted in the original Granato's model [14]. The quasi-equilibrium balance between all these clusters will be dependent on temperature and thermal prehistory defining the evolution of glass properties [83,84]. Some qualitative estimates of clustering kinetics are given elsewhere [85,86]. Further detailed work in this direction is challenging.

7. Comparison with Other Models

Quite a few models are suggested in the literature for a description of defects in metallic glasses. A common shortage in most of them is the lack of understanding of their nature. However, it turns out that many of these models are quite consistent with the IT, as sketched below.

Egami suggested that the main properties of MGs are determined by the defects, which create shear, hydrostatic compression and/or tension (so-called τ-, p- and n-defects) [8]. Meanwhile, the dumbbell interstitials are featured by the same combination of stresses. Figure 12 shows the changes of the Voronoi polyhedra volume (indicated by the numbers) for a dumbbell interstitial in a Cu crystal with respect to the ideal lattice. It is seen that the interstitial produces both positive and negative changes of the Voronoi polyhedra volume. In other words, the defect creates both hydrostatically compressed and hydrostatically tensioned regions. One can expect that the same should be applicable for an interstitial-type defect in the glass. Taking into account that shear stress field is one of its main characteristics, one can conclude that the defect demonstrates the properties similar to those of Egami's τ-, p- and n-defects.

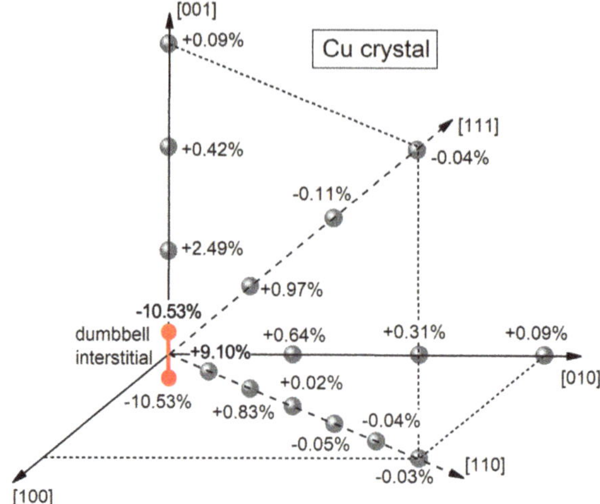

Figure 12. Dumbbell [001]-oriented interstitial (two red circles) and the relative changes of the volume of the Voronoi polyhedra $\Delta V/V$ (indicated by the numbers) as compared with the ideal lattice for the atoms (shaded circles) in a copper crystal. $\Delta V/V$-changes are shown along the dumbbell axis, perpendicular to it and in the [111] direction. It is seen that the defect creates both positive and negative changes of the Voronoi polyhedra volume (i.e., changes of the local density) [27]. With permission from JETP Letters, 2019.

It is often mentioned that supercooled liquids contain "strings" ("string-like" solitons), which become frozen in the solid glass [12,87–91]. It is also sometimes said that these defects resemble the signatures of dumbbell interstitials in crystals [33,88]. Meanwhile, the "string" idea is compatible with the IT since a split interstitial in the original Granato's model was considered as a string segment [13,14]. Similar arguments can be applied for the defects viewed as "shear transformation zones" [3,92], "soft spots" [70,76], "soft zones" [92], "flow units" [10,93], "liquid-like regions" [11], "geometricallly unfavored motifs" [76,94] and "regions with large non-affine displacements" [3,9,76]. Since the region of an interstitial-type defect is characterized by the big shear susceptibility, the shear deformation of the surrounding material upon action of the applied stress is bigger that of the matrix and contains a large non-affine component [41]. Naturally, the above terms ("soft zones", "liquid-like regions", etc.) can be applied to this region. It should be noted that these defects also contribute to the low-frequency part of the vibration spectrum of a glassy structure [76,93] similar to what is assumed by the IT.

Another approach to the understanding of defects in MGs is the "free volume" model, which was suggested long ago [1,95,96] and subjected to numerous modifications since then (e.g., Ref. [97]). In this approach, the defects are considered as regions of the reduced density (vacancy-like "free volume"), which affect relaxation and deformation phenomena in MGs. The premise for its popularity consists, on the one hand, in a decreased density of MGs (frozen-in "free volume") with respect to their maternal crystals and, on the other hand, in the existence of numerous correlations of MGs' properties with the amount of the "free volume" [98]. Although this model was repeatedly criticized in different directions [5,99], it nonetheless constitutes perhaps the most popular approach. Meanwhile, it is quite compatible with the IT. Indeed, one should recall that the volume change ΔV_v upon vacancy formation is $\Delta V_v / \Omega = r_v + 1$, where r_v is the corresponding relaxation volume and Ω is the volume per atom [56]. Taking into account the volume change ΔV_i occurring upon interstitial formation (see Section 5.2 above) and accepting, e.g., for Al, $r_i = 1.9$ and $r_v = -0.38$ [24], one can easily arrive to the ratio of the relative volume changes produced by interstitials and vacancies, $(\Delta V/V)_i / (\Delta V/V)_v = (r_i - 1)/(r_v + 1) = 1.45$. This means that the volume changes for interstitials and vacancies have the same sign and are quite comparable in the magnitude [100]. Thus, a decrease of the defect concentration within both the free volume model and the IT should lead to the densification of the glass by about the same amount. Moreover, since the free volume in both models is proportional to the number of the defects, one can naturally expect a correlation of material properties with the amount of the free volume. However, there are major differences between these approaches. In the former one, it is the free volume that constitutes the principal source for the property changes. The IT considers the features of the interstitial-type defect (large shear susceptibility, high formation entropy, and specific strain fields) to be of major importance while defect-related volume changes are of secondary relevance. It is also very important that vacancy-like free volume appears to have the spherical symmetry, which is why it *should not* interact with the external shear stress. Conversely, an interstitial defect is strongly asymmetric (Figure 1b) and constitutes an elastic dipole displaying strong sensitivity to the external shear load. Moreover, if a metallic glass is formed by melting of a loosely packed crystalline structure (related to the generation of interstitial-type defects), the creation of the free volume is not necessary at all. In this case, the density of glass can be even bigger than that of the maternal crystal (see Section 5.2 above) that cannot be understood within the free volume approach.

8. Concluding Remarks

The Interstitialcy Theory (IT) of condensed matter states thermodynamically predicts that melting of simple metallic crystals is related to a rapid increase of the concentration of the interstitial defects in their most stable dumbbell form just near the melting temperature. Recently, rather convincing (although indirect) experimental arguments confirming this hypothesis were obtained. Computer simulations showed that these defects remain identifiable structural objects in the liquid state. Rapid melt quenching freezes them in the solid glass. In the liquid and glassy states, these defects do not have any clear topological pattern as in crystals (two atoms trying to occupy the same lattice site) but nonetheless display all the properties characteristic of dumbbell interstitials in crystals, as confirmed by computer modeling of mono-atomic metallic systems. These properties include strong susceptibility to the applied shear stress, specific strain fields and low-frequency modes in the terahertz vibration spectrum. It is found that rather numerous relaxation phenomena can be understood by assuming that these interstitial-type defects indeed exist in real multicomponent metallic glasses (MGs).

The mathematical formalism of the IT is quite simple and based on the two relations linking the unrelaxed shear modulus, which constitutes the basic thermodynamical parameter of the IT (being the second derivative of the Gibbs free energy with respect to the shear strain), with the formation enthalpy of interstitial-type defects and their concentration. A thermoactivated change of the defect concentration (due to structural relaxation below T_g, in the supercooled liquid state or upon crystallization) leads to an alteration of the formation enthalpy and results in numerous heat effects, which are intrinsically related to the relaxation of the shear modulus, as verified by specially designed experiments. On the other hand, changes of the defect concentration lead to certain volume changes, which can also be monitored by the shear modulus relaxation. Accepting reasonable values of the material parameter for the interstitial-type volume relaxation (i.e., the relaxation volume), one can explain the kinetics and final volume changes occurring upon structural relaxation and crystallization.

The IT leads to the conclusion that the excess internal energy and enthalpy of the glassy structure with respect to the maternal crystal is mostly related to the elastic energy of the interstitial-type defect structure frozen-in from the melt upon glass production. The full dissipation of this elastic energy constitutes the heat of crystallization. On the other hand, since the total amount of the defects is determined by the melting, there arises a relationship between the latent heat absorbed upon melting and heat release occurring upon structural relaxation and crystallization, as verified experimentally.

Within the framework of the IT, the low temperature boson heat capacity peak originates from low-frequency resonance vibration modes of interstitial-type defects and, thus, its height should be proportional to the defect concentration. Specially performed experiments showed that this is indeed the case provided that the concentration is derived from measurements of the shear modulus, as assumed by the IT. The defect concentration determines the excess enthalpy of the glass and it is the origin, which defines the observed dependence of the boson peak height on the experimentally measured excess enthalpy.

In general, the obtained results convincingly demonstrate an intrinsic relationship of the shear modulus relaxation with the heat and volume effects occurring upon structural relaxation and crystallization of MGs. At that, the properties of the glass are tightly related with those of the maternal crystal, in line with the basic assumptions of the IT. It has also been shown that other models describing defects and properties of MGs are largely compatible with the IT. The ways for the development of the IT and its relation to other models on defects in MGs are discussed.

Author Contributions: The authors equally contributed to this work.

Funding: This research was supported by the Ministry of Science and Education of the Russian Federation under the grant 3.1310.2017/4.6.

Acknowledgments: The authors are grateful to their colleagues, Yu.P. Mitrofanov, R.A. Konchakov, A.S. Makarov, G.V. Afonin and E.V. Goncharova, for long-term fruitful cooperation.

Conflicts of Interest: The authors declare no conflict of interest.

References

1. Spaepen, F. A microscopic mechanism for steady state inhomogeneous flow in metallic glasses. *Acta Metall.* **1977**, *25*, 407–415. [CrossRef]
2. Argon, A.S. Plastic deformation in metallic glasses. *Acta Metall.* **1979**, *27*, 47–58. [CrossRef]
3. Falk, M.L.; Langer, J.S. Dynamics of viscoplastic deformation in amorphous solids. *Phys. Rev. E* **1998**, *57*, 7192–7205. [CrossRef]
4. Miracle, D.B. The efficient cluster packing model—An atomic structural model for metallic glasses. *Acta Mater.* **2006**, *54*, 4317–4336. [CrossRef]
5. Miracle, D.B.; Egami, T; Flores, K.M.; Kelton, K.F. Structural aspects of metallic glasses. *MRS Bull.* **2007**, *32*, 629–634. [CrossRef]
6. Miracle, D.B.; Greer, A.L.; Kelton, K.F. Icosahedral and dense random cluster packing in metallic glass structures. *J. Non-Cryst. Sol.* **2008**, *354*, 4049–4055. [CrossRef]
7. Cheng, Y.Q.; Ma, E. Atomic-level structure and structure–property relationship in metallic glasses. *Prog. Mater. Sci.* **2011**, *56*, 379–473. [CrossRef]
8. Egami, T. Atomic level stresses. *Prog. Mater. Sci.* **2011**, *56*, 637–653. [CrossRef]
9. Peng, H.L.; Li, M.Z.; Wang, W.H. Structural signature of plastic deformation in metallic glasses. *Phys. Rev. Lett.* **2011**, *106*, 135503. [CrossRef] [PubMed]
10. Wang, D.P.; Zhu, Z.G.; Xue, R.J.; Ding, D.W.; Bai, H.Y.; Wang, W.H. Structural perspectives on the elastic and mechanical properties of metallic glasses. *J. Appl. Phys.* **2013**, *114*, 173505. [CrossRef]
11. Z. Wang, Z.; Sun, B.A.; Bai, H.Y.; Wang, W.H. Evolution of hidden localized flow during glass-to-liquid transition in metallic glass. *Nat. Commun.* **2014**, *5*, 5823. [CrossRef]
12. Zhang, H.; Zhong, C.; Douglas, J.F.; Wang, X.; Cao, Q.; Zhang, D.; Jiang, J.-Z. Role of string-like collective atomic motion on diffusion and structural relaxation in glass forming Cu-Zr alloys. *J. Chem. Phys.* **2015**, *142*, 164506. [CrossRef]
13. Granato, A.V. Interstitialcy model for condensed matter states of face-centered-cubic metals. *Phys. Rev. Lett.* **1992**, *68*, 974–977. [CrossRef]
14. Granato, A.V. Interstitialcy theory of simple condensed matter. *Eur. J. Phys.* **2014**, *87*, 18. [CrossRef]

15. Khonik, V.A. Understanding of the structural relaxation of metallic glasses within the framework of the interstitialcy theory. *Metals* **2015**, *5* 504–529. [CrossRef]

16. Khonik, V.A. Interstitialcy theory of condensed matter states and its application to non-crystalline metallic materials. *Chin. Phys. B* **2017**, *26*, 016401. [CrossRef]

17. Holder, J.; Rehn, L.E.; Granato, A.V. Effect of self-interstitials on the elastic constants of copper. *Phys. Rev. B* **1974**, *10*, 363–375. [CrossRef]

18. Holder, J.; Granato, A.V.; Rehn, L.E. Experimental evidence for split interstitials in copper. *Phys. Rev. Lett.* **1974**, *32*, 1054–1057. [CrossRef]

19. Robrock, K.-H.; Schilling, W. Diaelastic modulus change of aluminum after low temperature electron irradiation. *J. Phys. F Metal Phys.* **1976**, *6*, 303–314. [CrossRef]

20. Seitz, F. On the theory of diffusion in metals. *Acta Cryst.* **1950**, *3*, 355–363. [CrossRef]

21. Nowick, A.S.; Berry, B.S. *Anelastic Relaxation in Crystalline Solids*; Academic Press: New York, NY, USA; London, UK, 1972.

22. Schilling, W. Self-interstitial atoms in metals. *J. Nucl. Mater.* **1978**, *69–70*, 465–489. [CrossRef]

23. Robrock, K.H. *Mechanical Relaxation of Interstitials in Irradiated Metals*; Springer: Berlin, Germany, 1990.

24. Wolfer, W.G. Fundamental properties of defects in metals. In *Comprehensive Nuclear Materials*; Konings, R.J.M., Ed.; Elsevier: New York, NY, USA, 2012.

25. Ma, P.-W.; Dudarev, S.L. Universality of point defect structure in body-centered cubic metals. *Phys. Rev. Mater.* **2019**, *3*, 013605. [CrossRef]

26. Ma, P.-W.; Dudarev, S.L. Symmetry-broken self-interstitial defects in chromium, molybdenum, and tungsten. *Phys. Rev. Mater.* **2019**, *3*, 043606. [CrossRef]

27. Konchakov, R.A.; Makarov, A.S.; Afonin, G.A.; Kretova, M.A.; Kovelev, N.P.; Khonik, V.A. Relation between the shear and dilatation energy of interstitial defects in metallic crystals. *J. Exp. Theor. Phys. Lett.* **2019**, *109*, 473–478.

28. Dederichs, P.H.; Lehman, C.; Schober, H.R.; Scholz, A.; Zeller, R. Lattice theory of point defects. *J. Nucl. Mater.* **1978**, *69–70*, 176–199. [CrossRef]

29. Born, M. Thermodynamics of crystals and melting. *J. Chem. Phys.* **1939**, *7*, 591–603. [CrossRef]

30. Forsblom, M.; Grimvall, G. How superheated crystals melt. *Nat. Mater* **2005**, *4*, 388–390. [CrossRef] [PubMed]

31. Nemilov, S.V. Kinetics of elementary processes in the condensed state. II. Shear relaxation and the equation of state for solids. *Russ. J. Phys. Chem.* **1968**, *42*, 726–729.

32. Dyre, J.C.; Olsen, N.B.; Christensen, T. Local elastic expansion model for viscous-flow activation energies of glass-forming molecular liquids. *Phys. Rev. B* **1996**, *53*, 2171–2174. [CrossRef]

33. Nordlund, K.; Ashkenazy, Y.; Averback, R.S.; Granato, A.V. Strings and interstitials in liquids, glasses and crystals. *Europhys. Lett.* **2005**, *71*, 625–631. [CrossRef]

34. Dyre, J.C. Dominance of shear elastic energy far from a point defect in a solid. *Phys. Rev. B* **2007**, *75*, 092102. [CrossRef]

35. Safonova, E.V.; Mitrofanov, Y.P.; Konchakov, R.A.; Vinogradov, A.Y.; Kobelev, N.P.; Khonik, V.A. Experimental evidence for thermal generation of interstitials in a metallic crystal near the melting temperature. *J. Phys. Condens. Matter* **2016**, *28*, 215401. [CrossRef]

36. Goncharova, E.V.; Makarov, A.S.; Konchakov, R.A.; Kobelev, N.P.; Khonik, V.A. Premelting generation of interstitial defects in polycrystalline indium. *J. Exp. Theor. Phys. Lett* **2017**, *106*, 35–39. [CrossRef]

37. Gottstein, G. *Physical Foundations of Materials Science*; Springer: Berlin, Germany, 2004.

38. Safonova, E.V.; Konchakov, R.A.; Mitrofanov, Y.P.; Kobelev, N.P.; Vinogradov, A.Yu.; Khonik V.A. Contribution of interstitial defects and anharmonicity to the premelting increase in the heat capacity of single-crystal aluminum. *J. Exp. Theor. Lett.* **2016**, *103* 765–768. [CrossRef]

39. Kobelev, N.P.; Khonik, V.A. On the enthalpy and entropy of point defect formation in crystals. *J. Exp. Theor. Phys.* **2018**, *126*, 340–346. [CrossRef]

40. Konchakov, R.A.; Kobelev, N.P.; Khonik, V.A.; Makarov, A.S. Elastic dipoles in the model of single-crystal and amorphous copper. *Phys. Sol. State* **2016**, *58*, 215–222. [CrossRef]

41. Goncharova, E.V.; Konchakov, R.A.; Makarov, A.S.; Kobelev, N.P.; Khonik, V.A. Identification of interstitial-like defects in a computer model of glassy aluminum. *J. Phys. Condens. Matter* **2017**, *29*, 305701. [CrossRef]

42. Kobelev, N.P.; Kolyvanov, E.L.; Khonik, V.A. Higher order elastic moduli of the bulk metallic glass $Zr_{52.5}Ti_5Cu_{17.9}Ni_{14.6}Al_{10}$. *Phys. Sol. State* **2007**, *49*, 1209–1215. [CrossRef]

43. Kobelev, N.P.; Kolyvanov, E.L.; Khonik, V.A. Higher-order elastic moduli of the metallic glass $Pd_{40}Cu_{30}Ni_{10}P_{20}$. *Phys. Sol. State* **2015**, *57*, 1483–1487. [CrossRef]

44. Konchakov, R.A.; Makarov, A.S.; Afonin, G.V.; Mitrofanov, Y.P.; Kobelev, N.P.; Khonik V.A. Estimate of the fourth-rank shear modulus in metallic glasses. *J. Alloys Compd.* **2017**, *714*, 168–171. [CrossRef]

45. Mitrofanov, Y.P.; Wang, D.P.; Makarov, A.S.; Wang, W.H.; Khonik, V.A. Towards understanding of heat effects in metallic glasses on the basis of macroscopic shear elasticity. *Sci. Rep.* **2016**, *6*, 23026. [CrossRef]

46. Makarov, A.S.; Afonin, G.V.; Mitrofanov, Y.P.; Kobelev, N.P.; Khonik, V.A. Heat effects occurring in the supercooled liquid state and upon crystallization of metallic glasses as a result of thermally activated evolution of their defect systems. *Phys. Stat. Sol. (A)* **2019**, submitted.

47. Kobelev, N.P.; Khonik, V.A. Theoretical analysis of the interconnection between the shear elasticity and heat effects in metallic glasses. *J. Non-Cryst. Sol.* **2015**, *427*, 184–190. [CrossRef]

48. Afonin, G.V.; Mitrofanov, Y.P.; Makarov, A.S.; Kobelev, N.P.; Wang, W.H.; Khonik, V.A. Universal relationship between crystallization-induced changes of the shear modulus and heat release in metallic glasses. *Acta Mater.* **2016**, *115*, 204–209. [CrossRef]

49. Makarov, A.S.; Mitrofanov, Y.P.; Afonin, G.V.; Kobelev, N.P.; Khonik, V.A. Shear susceptibility—A universal integral parameter relating the shear softening, heat effects, anharmonicity of interatomic interaction and "defect" structure of metallic glasses. *Intermetallics* **2017**, *87*, 1–5. [CrossRef]

50. Kobelev, N.P.; Khonik, V.A.; Makarov, A.S.; Afonin, G.V.; Mitrofanov, Yu.P. On the nature of heat effects and shear modulus softening in metallic glasses: A generalized approach. *J. Appl. Phys.* **2014**, *115*, 033513. [CrossRef]

51. Makarov, A.S.; Mitrofanov, Y.P.; Afonin, G.V.; Kobelev, N.P.; Khonik, V.A. Predicting temperature dependence of the shear modulus of metallic glasses using calorimetric data. *Scr. Mater.* **2019**, *168*, 10–13. [CrossRef]

52. Makarov, A.S.; Afonin, G.V.; Mitrofanov, Y.P.; Konchakov, R.A.; Kobelev, N.P.; Qiao, J.C.; Khonik, V.A. Relationship between the heat effects and shear modulus changes occurring upon heating of a metallic glass into the supercooled liquid state. *J. Non-Cryst. Sol.* **2018**, *500*, 129–132. [CrossRef]

53. Mitrofanov, Y.P.; Afonin, G.V.; Makarov, A.S.; Kobelev, N.P.; Khonik, V.A. A new understanding of the sub-T_g enthalpy relaxation in metallic glasses. *Intermetallics* **2018**, *101*, 116–122. [CrossRef]

54. Chen, H.S. On mechanisms of structural relaxation in a $Pd_{48}Ni_{32}P_{20}$ glass. *J. Non-Cryst. Sol.* **1981**, *46*, 289–305. [CrossRef]

55. Busch, R.; Johnson, W.L. The kinetic glass transition of the $Zr_{46.75}Ti_{8.25}Cu_{7.5}Ni_{10}Be_{27.5}$ bulk metallic glass former-supercooled liquids on a long time scale. *Appl. Phys. Lett.* **1998**, *72*, 2695–2697. [CrossRef]

56. Gordon, C.A.; Granato, A.V.; Simmons, R.O. Evidence for the self-interstitial model of liquid and amorphous states from lattice parameter measurements in krypton. *J. Non-Cryst. Sol.* **1996**, *205–207*, 216–220. [CrossRef]

57. Goncharova, E.V.; Konchakov, R.A.; Makarov, A.S.; Kobelev, N.P.; Khonik, V.A. On the nature of density changes upon structural relaxation and crystallization of metallic glasses. *J. Non-Cryst. Sol.* **2017**, *471*, 396–399. [CrossRef]

58. Harms, U.; Jin, O.; Schwarz, R.B. Effects of plastic deformation on the elastic modulus and density of bulk amorphous $Pd_{40}Cu_{30}Ni_{10}P_{20}$. *Non-Cryst. Sol.* **2003**, *317*, 200–205. [CrossRef]

59. Shen, T.D.; Harms, U.; Schwarz, R.B. Correlation between the volume change during crystallization and the thermal stability of supercooled liquids. *Appl. Phys. Lett.* **2003**, *83*, 4512–4514. [CrossRef]

60. Safarik, D.J.; Schwarz, R.B. Elastic constants of amorphous and single-crystal $Pd_{40}Cu_{40}P_{20}$. *Acta Mater.* **2007**, *55*, 5736–5746. [CrossRef]

61. Panova, G.K.; Chernoplekov, N.A.; Shikov, A.A.; Savel'ev, B.I.; Khlopkin, M.N. Effects of amorphization on the vibrational specific heat of metallic glasses. *Sov. Phys. JETP* **1985**, *61*, 595–598.

62. Zhao, Y.; Zhang, B.; Sato, K. Unusual volume change associated with crystallization in Ce-Ga-Cu bulk metallic glass. *Intermetallics* **2017**, *88*, 1–5. [CrossRef]

63. Makarov, A.S.; Mitrofanova, Y.P.; Konchakov, R.A.; Kobelev, N.P.; Csach, K.; Qiao, J.C.; Khonik, V.A. Density and shear modulus changes occurring upon structural relaxation and crystallization of Zr-based bulk metallic glasses: In situ measurements and their interpretation. *J. Non-Cryst. Sol.* **2019**, submitted.

64. Afonin, G.V.; Mitrofanov, Y.P.; Kobelev, N.P.; da Silva Pinto, M.W.; Wilde, G.; Khonik, V.A. Relationship between the enthalpies of structural relaxation, crystallization and melting in metallic glass-forming systems. *Scr. Mater.* **2019**, *166*, 6–9. [CrossRef]

65. Philips, W.A. *Amorphous Solids: Low Temperature Properties*; Springer: Berlin, Germany, 1981.

66. Gil, L.; Ramos, M.A.; Bringer, A.; Buchenau, U. Low-temperature specific heat and thermal conductivity of glasses. *Phys. Rev. Lett.* **1993**, *70*, 182–185. [CrossRef]

67. Li, Y.; Bai, H.Y.; Wang, W.H.; Samwer, K. Low-temperature specific-heat anomalies associated with the boson peak in CuZr-based bulk metallic glasses. *Phys. Rev. B* **2006**, *74*, 052201. [CrossRef]

68. Zorn, R. The boson peak demystified? *Physics* **2011**, *4*, 44. [CrossRef]

69. Shintani, H.; Tanaka, H. Universal link between the boson peak and transverse phonons in glass. *Nat. Mater.* **2008**, *7*, 870–877. [CrossRef]

70. Sheng, H.; Ma, E.; Kramer, M. Relating dynamic properties to atomic structure in metallic glasses. *JOM* **2012**, *64*, 856–881. [CrossRef]

71. Jakse, N.; Nassour, A.; Pasturel1, A. Structural and dynamic origin of the boson peak in a Cu-Zr metallic glass. *Phys. Rev. B* **2012**, *85*, 174201. [CrossRef]

72. Granato, A.V. Interstitial resonance modes as a source of the boson peak in glasses and liquids. *Physica B* **1996**, *219–220*, 270–272. [CrossRef]

73. Khonik, V.A.; Kobelev, N.P.; Mitrofanov, Y.P.; Zakharov, K.V.; Vasiliev A.N. Boson heat capacity peak in metallic glasses: Evidence of the same defect-induced heat absorption mechanism in structurally relaxed and partially crystallized states. *Phys. Stat. Sol. RRL* **2018**, *12*, 1700412. [CrossRef]

74. Mitrofanov, Y.P.; Makarov, A.S.; Afonin, G.V.; Zakharov, K.V.; Vasiliev, A.N.; Kobelev, N.P.; Wilde, G.; Khonik, V.A. Relationship between the boson heat capacity peak and the excess enthalpy of a metallic glass. *Phys. Stat. Sol. RRL* **2019**, 1900046. [CrossRef]

75. Brink, T.; Koch, L.; Albe, K. Structural origins of the boson peak in metals: From high-entropy alloys to metallic glasses. *Phys. Rev. B* **2016**, *94*, 224203. [CrossRef]

76. Ding, J.; Patinet, S.; Falk, M.L.; Cheng, Y.; Ma, E. Soft spots and their structural signature in a metallic glass. *Proc. Natl. Acad. Sci. USA* **2014**, *111*, 14052. [CrossRef]

77. Mitrofanov, Y.P.; Kobelev, N.P.; Khonik, V.A. On the relationship of the properties of metallic glasses and their maternal crystals. *Phys. Sol. State* **2019**, *61*, 962–968.

78. Granato, A.V. Self-interstitials as basic structural units if liquids and glasses . *J. Phys. Chem. Sol.* **1994**, *55*, 931–939. [CrossRef]

79. Wakeda, M.; Saida, J.; Li, J.; Ogata, S. Controlled rejuvenation of amorphous metals with thermal processing. *Sci. Rep.* **2015**, *5*, 10545. [CrossRef]

80. Guo, W.; Yamada, R.; Saida J.; Lü, S.; Wu, S. Thermal rejuvenation of a heterogeneous metallic glass. *J. Non-Cryst. Sol.* **2018**, *498*, 8–13. [CrossRef]

81. Ingle, W.; Perrin, R.C.; Schober, H.R. Interstitial cluster in FCC metals. *J. Phys. F Met. Phys.* **1981**, *11*, 1161–1173. [CrossRef]

82. Greer, A.L. Metallic glasses. In *Physical Metallurgy*; Laughlin, D.E., Hono, K., Eds.; Elsevier: Oxford, UK, 2014; Volume I, pp. 305–385.

83. Cheng, Y.Q.; Cao, A.J.; Ma, E. Correlation between the elastic modulus and the intrinsic plastic behavior of metallic glasses: The roles of atomic configuration and alloy composition. *Acta Mater.* **2009**, *57*, 3253–3267. [CrossRef]

84. Ding J.; Cheng Y.Q.; Ma, E. Full icosahedra dominate local order in $Cu_{64}Zr_{34}$ metallic glass and supercooled liquid. *Acta Mater.* **2014**, *69*, 343–354. [CrossRef]

85. Mitrofanov, Y.P.; Kobelev, N.P.; Khonik, V.A. Different metastable equilibrium states in metallic glasses occurring far below and near the glass transition. *J. Non-Cryst. Sol.* **2018**, *497*, 48–55. [CrossRef]

86. Makarov, A.S.; Afonin, G.V.; Mitrofanov, Y.P.; Konchakov, R.A.; Kobelev, N.P.; Qiao, J.C.; Khonik, V.A. Evolution of the activation energy spectrum and defect concentration upon structural relaxation of a metallic glass determined using calorimetry and shear modulus data. *J. Alloys Comp.* **2018**, *745*, 378–384. [CrossRef]

87. Donati, C.; Glotzer, S.C.; Poole P.H.; Kob, W.; Plimpton, S.J. Spatial correlations of mobility and immobility in a glass-forming Lennard-Jones liquid. *Phys. Rev. E* **1999**, *60*, 3107–3119. [CrossRef]

88. Oligschleger, C.; Schober, H.R. Collective jumps in a soft-sphere glass. *Phys. Rev. B* **1999**, *59*, 811–821. [CrossRef]

89. Schober, H.R. Collectivity of motion in undercooled liquids and amorphous solids. *J. Non-Cryst. Sol.* **2002**, *307–310*, 40–49. [CrossRef]

90. Betancourt, B.A.P.; Douglas, J.F.; Starr, F.W. String model for the dynamics of glass-forming liquids. *J. Chem. Phys.* **2014**, *140*, 204509. [CrossRef]

91. Wang, Y.-J.; Du, J.-P; Shinzato, S; Dai, L.-H.; Ogata, S. A free energy landscape perspective on the nature of collective diffusion in amorphous solids. *Acta Mater.* **2018**, *157*, 165–173. [CrossRef]

92. Li, W.; Gao, Y.; Bei, H. On the correlation between microscopic structural heterogeneity and embrittlement behavior in metallic glasses. *Sci. Rep.* **2015**, *5*,14786. [CrossRef]

93. Liu, S.T.; Li, F.X.; Li, M.Z.; Wang, W.H. Structural and dynamical characteristics of flow units in metallic glasses. *Sci. Rep.* **2017**, *7*, 11558. [CrossRef]

94. Ma, E. Tuning order in disorder. *Nat. Mater.* **2015**, *14*, 547–552. [CrossRef]

95. van den Beukel, A.; Radelaar, S. On the kinetics of structural relaxation in metallic glasses. *Acta Metall.* **1983**, *31*, 419–427. [CrossRef]

96. van den Beukel, A.; Sietsma, J. The glass transition as a free volume related kinetic phenomenon. *Acta Met. Mater.* **1990**, *38*, 383–389. [CrossRef]

97. Spaepen, F. Homogeneous flow of metallic glasses: A free volume perspective. *Scr. Mater.* **2006**, *54*, 363–367. [CrossRef]

98. Wang, W.H. The elastic properties, elastic models and elastic perspectives of metallic glasses. *Prog. Mater. Sci.* **2012**, *57*, 487–656. [CrossRef]

99. Cheng, Y.Q.; Ma, E. Indicators of internal structural states for metallic glasses: Local order, free volume, and configurational potential energy. *Appl. Phys. Lett.* **2008**, *93*, 051910. [CrossRef]

100. Khonik, V.A.; Kobelev, N.P. Alternative understanding for the enthalpy vs. volume change upon structural relaxation of metallic glasses. *J. Appl. Phys.* **2014**, *115*, 093510. [CrossRef]

Article

Heat Effects Occurring in the Supercooled Liquid State and Upon Crystallization of Metallic Glasses as a Result of Thermally Activated Evolution of Their Defect Systems

Andrei Makarov [1,†], Gennadii Afonin [1,*,†], Yurii Mitrofanov [1,†] and Nikolai Kobelev [2,†] and Vitaly Khonik [1,†]

1   Department of General Physics, State Pedagogical University, Lenin St. 86, Voronezh 394043, Russia; a.s.makarov.vrn@gmail.com (A.M.); mitrofanovyup@gmail.com (Y.M.); v.a.khonik@vspu.ac.ru (V.K.)
2   Institute of Solid State Physics, Russian Academy of Sciences, Chernogolovka, Moscow 142432, Russia; kobelev@issp.ac.ru
*   Correspondence: afoningv@gmail.com
†   These authors contributed equally to this work.

Received: 20 February 2020; Accepted: 19 March 2020; Published: 24 March 2020

Abstract: We show that the kinetics of endothermal and exothermal effects occurring in the supercooled liquid state and upon crystallization of metallic glasses can be well reproduced using temperature dependences of their shear moduli. It is argued that the interrelation between the heat effects and shear modulus relaxation reflects thermally activated evolution of interstitial-type defect system inherited from the maternal melt.

Keywords: metallic glasses; heat effects; shear modulus; defects

---

1. Introduction

The non-crystallinity of glasses necessarily leads to the heat release or heat absorption upon heat treatment. This applies in full to metallic glasses (MGs) [1]. Exothermal effects in MGs take place upon structural relaxation below the glass transition temperature T_g as well as upon crystallization. Strong endothermal reaction is observed in the supercooled liquid state, i.e., at temperatures $T_g \leq T < T_x$, where T_x is the crystallization onset temperature [1]. Besides that, any physical impact (thermal cycling, plastic deformation, irradiation, etc.) leads to certain additional heat effects in MGs [2–4].

Current literature most often relates the heat effects in MGs to the changes of the free volume frozen-in upon glass production, as was originally suggested in the 1980s [5]. Later, the enthalpy relaxation in MGs was linked to the amount of the free volume in the simplest linear form [6] and subsequent experiments seemed to confirm this relationship [7,8]. As a matter of fact, heat effects in MGs are often considered as an indirect measure of the free volume since then [8].

However, it was repeatedly mentioned that the free volume concept has definite theoretical shortcomings [9] and the enthalpy release observed during relaxation cannot be simply interpreted as a measure of the free volume change [10]. On the other hand, exothermal reaction occurring upon crystallization is currently considered solely from a general viewpoint relating it to the appearance of crystalline phases, without discussing any specific details that might lead to the release of heat. We are unaware of any physical model, which could explain all heat effects taking place in the glassy state as well as upon crystallization within a unified physical concept, with the only exception discussed below.

Heat effects can be naturally explained if the defect system of glass is considered within the framework of the Interstitialcy theory (IT) suggested by Granato [11,12]. The IT argues that melting of metals is associated with the rapid increase of the concentration of interstitial defects in their most stable dumbbell form, in line with recent experimental observations [13,14]. These "defects" remain identifiable structural units in the liquid state [15] while rapid melt quenching freezes them in the solid glass. They retain all basic properties of dumbbell interstitials in crystals—high sensitivity to the applied shear stress, specific shear strain fields as well as characteristic low- and high-frequency vibration modes—although do not have any characteristic geometrical image like in crystals (two atoms trying to occupy the same lattice site) [16]. These entities can be considered as interstitial-type "defects" (quotation marks are omitted hereafter). Heat effects can be then interpreted in terms of the changes of the defect concentration. This approach provides quantitative explanations for quite a few other relaxation phenomena in MGs [17].

A description of the heat effects within the framework of the IT is based on an expression for the formation enthalpy of interstitial-type defects [11,12], $H = \alpha\Omega G$, where the dimensionless $\alpha \approx 1$ is related to the defect strain field [18], Ω is the volume per atom and G is the unrelaxed shear modulus. The latter is related to the defect concentration c as $G = \mu\, exp(-\alpha\beta c)$, where μ is the shear modulus of the maternal crystal (i.e., the one, which was melted and then used for glass production by melt quenching) measured at the same temperature and the dimensionless shear susceptibility β characterises the anharmonicity of the interatomic potential and by the definition equals the ratio of the fourth-order non-linear shear modulus to the second-order (i.e., "usual") shear modulus [17]. Typically, $\beta = 15 - 20$ depending on MGs' chemical composition [17].

Thus, any change of the defect concentration leads to heat release/absorption depending on the c-change sign. For the glassy state, this mechanism was first suggested in Ref. [19]. It was later found that the whole excess enthalpy (heat content) of glass with respect to the maternal crystal (i.e., the difference between the heat contents of the glassy and crystalline states) within the IT framework is related to the elastic energy of interstitial-type defect system and this energy is fully released as heat when the defect concentration drops down to zero as a result of crystallization [20]. Thus, all exo- and endothermal heat effects occurring upon structural relaxation and crystallization can be considered as different sides of the same process—a change of interstitial-type defect concentration. In this case, the heat flow upon structural relaxation and crystallization of glass should be described by the same kinetic law, which relates the heat effects with the defect concentration. At that, the latter can be monitored by shear modulus measurements.

The two main equations of the IT given above lead to the difference in the heat flow of the glassy and crystalline states (per unit mass) conditioned by this relaxation mechanism [18]:

$$\Delta W(T) = \frac{1}{m}\frac{dH_g(T)}{dt} = \frac{\dot{T}}{\beta\rho}\left[\frac{G(T)}{\mu(T)}\frac{d\mu(T)}{dT} - \frac{dG(T)}{dT}\right], \tag{1}$$

where H_g is the enthalpy of glass, $\dot{T}$ is the heating rate, ρ is the density and m is the sample's mass. It is seen that the heat flow ΔW is fully controlled by the shear moduli of glass and maternal crystal, $G(T)$ and $\mu(T)$. All other quantities are constants while any fitting parameters are absent. Equation (1) was successfully tested for the heat effects occurring in the glassy state [18] (A similar heat flow law was first derived in Ref. [19]. It differs from Equation (1) by slightly another expression in the square brackets: $\left[\frac{G_{rt}}{\mu_{rt}}\frac{d\mu(T)}{dT} - \frac{dG(T)}{dT}\right]$, where G_{rt} and μ_{rt} are the shear moduli of glass and maternal crystal at room temperature, respectively. This heat flow law provides the results very close to those given by Equation (1). The latter, however, is more general [18]). However, as mentioned above, the same mechanism should be valid for the crystallization as well, as first argued in Ref. [18] and recently confirmed experimentally on a particular metallic glass [21]. In this work, we show that Equation (1) is valid both for the relaxation in the supercooled liquid state and the whole crystallization kinetics of four Zr-based MGs displaying diverse physical properties.

2. Experimental

X-ray non-crystalline $Zr_{47.5}Cu_{47.5}Al_5$, $Zr_{65}Cu_{15}Al_{10}Ni_{10}$, $Zr_{47}Cu_{45}Al_7Fe_1$ and $Zr_{52.5}Cu_{17.9}Ni_{14.6}Al_{10}Ti_5$ (at.%, labelled as $ZrAl_5$, $ZrNi_{10}$, $ZrFe_1$ and $ZrTi_5$ hereafter) produced as $5 \times 2 \times (40-50)$ mm^3 bars were used for the investigation. The choice of these glasses was conditioned by the following reasons. First, all above Zr-based glasses (either in the initial or preannealed states) display a level of the mechanical damping in the supercooled liquid state (i.e., above T_g), which is small enough to ensure automatic measurements of the shear modulus (see below). As noted earlier, a high enough damping level above T_g results in the loss of automatic resonant frequency tracking [21], which strongly limits the range of possible glass compositions. Second, these glasses have diverse and, in some sense, particular properties. Specifically, the $ZrNi_{10}$ glass is very resistant to oxidation [22] that strongly favours precise calorimetric measurements on relatively small samples and their subsequent comparison with calculation results. The glasses $ZrAl_5$ and $ZrFe_1$ display, respectively, big and small compression plasticity [23] that could potentially affect their high temperature behavior. The $ZrTi_5$ glass is being produced industrially for different applications and demonstrates enhanced corrosion resistance [24]. Besides that, the choice of the above MGs is conditioned by the fact that they can be fully crystallized (without any additional phase transformations) below the maximal temperature ($\approx$900 K) achievable upon standard calorimetric measurements. Thus, it is important to check the IT approach sketched above on MGs displaying diverse physical properties.

Heat effects were measured by a Hitachi DSC 7020 instrument in flowing N_2 (99.999%). The mass of the samples was 50–70 mg. Every DSC run on a glassy sample was taken up to the temperature of the full crystallization. It was next followed by the 2nd run on the same sample and the difference between the two runs was then calculated. It is this difference, which is shown below in Figure 2 and compared with ΔW-calculations performed using Equation (1).

The electromagnetic acoustic transformation (EMAT) method (see Ref. [25] for the method's details) was applied to measure the transverse resonant frequencies f (500–700 kHz) of $5 \times 5 \times 2$ mm^3 samples at temperatures of up to 810 K in a vacuum of $\approx$0.01 Pa. For this purpose, frequency scanning was automatically performed every 10–15 s and the resonant frequency was determined as a maximal signal response received by the pick-up coil upon scanning. The half-width of the resonant curve was used to calculate the mechanical quality factor (equal to the inverse damping) and the latter was used to estimate the precision of resonant frequency determination. The shear modulus was then calculated as $G(T) = G_{rt}f^2(T)/f_{rt}^2$, where f_{rt} and G_{rt} are the vibration frequency and shear modulus at room temperature, respectively. This way of G-calculation ignores possible density changes that can occur upon heating (usually less than 1%). The errors for the absolute G_{rt}-values were accepted to be 1–2%. Then, the errors in the absolute $G(T)$ data are about the same while the error in the measurements of $G(T)$-changes was estimated to be 5 ppm near room temperature and about 100 ppm near T_g. A heating/cooling rate of 3 K/min was accepted in all shear modulus and DSC measurements.

Three of the MGs under investigation ($ZrAl_5$, $ZrNi_{10}$ and $ZrFe_1$) were tested in the initial state. This was impossible for the fourth glass ($ZrTi_5$) because of the large damping near T_g, which results in the loss of EMAT automatic signal tracking. To avoid this effect, this MG was first annealed by heating into the supercooled liquid region and cooling back to room temperature. This MG is labelled as "relaxed" below. Both initial and relaxed MGs were first tested from room temperature up to temperatures of 780–810 K, which in all cases lead to the full crystallization. The second run for each glass was performed on the same sample in order to measure shear modulus $\mu(T)$ in the crystalline state.

3. Results and Discussion

Figure 1a–d shows temperature dependences of the shear modulus $G(T)$ for all MGs under investigation. The room-temperature shear moduli G_{rt} listed in this Figure are taken from Ref. [26]. It is seen that $G(T)$-patterns for all MGs are quite similar. Heating of glassy samples leads to a monotonous decrease of G while the slope $|dG/dT|$ decreases by several times near T_g, which constitutes a typical behavior upon high-frequency G-measurements [17]. Heating by 50–70 K above T_g results in the beginning of crystallization, which leads to an increase of the shear modulus by 22% ($ZrFe_1$) to 40% ($ZrNi_{10}$). In the crystalline state (2nd run), the shear modulus $\mu(T)$ demonstrates a featureless decrease with temperature.

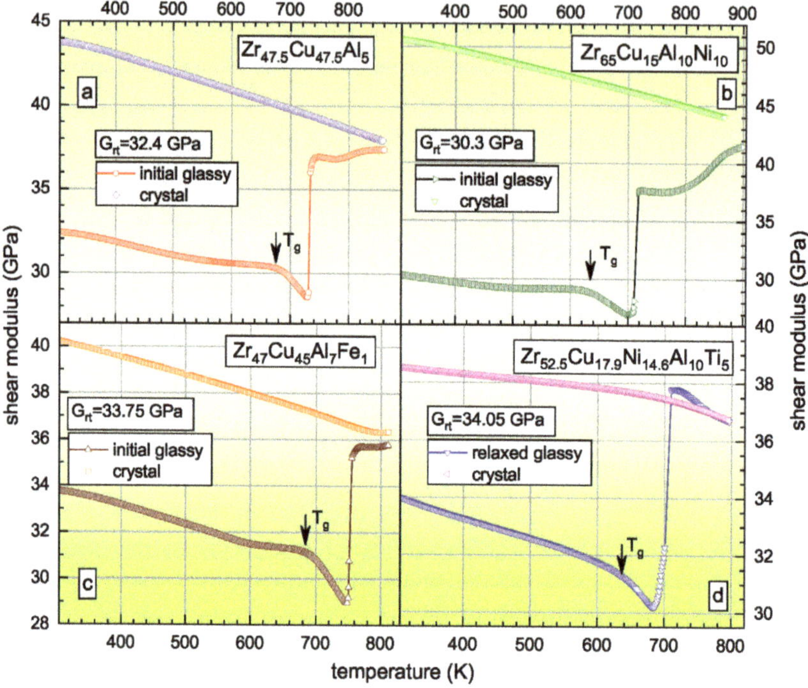

Figure 1. Temperature dependences of the shear modulus measured at 3 K/min for the glasses under investigation in the initial state (**a–c**), after relaxation (**d**) and after full crystallization (**a–d**). The calorimetric T_g's are indicated by the arrows. The room-temperature shear moduli are taken from Ref. [26].

Figure 2 shows DSC traces of the MGs under investigation at the same heating rate of 3 K/min. Initial glasses ($ZrAl_5$, $ZrNi_{10}$ and $ZrFe_1$) demonstrate (*i*) exothermal reaction below T_g (not seen in the scale of Figure 2a–c), (*ii*) heat absorption in the supercooled liquid state (i.e., at temperatures $T_g \leq T < T_x$), which is bigger compared with that of the glassy and crystalline phases, and (*iii*) large heat release due to the crystallization. In the relaxed MG ($ZrTi_5$), the feature (*i*) is absent, as one would expect.

Figure 2 also gives the heat flow $\Delta W(T)$-curves calculated with Equation (1) at $\dot{T} = 3$ K/min using the corresponding temperature dependences of the shear moduli $G(T)$ and $\mu(T)$ given in Figure 1. The way of the determination of the shear susceptibility β from the experimental data and its values together with the densities for the MGS under investigation are given in Ref [26]. The insets in Figure 2 show experimental and calculated $\Delta W(T)$-curves in the supercooled liquid and crystallization temperature regions on enlarged scales. Figure 2 in general demonstrates that the calculation provides a rather good quantitative reproduction of the experimental $\Delta W(T)$-curves. First of all, this applies to the heat absorption for all glasses in the supercooled liquid state. Temperature position of the crystallization exothermal peak is reproduced within $\approx$10 K for $ZrNi_{10}$ and $\leq$6 K for other glasses. The height of this peak is reproduced within 10%–15% accuracy with the exception of

$ZrFe_1$ for which the calculation gives about 60% of the experimental height (not shown in Figure 2). It should be emphasized that the density of data points provided by our EMAT system in the range of fast crystallization (i.e., upon a rapid change of the shear modulus) is significantly smaller than that for temperatures $T < T_x$ (see Figure 1) and this constitutes a significant source for the calculation errors when using Equation (1), which contains both derivatives of the shear moduli and their difference. Nonetheless, one can conclude that this equation provides a good description of the heat effects on the basis of shear modulus data. It is interesting to note that the calculation reproduces the fine details of the crystallization kinetics as exemplified by the data around $T = 700$ K shown in the inset of Figure 2d.

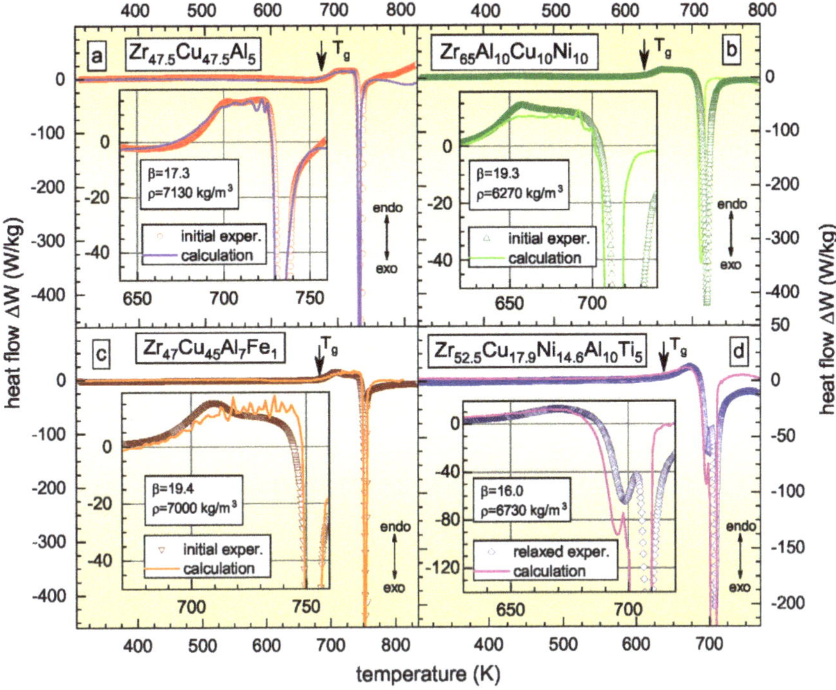

Figure 2. Experimental and calculated using Equation (1) DSC traces for the MGs under investigation in the initial (**a–c**) and relaxed states (**d**). The insets show the heat flow in the supercooled liquid state and upon crystallization on an enlarged scale. The shear susceptibilities β and the densities ρ are taken from Ref. [26].

The obtained results confirm the basic idea of the IT sketched above: the origin of the heat absorption and heat release occurring in the glass transition range as well as upon crystallization can be understood as a result of the change of the concentration of interstitial-type defects frozen-in from the melt upon glass production. In particular, structural relaxation below T_g provides relatively small decrease of the defect concentration and constitutes the reason of moderate exothermal heat effect (feature (*i*) mentioned above). The defect concentration above T_g increases with temperature leading to the heat absorption (feature (*ii*)). Finally, fast crystallization at $T > T_x$ results in a rapid drop of the defect concentration down to zero providing quick dissipation of the defect elastic energy into heat, which is released as a strong crystallization peak (feature (*iii*)) [20].

As mentioned above, the interstitial defects in crystalline structures are unambiguously geometrically defined as two atoms trying to occupy the same lattice site. However, the situation in metallic non-crystalline structures is more complex. Molecular static simulation performed on a monoatomic glassy structure revealed localized nano-regions, which display the properties similar to those of dumbbell interstitials in simple metals [16]. These findings imply that interstitial-type defects can indeed exist in a monatomic non-crystalline structure. The issue whether such structural entities can be found in polyatomic glassy structures constitutes a major challenge for further research in this direction.

4. Conclusions

The investigation of four Zr-based glasses displaying diverse physical properties showed that the heat effects occurring in the supercooled liquid state and upon crystallization can well reproduced with Equation (1) using experimentally determined temperature dependences of shear moduli of the glass and maternal crystal. This equation is derived on the basis of the Interstitialcy theory and assumes that the heat effects in all temperature ranges take place due to the thermally activated change of the concentration of interstitial-type defects frozen-in from the melt upon glass production. A change of the defect concentration leads to a release or absorption of the total defect formation enthalpy constituting the physical reason for the heat effects both in the glassy state and upon crystallization.

Author Contributions: All authors equally contributed to this work. All authors have read and agreed to the published version of the manuscript.

Funding: This research was supported by the Russian Science Foundation under the grant 20-62-46003.

Conflicts of Interest: The authors declare no conflict of interest.

References

1. Chen, H.S. Glassy metals. *Rep. Prog. Phys.* **1980**, *43*, 353. [CrossRef]
2. Saida, J.; Yamada, R.; Wakeda, M.; Ogata, S. Thermal rejuvenation in metallic glasses. *Sci. Technol. Adv. Mater.* **2017**, *18*, 152–162. [CrossRef] [PubMed]
3. Ebner, C.; Escher, B.; Gammer, C.; Eckert, J.; Pauly, S.; Rentenberger, C. Structural and mechanical characterization of heterogeneities in a CuZr-based bulk metallic glass processed by high pressure torsion. *J. Alloys Compd.* **2018**, *160*, 147–157. [CrossRef]
4. Pan, J.; Wang, Y.X.; Guo, Q.; Zhang, D.; Greer, A.L.; Li, Y. Extreme rejuvenation and softening in a bulk metallic glass. *Nat. Commun.* **2018**, *9*, 560. [CrossRef] [PubMed]
5. van den Beukel, A.; Radelaar, S. On the kinetics of structural relaxation in metallic glasses. *Acta Metall.* **1983**, *31*, 419–427. [CrossRef]
6. van den Beukel, A.; Sietsma, J. The glass transition as a free volume related kinetic phenomenon. *Acta Met. Mater.* **1990**, *38*, 383–389. [CrossRef]
7. Slipenyuk, A.; Eckert, J. Correlation between enthalpy change and free volume reduction during structural relaxation of $Zr_{55}Cu_{30}Al_{10}Ni_5$ metallic glass. *Scr. Mater.* **2004**, *50*, 39–44. [CrossRef]
8. Haruyama, O.; Nakayama, Y.; Wada, R.; Tokunaga, H.; Okada, J.; Ishikawa, T.; Yokoyama, Y. Volume and enthalpy relaxation in $Zr_{55}Cu_{30}Ni_5Al_{10}$ bulk metallic glass. *Acta Mater.* **2010**, *58*, 1829–1836. [CrossRef]
9. Miracle, D.B.; Egami, T.; Flores, K.M.; Kelton, K.F. Structural Aspects of Metallic Glasses. *MRS Bull.* **2007**, *32*, 629–634. [CrossRef]
10. Cheng, Y.Q.; Ma, E. Indicators of internal structural states for metallic glasses: Local order, free volume, and configurational potential energy. *Appl. Phys. Lett.* **2008**, *93*, 051910. [CrossRef]
11. Granato, A.V. Interstitialcy model for condensed matter states of face-centered-cubic metals. *Phys. Rev. Lett.* **1992**, *68*, 974–977. [CrossRef] [PubMed]
12. Granato, A.V. Interstitialcy theory of simple condensed matter. *Eur. J. Phys.* **2014**, *87*, 18. [CrossRef]
13. Safonova, E.V.; Mitrofanov, Y.P.; Konchakov, R.A.; Vinogradov, A.Y.; Kobelev, N.P.; Khonik, V.A. Experimental evidence for thermal generation of interstitials in a metallic crystal near the melting temperature. *J. Phys. Cond. Matter* **2016**, *28*, 215401. [CrossRef] [PubMed]
14. Goncharova, E.V.; Makarov, A.S.; Konchakov, R.A.; Kobelev, N.P.; Khonik, V.A. Premelting generation of interstitial defects in polycrystalline indium. *J. Exp. Theor. Phys. Lett.* **2017**, *106*, 35–39. [CrossRef]

15. Nordlund, K.; Ashkenazy, Y.; Averback, R.S.; Granato, A.V. Strings and interstitials in liquids, glasses and crystals. *Europhys. Lett.* **2005**, *71*, 625–631. [CrossRef]

16. Goncharova, E.V.; Konchakov, R.A.; Makarov, A.S.; Kobelev, N.P.; Khonik, V.A. Identification of interstitial-like defects in a computer model of glassy aluminum. *J. Phys. Condens. Matter* **2017**, *29*, 305701. [CrossRef]

17. Khonik, V.A.; Kobelev, N.P. Metallic glasses: A new approach to the understanding of the defect structure and physical properties. *Metals* **2019**, *9*, 605. [CrossRef]

18. Kobelev, N.P.; Khonik, V.A. Theoretical analysis of the interconnection between the shear elasticity and heat effects in metallic glasses. *J. Non-Cryst. Sol.* **2015**, *427*, 184–190. [CrossRef]

19. Mitrofanov, Y.P.; Makarov, A.S.; Khonik, V.A.; Granato, A.V.; Joncich, D.M.; Khonik, S.V. On the nature of enthalpy relaxation below and above the glass transition of metallic glasses. *Appl. Phys. Lett.* **2012**, *101*, 131903. [CrossRef]

20. Afonin, G.V.; Mitrofanov, Y.P.; Makarov, A.S.; Kobelev, N.P.; Wang, W.H.; Khonik, V.A. Universal relationship between crystallization-induced changes of the shear modulus and heat release in metallic glasses. *Acta Mater.* **2016**, *115*, 204–209. [CrossRef]

21. Mitrofanov, Y.P.; Wang, D.P.; Makarov, A.S.; Wang, W.H.; Khonik, V.A. Towards understanding of heat effects in metallic glasses on the basis of macroscopic shear elasticity. *Sci. Rep.* **2016**, *6*, 23026. [CrossRef] [PubMed]

22. Kim, C.W.; Jeong, H.G.; Lee, D.B. Oxidation of $Zr_{65}Al_{10}Ni_{10}Cu_{15}$ bulk metallic glass. *Mater. Lett.* **2008**, *62*, 584–586. [CrossRef]

23. Wang, D.P.; Zhao, D.Q.; Ding, D.W.; Bai, H.Y.; Wang, W.H. Understanding the correlations between Poisson's ratio and plasticity based on microscopic flow units in metallic glasses. *J. Appl. Phys.* **2014**, *115*, 123507. [CrossRef]

24. Chieh, T.C.; Chu, J.; Liu, C.T.; Wu, J.K. Corrosion of $Zr_{52.5}Cu_{17.9}Ni_{14.6}Al_{10}Ti_5$ bulk metallic glasses in aqueous solutions. *Mater. Lett.* **2003**, *57*, 3022–3025. [CrossRef]

25. Vasil'ev, A.N.; Gaidukov, Y.P. Electromagnetic excitation of sound in metals. *Soviet Physics Uspekhi* **1983** *26*, 952. [CrossRef]

26. Afonin, G.V.; Mitrofanov, Y.P.; Makarov, A.S.; Kobelev, N.P.; Khonik, V.A. On the origin of heat effects and shear modulus changes upon structural relaxation and crystallization of metallic glasses. *J. Non-Cryst. Sol.* **2017**, *475*, 48–52. [CrossRef]

Article

Dynamic Mechanical Relaxation in LaCe-Based Metallic Glasses: Influence of the Chemical Composition

Minna Liu [1], Jichao Qiao [1,*], Qi Hao [1], Yinghong Chen [1], Yao Yao [1], Daniel Crespo [2] and Jean-Marc Pelletier [3]

[1] School of Mechanics, Civil Engineering Architecture, Northwestern Polytechnical University, Xi'an 710072, China; liuminna@mail.nwpu.edu.cn (M.L.); haoq@mail.nwpu.edu.cn (Q.H.); cyhongc@mail.nwpu.edu.cn (Y.C.); yaoy@nwpu.edu.cn (Y.Y.)

[2] Departament de Física, Barcelona Research Center in Multiscale Science and Technology & Institut de Tècniques Energètiques, Universitat Politècnica de Catalunya, 08930 Barcelona, Spain; Daniel.Crespo@upc.edu

[3] MATEIS, UMR CNRS5510, Bat. B. Pascal, INSA-Lyon, F-69621 Villeurbanne CEDEX, France; jean-marc.pelletier@insa-lyon.fr

* Correspondence: qjczy@nwpu.edu.cn

Received: 13 August 2019; Accepted: 14 September 2019; Published: 17 September 2019

Abstract: The mechanical relaxation behavior of the $(La_{0.5}Ce_{0.5})_{65}Al_{10}(Co_xCu_{1-x})_{25}$ at% (x = 0, 0.2, 0.4, 0.6, and 0.8) metallic glasses was probed by dynamic mechanical analysis. The intensity of the secondary β relaxation increases along with the Co/Cu ratio, as has been reported in metallic glasses where the enthalpy of mixing for all pairs of atoms is negative. Furthermore, the intensity of the secondary β relaxation decreases after physical aging below the glass transition temperature, which is probably due to the reduction of the atomic mobility induced by physical aging.

Keywords: metallic glasses; mechanical spectroscopy; mixing enthalpy; Kohlrausch–Williams–Watts equation; structural heterogeneity

---

1. Introduction

Metallic glasses (MGs) have been extensively studied for several decades because they exhibit unique physical, chemical and mechanical properties and have no crystal defects (i.e., dislocations, grain boundaries, and vacancies) [1–3]. Compared to other glassy materials (i.e., amorphous polymers, oxide glasses, and other non-crystalline solids), metallic glasses show high yield strength and resilience, large fracture toughness, and attractive corrosion resistance [4–8]. It is well known that the mechanical and physical properties of metallic glasses, i.e., plasticity, glass transition behavior, and diffusion phenomena, are bound up with their mechanical relaxation modes [1,6,9–11]. Nevertheless, below the glass transition temperature, metallic glasses have insufficient ductility due to shear band instability during plastic deformation, which dramatically reduces the use in structural applications [12]. Compared to crystalline metals, metallic glasses are basically characterized by brittleness at room temperature. One way to overcome the macroscopic brittle behavior of metallic glass is to reduce the size. Previous studies have shown that brittle behavior can be mitigated when the sample size is reduced to sub-micron levels, thereby reducing the effects of instability on material behavior [13,14]. Below the glass transition, metallic glasses are thermodynamically in a non-equilibrium state, as there is a large enthalpy difference from the crystallized state [15]. However, the topological structure as well as the physical mechanism of relaxation in glassy materials remain unresolved issues [16–18].

In the supercooled liquid phase region of metallic glasses, relaxation processes drive the glass towards more stable states. The primary α relaxation and secondary β relaxation are considered the elementary relaxation processes [19–21]. Johari et al. [22] proposed that glasses and glass-forming liquids have two relaxation modes: (i) The primary (α) relaxation, which is a global, structural atomic or molecular rearrangement observed at relatively high temperatures and closely related to the glass transition phenomenon; (ii) The secondary (β) relaxation, a low energy process which is observed under the glass transition temperature T_g. While the α relaxation process shows a complex dependence on temperature, the secondary β relaxation generally submits to an Arrhenius temperature dependence rule. The β relaxation shows up as an over wing in the high-frequency tail or a side shoulder at low temperature of α relaxation [23,24]. Contrary to the main relaxation, β relaxation is associated with the motion of atoms or molecules inside a glass material without topological rearrangement. The study on the connection between the diffusion behavior of the amorphous alloy, plastic deformation, and glass transition on the relaxation process of amorphous alloy is of great significance for the assessment of the potential applications of metallic glasses.

Literature results show that La-based metallic glasses display a conspicuous secondary relaxation [25]. Therefore, La-based metallic glasses are an ideal model system to investigate the relaxation process. In the present study, the dynamic mechanical properties of emblematic LaCe-based metallic glasses were investigated by mechanical spectroscopy. The physical mechanism of mechanical relaxation process was analyzed relied on the Kohlrausch–Williams–Watts (KWW) equation.

2. Experimental Procedure

The $(La_{0.5}Ce_{0.5})_{65}Al_{10}(Co_xCu_{1-x})_{25}$ at% (x = 0, 0.2, 0.4, 0.6 and 0.8) precursor alloy was produced by arc melting in high-purity argon atmosphere, after titanium melting for the removal of residual oxygen. The alloy was re-melted at least four times to ensure chemical homogeneity. Suction casting copper method was eventually used to produce plates of 2 mm thickness.

The structural properties of the samples were checked by X-ray diffraction (XRD, Philips PW 3830, Amsterdam, Netherlands) using monochromatic Cu-Kα radiation. The glass transition T_g and crystallization onset T_x temperatures were determined by differential scanning calorimeter (DSC, NEZTCH 404 C, Bavaria, Germany) at a heating rate of 10 K/min. Dynamical mechanical analysis (DMA, TA Q800, USA) was used to monitor the mechanical relaxation behavior. Samples of 30 mm (length) × 2 mm (width) × 1 mm (thickness) for DMA analysis were produced on a precise water-cooled low-speed cutting machine. DMA measurements were performed on single cantilever, at a 3 K/min heating rate; the complex elastic modulus is denoted as E (storage modulus E' and the loss modulus E'').

3. Experimental Results and Discussions

3.1. Structural and Thermal Properties

The amorphous nature of $(La_{0.5}Ce_{0.5})_{65}Al_{10}(Co_xCu_{1-x})_{25}$ at% (x = 0, 0.2, 0.4, 0.6 and 0.8) was confirmed by XRD. XRD patterns of $(La_{0.5}Ce_{0.5})_{65}Al_{10}(Co_xCu_{1-x})_{25}$ at% (x = 0, 0.2, 0.4, 0.6, and 0.8) bulk metallic glasses, as presented in Figure 1a, exhibit broad diffraction peaks, and no traces of crystalline phases are detected. Therefore, the glassy nature of the $(La_{0.5}Ce_{0.5})_{65}Al_{10}(Co_xCu_{1-x})_{25}$ at% (x = 0, 0.2, 0.4, 0.6, and 0.8) bulk metallic glasses (BMG) was verified.

DSC curves of the $(La_{0.5}Ce_{0.5})_{65}Al_{10}(Co_xCu_{1-x})_{25}$ at% (x = 0, 0.2, 0.4, 0.6and 0.8) bulk metallic glasses are presented in Figure 1b. The main thermal events are the glass transition and subsequent crystallization. The glass transition temperatures T_g, indicated by arrows in the figure, increase almost linearly with the substitution of Copper by Cobalt. While the Co-free $((La_{0.5}Ce_{0.5})_{65}Al_{10}Cu_{25})$ alloy has a glass transition temperature of 372 K, the T_g of $(La_{0.5}Ce_{0.5})_{65}Al_{10}(Co_{0.8}Cu_{0.2})_{25}$ alloy increases up to 404 K. Crystallization temperatures T_x show a less predictable behavior. Given that the difference between the crystallization and glass transition temperatures is a parameter largely related to the glass stability, it is observed that $(La_{0.5}Ce_{0.5})_{65}Al_{10}(Co_{0.2}Cu_{0.8})_{25}$ metallic glass is the most stable glass. The XRD patterns and DSC curves corroborate the amorphous properties of the studied alloys.

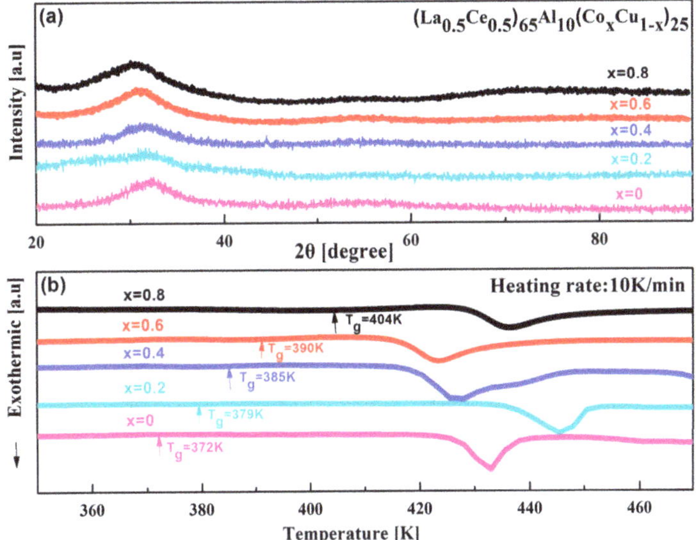

Figure 1. (**a**) XRD patterns of the $(La_{0.5}Ce_{0.5})_{65}Al_{10}(Co_xCu_{1-x})_{25}$ at% (x = 0, 0.2, 0.4, 0.6, and 0.8), as-cast state. (**b**) Differential scanning calorimeter (DSC) curves of the $(La_{0.5}Ce_{0.5})_{65}Al_{10}(Co_xCu_{1-x})_{25}$ at% (x = 0, 0.2, 0.4, 0.6, and 0.8) at a heating rate of 10 K/min. The glass transition temperature T_g of the different metallic glasses is pointed out in the figure.

3.2. Dynamic Mechanical Analysis

3.2.1. Constant Frequency Measurements

Dynamic mechanical analysis is very sensitive to the atomic and molecular mobility of amorphous materials. The $(La_{0.5}Ce_{0.5})_{65}Al_{10}(Co_{0.4}Cu_{0.6})_{25}$ metallic glass dynamic mechanical response was measured from room temperature to 550 K with a heating rate of 3 K/min and a driving frequency of 1 Hz. Figure 2 exhibits the normalized storage modulus E' and loss modulus E'' as a function of temperature, where E_u represents the unrelaxed modulus at room temperature. The temperature dependence of E' and E'' is very similar to most other BMGs [9,26,27]. It is worth noticing that when the temperature is below the T_g, loss modulus curve showed no apparent β relaxation [9,28,29]. The storage modulus and loss modulus of the LaCe-based metallic glass vary with temperature, and the process can be segmented into three different regions:

Region (I): In the low temperature region, i.e., beneath 350 K, the normalized storage modulus is large and close to unity. Contrarily, the loss modulus E'' is negligible. Therefore, the glassy material mainly exhibits elastic deformation in this temperature range, and the viscoelastic component can be neglected.

Region (II): The medium temperature region comprises the temperature range of 390 K to 460 K. A large increase of the loss modulus E'' is observed reaching a peak at the temperature T_α, denoting the α relaxation characteristic of amorphous materials. This process is the dynamic glass transition. The storage modulus E' starts to diminish while the loss modulus E'' boosts. This temperature range falls within the super-cooled liquid region of metallic glasses.

Region (III): The high temperature region starts at 460 K. The storage modulus E' increases again reaching a value similar to that at room temperature, due to crystallization. The loss modulus E'' falls to a relatively low value.

As shown in Figure 2, the β relaxation is not observed below T_g [22,23]. β relaxation in this glass should be found in the temperature range from 300 to 400 K.

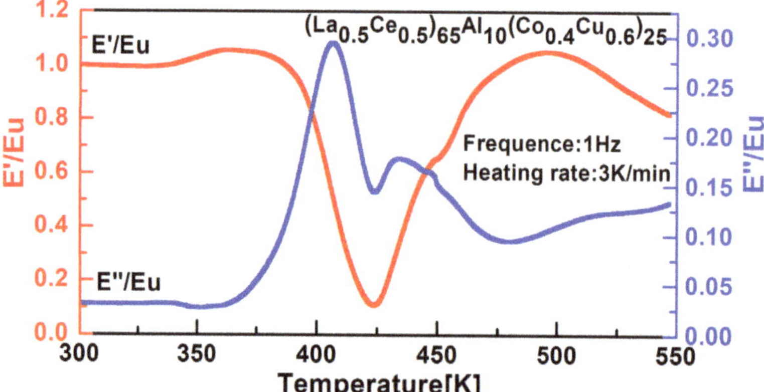

Figure 2. Thermal dependence of the normalized storage E'/E_u and loss modulus E''/E_u of (La$_{0.5}$Ce$_{0.5}$)$_{65}$Al$_{10}$(Co$_{0.4}$Cu$_{0.6}$)$_{25}$ metallic glass. Measurement was carried out at a fixed frequency of 1 Hz and a heating rate of 3 K/min. E_u is the unrelaxed modulus, which equals the value of E' at room temperature.

Figure 3 exhibits the normalized dynamic loss modulus E'' as a function of temperature at a constant frequency (1 Hz) in (La$_{0.5}$Ce$_{0.5}$)$_{65}$Al$_{10}$(Co$_x$Cu$_{1-x}$)$_{25}$ at% (x = 0, 0.2, 0.4, 0.6, and 0.8) glass alloys. The data are normalized to the values of temperature and loss modulus at the peak of the α relaxation, namely T_α and E''_{max}. Interestingly, the secondary relaxation process depends significantly on the chemical composition of the glass. The intensity of β relaxation of (La$_{0.5}$Ce$_{0.5}$)$_{65}$Al$_{10}$(Co$_x$Cu$_{1-x}$)$_{25}$ at% (x = 0, 0.2, 0.4, 0.6, and 0.8) decreases with the increase of the Cu content. The behavior of the loss module reveals a noticeable change in the features of the β relaxation around 0.8 T_g for the different compositions. For (La$_{0.5}$Ce$_{0.5}$)$_{65}$Al$_{10}$Cu$_{25}$ metallic glass, the β relaxation merely declares as a weak shoulder. It is significant that in several metallic glasses that exhibit an evident β relaxation, this has been correlated to plasticity [30]. It has also been proven that the mechanical relaxation process, especially the β relaxation, is sensitive to the micro-alloying in metallic glasses.

Previous works indicate that the β relaxation reflects the inherent structural heterogeneities in metallic glasses, described for instance as soft domains, liquid-like regions, local topological structure of loose packing regions, and flow units [1]. In the current study, minor addition of Cobalt in the (La$_{0.5}$Ce$_{0.5}$)$_{65}$Al$_{10}$(Co$_x$Cu$_{1-x}$)$_{25}$ at% (x = 0, 0.2, 0.4, 0.6, and 0.8) bulk metallic glasses is very important and reshapes the relaxation mode (i.e., β relaxation).

In order to associate the different behavior of β relaxation and the deformability of metallic glass with its structure, transmission electron microscopy (TEM) was carried out to reveal the microstructural characteristics of metallic glass [31]. The most notable structural feature is that the metallic glass is composed of two types of regions: Light regions with typical sizes ranging from 50 to 200 nm are enveloped by dark boundary regions, which are about 5–20 nm in width. It was further confirmed that both of the two regions are of a glassy nature. It is proposed that the local atomic motions of soft regions are responsible for β relaxations, and the heterogeneous structure improves the plasticity of metallic glasses though the formation of multiple shear bands [32].

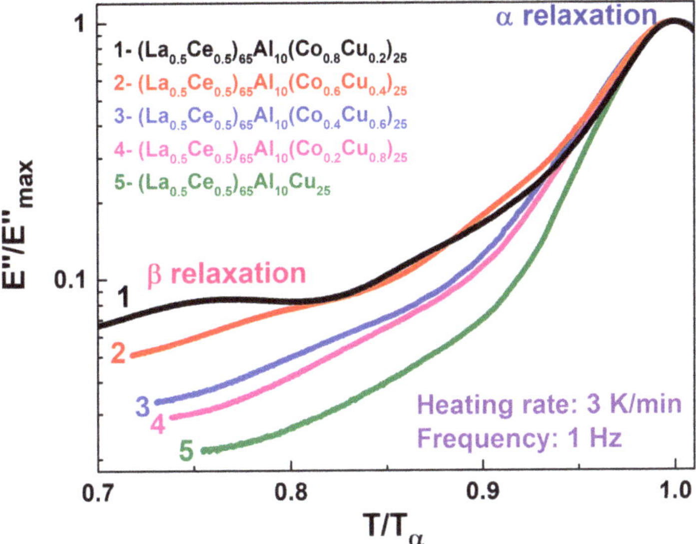

Figure 3. Temperature dependence of the loss modulus E''/E''_{max} in the $(La_{0.5}Ce_{0.5})_{65}Al_{10}(Co_xCu_{1-x})_{25}$ at% (x = 0, 0.2, 0.4, 0.6, 0.8) metallic glass (Heating rate: 3 K/min; frequency: 1 Hz).

The relationship between enthalpy of mixing and β relaxation can be used to qualitatively anticipate the intensity β relaxation of some metallic glasses [33]. An empirical rule on β relaxation has been established [34]: Pronounced β relaxation is associated with alloys where all the atomic pairs have larger and similar negative values of the mixing enthalpy. On the other hand, positive or large fluctuations in the values of mixing enthalpy reduce and even suppress β relaxation [33].

Regarding the β relaxation in $(La_{0.5}Ce_{0.5})_{65}Al_{10}(Co_xCu_{1-x})_{25}$ at% metallic glasses, the features of the mixing enthalpy between the constituent atoms correspond to the apparent β relaxation. Figure 4 shows the mixing enthalpy of the constituents of the $(La_{0.5}Ce_{0.5})_{65}Al_{10}(Co_xCu_{1-x})_{25}$ at% (x = 0, 0.2, 0.4, 0.6, and 0.8) metallic glasses (The data of mixing enthalpy are derived from the reference [35]). The mixing enthalpy ΔHm of the "solvent" atoms, La/Ce, with the "solute" atoms are almost identical: ΔHm (La/Ce-Al) = −38 kJ/mol, ΔHm (La/Ce-Cu) = −21 kJ/mol, ΔHm (La-Co) = −17 kJ/mol and ΔHm (La-Cu) = −18 kJ/mol, reflecting the chemical similar chemistry of the rare-earth elements. As for the "solute" atoms, the mixing enthalpy of Cu-Co is positive, 6 kJ/mol, and the main difference appears when comparing the mixing enthalpies of Al-Cu, −1 kJ/mol, to that of Al-Co, −18 kJ/mol. Based on the empirical rules to determine ΔH_{mix}, the mixing enthalpy of $(La_{0.5}Ce_{0.5})_{65}Al_{10}(Co_xCu_{1-x})_{25}$ at% (x = 0, 0.2, 0.4, 0.6, 0.8) metallic glasses is given in Figure 4. We observed an actual decrease of the enthalpy of mixing as the concentration of Co increases. However, due to substitution of Cu by Co, the fluctuation on mixing enthalpies decreases, as the Al-Co mixing enthalpy is substantially more negative than that of Al-Cu and similar to those of La/Ce-Cu. According to the literature, this reduction on the mixing enthalpy fluctuation enhances the β relaxation [27].

Previous works proved that relaxation is connected to dynamic heterogeneity in glasses and related to the local movement of "weak spots" [36]. In particular, the microstructure inhomogeneity of metallic glass has been proven by means of microscopy and simulation [37].

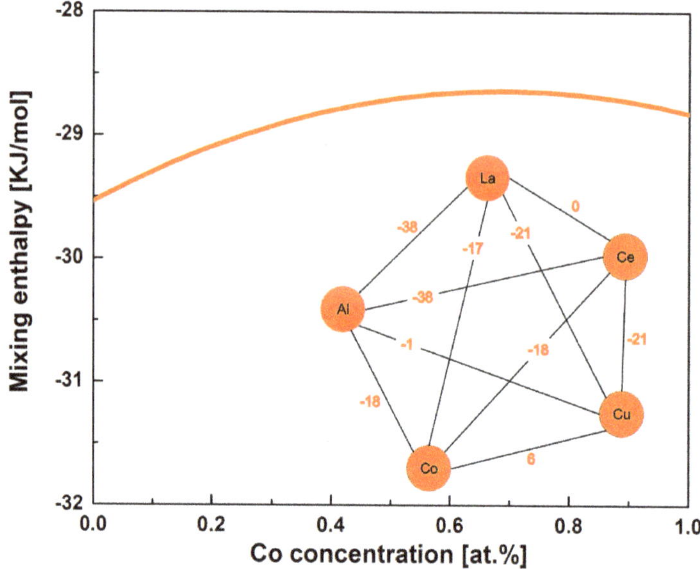

Figure 4. Mixing enthalpy of the $(La_{0.5}Ce_{0.5})_{65}Al_{10}(Co_xCu_{1-x})_{25}$ at% (x = 0, 0.2, 0.4, 0.6, and 0.8) metallic glasses. The inset displays the mixing enthalpy of constituent atoms (the data are taken from the reference [35]).

3.2.2. Physical Aging on the Secondary Relaxation of LaCe-Based Metallic Glass

From the thermodynamics point of view, annealing below the T_g drives the glassy state towards a more stable state of lower energy. Figure 5 presents the storage and loss factor of $(La_{0.5}Ce_{0.5})_{65}Al_{10}(Co_{0.8}Cu_{0.2})_{25}$ metallic glass after annealing at 363 K for 24 h, which certainly illustrates that the intensity of the β relaxation reduces by physical aging below T_g.

As proposed in previous works, the β relaxation of metallic glasses is ascribed to the structural heterogeneity or local motion of the "defects" [1,33]. These defects are denominated as flow units [33,38], quasi-point defects (QPDs) [39], liquid-like sites [40], weakly bonded zones or loose packing regions [41]. According to Figure 5, annealing below the glass transition temperature can lead to disappearance of "defects" in metallic glasses. Physical aging causes rearrangement of atoms, resulting in an increase of density and elastic modulus. In the metallic glass, the mobility of atoms is closely related to "defects" concentration. Annealing causes the metallic glass to evolve towards a higher density state with a consequent reduction of the local "free volume" available for atomic rearrangement. Subsequent cooling after the annealing does not alter the glassy state, since the cooling rate is much lower than in the initial production of the glass.

In addition, physical aging below the glass transition temperature T_g leads to enthalpy relaxation of glassy materials. Figure 6 shows the DSC trace of the $(La_{0.5}Ce_{0.5})_{65}Al_{10}(Co_{0.8}Cu_{0.2})_{25}$ metallic glass annealed at 363 K for 24 h. Comparison to the as-produced sample allows the identification of a notable enthalpy recovery, a consequence of the glass relaxation during annealing.

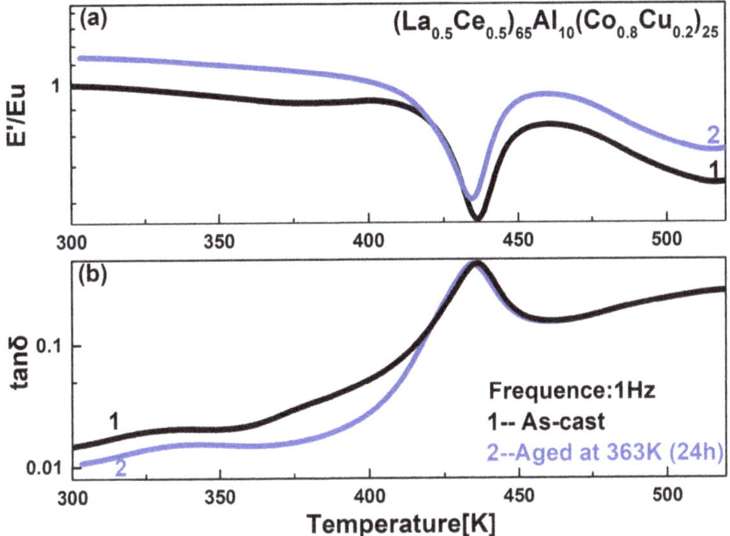

Figure 5. Temperature dependence of the normalized storage modulus (**a**) and loss factor (**b**) of $(La_{0.5}Ce_{0.5})_{65}Al_{10}(Co_{0.8}Cu_{0.2})_{25}$ metallic glass (heating rate: 3 K/min and frequency: 0.3 Hz) (1) As-cast (2) annealed one (annealing temperature: 363 K and annealing time: 24 h).

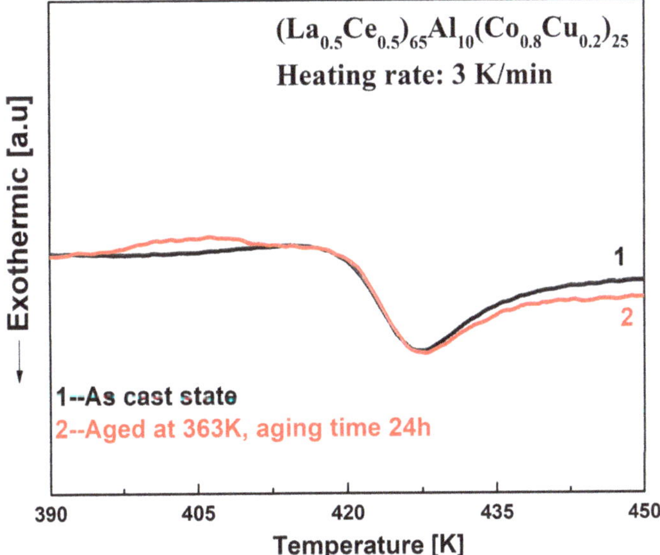

Figure 6. Enthalpy relaxation in $(La_{0.5}Ce_{0.5})_{65}Al_{10}(Co_{0.8}Cu_{0.2})_{25}$ metallic glass bulk metallic glasses after aging at 363 K.

The structural relaxation observed in the DMA test can be analyzed by the growth of the loss factor ($tan\ \delta = E''/E'$) [42,43]. Figure 7 shows the loss factor ($tan\ \delta$) evolution versus annealing time in the $(La_{0.5}Ce_{0.5})_{65}Al_{10}(Co_{0.8}Cu_{0.2})_{25}$ bulk metallic glass at 363 K. Previous works have characterized the structural relaxation below T_g in amorphous materials, particularly in bulk metallic glasses, by using the Kohlrausch–Williams–Watts (KWW) equation [9,44].

$$\tan \delta(t_a) - \tan \delta(t_a = 0) = A\{1 - e^{[-(\frac{t_a}{\tau})^{\beta_{aging}}]}\}$$

(1)

where $A = \tan \delta(t_a \to \infty) - \tan \delta(t_a = 0)$ is the maximum value of the dynamic relaxation. τ is the relaxation time, and β_{aging} is the Kohlrausch exponent with values between 0 and 1. The best fit of Equation (1) to the data on the loss factor of the $(La_{0.5}Ce_{0.5})_{65}Al_{10}(Co_{0.8}Cu_{0.2})_{25}$ metallic glass was obtained with τ = 10,594 s and β_{aging} = 0.4.

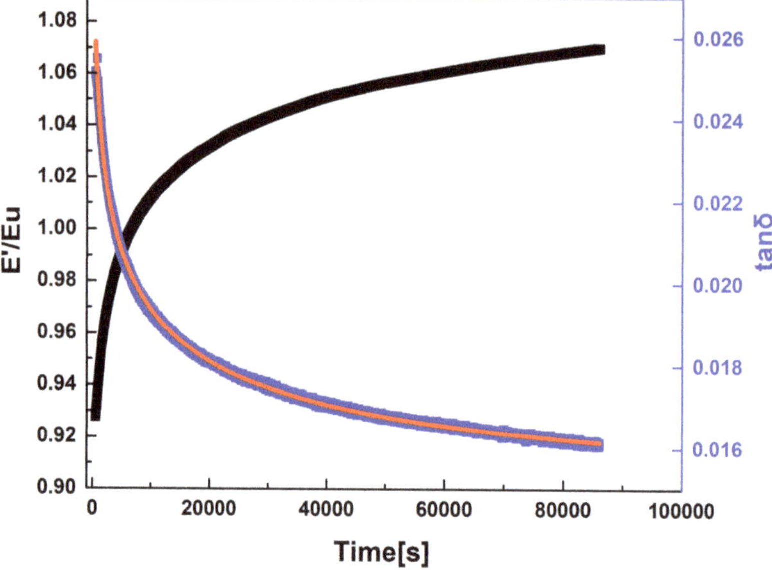

Figure 7. Evolution of the storage modulus E' and loss factor $\tan\delta$ for $(La_{0.5}Ce_{0.5})_{65}Al_{10}(Co_{0.8}Cu_{0.2})_{25}$ metallic glass with the annealing time. The aging temperature is T_α = 363 K, and the driving frequency is 1 Hz. The red curve is the best fit from Equation (1) obtained for the parameters τ = 10,594 s and β_{aging} = 0.4.

The Kohlrausch exponent β_{KWW} reveals the presence of a broad distribution of relaxation times in the glass, with β_{KWW} aging = 1 corresponding to a single Debye relaxation time.

Experimental values of the parameter β_{KWW} in amorphous alloys are in the range of 0.24–1 [45]. For amorphous polymers, values extend from 0.24 (polyvinyl chloride) to 0.55 (polyisobutylene), for alcohols from 0.45 to 0.75, while for orientational glasses and networks values are up to 1. In bulk metallic glasses, it appears that β_{KWW} is related to the fragility of the amorphous materials [42]. Values of β_{KWW} close to 1 indicate that the system is a strong glass former while values less than 0.5 suggest that the glass is a fragile glass [46,47]. According to the available literature, no defined trend characterizes the stretching parameters β_{KWW} in bulk metallic glasses, as it is either temperature dependent or temperature independent [48–50]. For example, $Pd_{42.5}Ni_{7.5}Cu_{30}P_{20}$ bulk metallic glass has very similar fragility parameters, 59 < m < 67, and similar stretched exponents, 0.59 < β_{KWW} < 0.6 [44]. However, experimental data indicate that the Kohlrausch exponent β_{aging} is around 0.4 for temperatures close to T_g [44], as found in the present case.

As proposed by Wang et al. [51], the kinetic parameter β_{KWW} is associated with the dynamic heterogeneity. The β_{KWW} parameter takes low values when the temperature is below the β relaxation peak and increases dramatically when the temperature surpasses the glass transition temperature. The β relaxation in metallic glasses is related to the reversible displacement of the "defects". When the stress relaxation is performed around the β relaxation temperature, only a small fraction of atoms are allowed to move. Thus, it can be concluded that lower β_{KWW} values around the β relaxation are ascribed to reversible "defects".

4. Conclusions

The dynamic mechanical properties of LaCe-based bulk metallic glasses were investigated by dynamic mechanical analysis. The experimental results reveal that the mechanical and thermal properties strongly depend on chemical composition. Substitution of Cu by Co enhances β relaxation due to the reduction of the enthalpy of mixing fluctuation. Physical aging below the glass transition temperature T_g reduces the concentration of local "defects" in metallic glasses and causes a decrease of the intensity for the β process. The curves of loss modulus can be well fitted by the KWW model, the fitting parameter β_{KWW} is approximately 0.4, indicating that the plastic deformation of the $(La_{0.5}Ce_{0.5})_{65}Al_{10}(Co_{0.8}Cu_{0.2})_{25}$ metallic glass is related to the microstructural heterogeneity.

Author Contributions: M.L., Q.H. and Y.C. conducted the experiments and analyzed the experimental results. Q.H. and M.L. contributed to the sample preparation. M.L., J.Q., D.C., Y.Y. and J.-M.P. proposed the idea of the work, originated the experiment and write the paper.

Funding: This research was funded by the Fundamental Research Funds for the Central Universities (Nos. 3102018ZY010, 3102019ghxm007 and 3102017JC01003), Astronautics Supporting Technology Foundation of China (2019-HT-XG) and the Natural Science Foundation of Shaanxi Province (No. 2019JM-344). The investigation of Y.H. Chen and M.N. Liu sponsored by the Seed Foundation of Innovation and Creation for Graduate Students in Northwestern Polytechnical University (No. ZZ2019014). D. Crespo acknowledges financial support from MINECO, Spain, grant FIS2017-82625-P, and Generalitat de Catalunya, grant 2017SGR0042.

References

1. Qiao, J.C.; Wang, Q.; Pelletier, J.M.; Kato, H.; Casalini, R.; Crespo, D.; Pineda, E.; Yao, Y.; Yang, Y. Structural heterogeneities and mechanical behavior of amorphous alloys. *Prog. Mater. Sci.* **2019**, *104*, 250–329. [CrossRef]
2. Hufnagel, T.C.; Schuh, C.A.; Falk, M.L. Deformation of metallic glasses: recent developments in theory, simulations, and experiments. *Acta Mater.* **2016**, *109*, 375–393. [CrossRef]
3. Wang, W.H. The elastic properties, elastic models and elastic perspectives of metallic glasses. *Prog. Mater. Sci.* **2012**, *57*, 487–656. [CrossRef]
4. Inoue, A. Stabilization of metallic supercooled liquid and bulk amorphous alloys. *Acta Mater.* **2000**, *48*, 279–306. [CrossRef]
5. Sun, B.A.; Wang, W.H. The fracture of bulk metallic glasses. *Prog. Mater. Sci.* **2015**, *74*, 211–307. [CrossRef]
6. Qiao, J.C.; Chen, Y.H.; Casalini, R.; Pelletier, J.M.; Yao, Y. Main relaxation and slow relaxation processes in a $La_{30}Ce_{30}Al_{15}Co_{25}$ metallic glass. *J. Mater. Sci. Technol.* **2019**, *35*, 982–986. [CrossRef]
7. Zhang, L.-C.; Jia, Z.; Lyu, F.; Liang, S.-X.; Lu, J. A review of catalytic performance of metallic glasses in wastewater treatment: Recent progress and prospects. *Prog. Mater. Sci.* **2019**, *105*, 100576. [CrossRef]
8. Jia, Z.; Wang, Q.; Sun, L.; Wang, Q.; Zhang, L.-C.; Wu, G.; Luan, J.-H.; Jiao, Z.-B.; Wang, A.; Liang, S.-X.; et al. Attractive in situ self-reconstructed hierarchical gradient structure of metallic glass for high efficiency and remarkable stability in catalytic performance. *Adv. Funct. Mater.* **2019**, *29*, 1807857. [CrossRef]
9. Qiao, J.C.; Pelletier, J.M. Dynamic mechanical relaxation in bulk metallic glasses: A review. *J. Mater. Sci. Technol.* **2014**, *30*, 523–545. [CrossRef]
10. Schuh, C.A.; Hufnagel, T.C.; Ramamurty, U. Mechanical behavior of amorphous alloys. *Acta Mater.* **2007**, *55*, 4067–4109. [CrossRef]
11. Wang, Q.; Liu, J.J.; Ye, Y.F.; Liu, T.T.; Wang, S.; Liu, C.T.; Lu, J.; Yang, Y. Universal secondary relaxation and unusual brittle-to-ductile transition in metallic glasses. *Mater. Today* **2017**, *20*, 293–300. [CrossRef]
12. Ghidelli, M.; Gravier, S.; Blandin, J.J.; Djemia, P.; Mompiou, F.; Abadias, G.; Raskin, J.P.; Pardoen, T. Extrinsic mechanical size effects in thin ZrNi metallic glass films. *Acta Mater.* **2015**, *90*, 232–241. [CrossRef]

13. Ghidelli, M.; Idrissi, H.; Gravier, S.; Blandin, J.-J.; Raskin, J.-P.; Schryvers, D.; Pardoen, T. Homogeneous flow and size dependent mechanical behavior in highly ductile $Zr_{65}Ni_{35}$ metallic glass films. *Acta Mater.* **2017**, *131*, 246–259. [CrossRef]

14. Han, Z.; Wu, W.F.; Li, Y.; Wei, Y.J.; Gao, H.J. An instability index of shear band for plasticity in metallic glasses. *Acta Mater.* **2009**, *57*, 1367–1372. [CrossRef]

15. Leuzzi, L.; Nieuwenhuizen, T.M. Thermodynamics of the glassy state. *J. Stat. Phys.* **2008**, *133*, 1185–1186.

16. Dixon, P.K.; Lei, W.; Nagel, S.R.; Williams, B.D.; Carini, J.P. Scaling in the relaxation of supercooled liquids. *Phys. Rev. Lett.* **1990**, *65*, 1108–1111. [CrossRef]

17. Ngai, K.; Wang, Z.; Gao, X.; Yu, H.; Wang, W. A connection between the structural α-relaxation and the β-relaxation found in bulk metallic glass-formers. *J. Chem. Phys.* **2013**, *139*, 014502. [CrossRef]

18. Qiao, J.C.; Wang, Q.; Crespo, D.; Yang, Y.; Pelletier, J.M. Amorphous physics and materials: Secondary relaxation and dynamic heterogeneity in metallic glasses: A brief review. *Chin. Phys. B* **2017**, *26*, 016402. [CrossRef]

19. Angell, C.A. Formation of glasses from liquids and biopolymers. *Science* **1995**, *267*, 1924–1935. [CrossRef]

20. Debenedetti, P.G.; Stillinger, F.H. Supercooled liquids and the glass transition. *Nature* **2001**, *410*, 259–267. [CrossRef]

21. Zhao, Z.; Wen, P.; Shek, C.; Wang, W. Measurements of slow β-relaxations in metallic glasses and supercooled liquids. *Phys. Rev. B* **2007**, *75*, 174201. [CrossRef]

22. Johari, G.P.; Goldstein, M. Viscous liquids and the glass transition. II. Secondary relaxations in glasses of rigid molecules. *J. Chem. Phys.* **1970**, *53*, 2372–2388.

23. Ngai, K.L.; Rendell, R.W. Cooperative dynamics in relaxation: A coupling model perspective. *J. Mol. Liq.* **1993**, *56*, 199–214. [CrossRef]

24. Kudlik, A.; Benkhof, S.; Blochowicz, T.; Tschirwitz, C.; Rössler, E. The dielectric response of simple organic glass formers. *J. Mol. Struct.* **1999**, *479*, 201–218. [CrossRef]

25. Qiao, J.C.; Wang, Y.-J.; Pelletier, J.M.; Keer, L.M.; Fine, M.E.; Yao, Y. Characteristics of stress relaxation kinetics of $La_{60}Ni_{15}Al_{25}$ bulk metallic glass. *Acta Mater.* **2015**, *98*, 43–50. [CrossRef]

26. Lyu, G.J.; Qiao, J.C.; Pelletier, J.M.; Yao, Y. The dynamic mechanical characteristics of Zr-based bulk metallic glasses and composites. *Mater. Sci. Eng., A* **2018**, *711*, 356–363. [CrossRef]

27. Qiao, J.C.; Yao, Y.; Pelletier, J.M.; Keer, L.M. Understanding of micro-alloying on plasticity in $Cu_{46}Zr_{47-x}Al_7Dy_x$ ($0 \leq x \leq 8$) bulk metallic glasses under compression: Based on mechanical relaxations and theoretical analysis. *Int. J. Plast.* **2016**, *82*, 62–75. [CrossRef]

28. Yao, Z.F.; Qiao, J.C.; Pelletier, J.M.; Yao, Y. Characterization and modeling of dynamic relaxation of a Zr-based bulk metallic glass. *J. Alloys Compd.* **2017**, *690*, 212–220. [CrossRef]

29. Qiao, J.C.; Casalini, R.; Pelletier, J.M.; Yao, Y. Dynamics of the strong metallic glass $Zn_{38}Mg_{12}Ca_{32}Yb_{18}$. *J. Non-Cryst. Solids* **2016**, *447*, 85–90. [CrossRef]

30. Yu, H.B.; Wang, W.H.; Bai, H.Y.; Samwer, K. The β-relaxation in metallic glasses. *Natl. Sci. Rev.* **2014**, *1*, 429–461. [CrossRef]

31. Yu, H.B.; Shen, X.; Wang, Z.; Gu, L.; Wang, W.H.; Bai, H.Y. Tensile plasticity in metallic glasses with pronounced β relaxations. *Phys. Rev. Lett.* **2012**, *108*, 015504. [CrossRef]

32. Ichitsubo, T.; Matsubara, E.; Yamamoto, T.; Chen, H.S.; Nishiyama, N.; Saida, J.; Anazawa, K. Microstructure of fragile metallic glasses inferred from ultrasound-accelerated crystallization in Pd-based metallic glasses. *Phys. Rev. Lett.* **2005**, *95*, 245501. [CrossRef]

33. Wang, W.H. Dynamic relaxations and relaxation-property relationships in metallic glasses. *Prog. Mater. Sci.* **2019**, *106*, 100561. [CrossRef]

34. Yu, H.B.; Samwer, K.; Wang, W.H.; Bai, H.Y. Chemical influence on β-relaxations and the formation of molecule-like metallic glasses. *Nat. Commun.* **2013**, *4*, 1345–1346.

35. Takeuchi, A.; Inoue, A. Classification of bulk metallic glasses by atomic size difference, heat of mixing and period of constituent elements and its application to characterization of the main alloying element. *Mater. Trans.* **2005**, *46*, 2817–2829. [CrossRef]

36. Liu, S.T.; Wang, Z.; Peng, H.L.; Yu, H.B.; Wang, W.H. The activation energy and volume of flow units of metallic glasses. *Scr. Mater.* **2012**, *67*, 9–12. [CrossRef]

37. Liu, Y.H.; Wang, D.; Nakajima, K.; Zhang, W.; Hirata, A.; Nishi, T.; Inoue, A.; Chen, M.W. Characterization of nanoscale mechanical heterogeneity in a metallic glass by dynamic force microscopy. *Phys. Rev. Lett.* **2011**, *106*, 125504. [CrossRef]

38. Liu, S.T.; Jiao, W.; Sun, B.A.; Wang, W.H. A quasi-phase perspective on flow units of glass transition and plastic flow in metallic glasses. *J. Non-Cryst. Solids* **2013**, *376*, 76–80. [CrossRef]

39. Perez, J.; Etienne, S.; Tatibouät, J. Determination of glass transition temperature by internal friction measurements. *Phys. Status Solidi a* **1990**, *121*, 129–138. [CrossRef]

40. Egami, T. Mechanical failure and glass transition in metallic glasses. *J. Alloys Compd.* **2011**, *509*, S82–S86. [CrossRef]

41. Qiao, J.C.; Pelletier, J.M. Kinetics of structural relaxation in bulk metallic glasses by mechanical spectroscopy: Determination of the stretching parameter β_{KWW}. *Intermetallics* **2012**, *28*, 40–44. [CrossRef]

42. Qiao, J.; Pelletier, J.M.; Casalini, R. Relaxation of bulk metallic glasses studied by mechanical spectroscopy. *J. Phys. Chem. B* **2013**, *117*, 13658–13666. [CrossRef]

43. Pelletier, J.M. Influence of structural relaxation on atomic mobility in a $Zr_{41.2}Ti_{13.8}Cu_{12.5}Ni_{10.0}Be_{22.5}$ (Vit1) bulk metallic glass. *J. Non-Cryst. Solids* **2008**, *354*, 3666–3670.

44. Qiao, J.; Casalini, R.; Pelletier, J.M.; Kato, H. Characteristics of the structural and Johari–Goldstein Relaxations in Pd-based metallic glass-forming liquids. *J. Phys. Chem. B* **2014**, *118*, 3720–3730. [CrossRef]

45. Böhmer, R.; Ngai, K.L.; Angell, C.A.; Plazek, D.J. Nonexponential relaxations in strong and fragile glass formers. *J. Chem. Phys.* **1993**, *99*, 4201–4209. [CrossRef]

46. Angell, A.C. Relaxation in liquids, polymers and plastic crystals—Strong/fragile patterns and problems. *J. Non-Cryst. Solids* **1991**, *131–133*, 13–31. [CrossRef]

47. Raghavan, R.; Murali, P.; Ramamurty, U. Influence of cooling rate on the enthalpy relaxation and fragility of a metallic glass. *Metall. Mater. Trans. A* **2008**, *39*, 1573–1577. [CrossRef]

48. Zhang, Y.; Hahn, H. Study of the kinetics of free volume in $Zr_{45.0}Cu_{39.3}Al_{7.0}Ag_{8.7}$ bulk metallic glasses during isothermal relaxation by enthalpy relaxation experiments. *J. Non-Cryst. Solids* **2009**, *355*, 2616–2621.

49. Zhang, T.; Ye, F.; Wang, Y.; Lin, J. Structural Relaxation of $La_{55}Al_{25}Ni_{10}Cu_{10}$ bulk metallic glass. *Metall. Mater. Trans. A* **2008**, *39*, 1953–1957. [CrossRef]

50. Gallino, I.; Shah, M.B.; Busch, R. Enthalpy relaxation and its relation to the thermodynamics and crystallization of the $Zr_{58.5}Cu_{15.6}Ni_{12.8}Al_{10.3}Nb_{2.8}$ bulk metallic glass-forming alloy. *Acta Mater.* **2007**, *55*, 1367–1376.

51. Wang, Z.; Sun, B.A.; Bai, H.Y.; Wang, W.H. Evolution of hidden localized flow during glass-to-liquid transition in metallic glass. *Nat. Commun.* **2014**, *5*, 5823. [CrossRef]

Article

Preparation and Characterization of Mg-RE Alloy Sheets and Formation of Amorphous/Crystalline Composites by Twin Roll Casting for Biomedical Implant Application

Haijian Wang [1], Dongying Ju [2,3,*] and Haiwei Wang [4,5]

[1]   Department of High-Tech Research Center, University of Saitama Institute of Technology, Fusaiji 1690, Fukaya 369-0293, Japan; whaijian@yahoo.com
[2]   High-Tech Research Center, Saitama Institute of Technology, Fusaiji 1690, Fukaya 369-0293, Japan
[3]   School of Materials and Metallurgy, University of Science and Technology Liaoning, Anshan 114051, China
[4]   Institute of Metal Research, Chinese Academy of Sciences, Shenyang 110016, China; hwwang17b@imr.ac.cn
[5]   School of Materials Science and Engineering, University of Science and Technology of China, Hefei 230026, China
*   Correspondence: dyju@sit.ac.jp; Tel.: +81-485-956-826

Received: 27 August 2019; Accepted: 1 October 2019; Published: 3 October 2019

Abstract: A new type of Mg-based metallic glass has attracted extensive attention due to its excellent corrosion resistance and favorable biocompatibility. In this study, an amorphous/crystalline composite Mg-RE alloy sheet was prepared by a vertical type twin roll caster (VTRC) method, and its microstructure was characterized by scanning electron microscopy (SEM), X-ray diffraction (XRD), and electron probe micro-analysis (EPMA) and transmission electron microscopy (TEM); furthermore, the corrosion behaviors of the Mg-RE alloy sheet were investigated in PBS solution using electrochemical techniques and immersion testing in a simulated physiological condition. Furthermore, it was implanted into the femur of rats to explore its prospect as biological transplantation material. Its microscopic characterization experiments show that the crystal structure is crystalline phase containing amorphous phase. Electrochemical experiments and immersion testing both showed that Mg-RE(La,Ce) sheet with VTRC has a better corrosion resistance than master alloy, and a uniform corrosion layer on the surface. In vivo, as an implant material, tests show that Mg-RE alloy sheets have better biocompatibility and induce new bone formation, and they can be expected to be utilized as implant materials in the future.

Keywords: metallic glass; twin roll casting; GFA; MicroCT

1. Introduction

Metallic glasses present excellent mechanical and chemical properties that are distinctive among solid metals, and are becoming a research hotspot for current studies in the field of metallic materials [1]. In the field of biological health, biomaterials are developing rapidly and improving people's life quality. Among biomaterials, bioinert metals have been found to be mainly used in cardiovascular scaffolds, orthopedics and dental implants [2–4]. However, the characteristics of these crystalline alloys, instance of high elastic modulus, relative low abrasion resistance, and stress corrosion cracking lead to bone stress shielding. Compared with traditional crystalline metals, metallic glass has an amorphous structure, higher strength, lower Young's modulus, better wear resistance, higher corrosion resistance, and anti-fatigue performance for some Ti-, Zr-, Fe-base systems [5,6]. Over the course of the decades, many metallic glassy alloys have been developed using a wide range of components, including Pd-,

Pt-, Zr-, Mg-, Ti-, Co-, and Au-base systems. Among the various different compositions of metallic glasses, Mg-based metallic glass has been widely studied for biomedical applications.

Many amorphization techniques have been developed for metallic glasses, including gun and splat quenching [7], melt spinning [8], high-pressure die casting [9,10], copper mold casting [11], and twin roll casting (TRC). TRC has the advantages of a shorter production cycle, low production cost and lower capital investment compared with conventional techniques [12]. In 1970, following Duvez's seminal discoveries, Chen and Miller developed a TRC technique for producing metastable uniform sheets [13]. Until now, the technique for producing metallic glass strips has been almost solely confined to laboratory-scale research [14–20].

Previous studies have shown that TRC is a useful technique for preparing amorphous alloy sheets with an extensive cooling speed. However, the major research until now has been based on horizontal double roll casters. It is turns out that the heat transfer efficiency of vertical-type twin roll casting (VTRC) is higher than that of horizontal double roll casting (HTRC), and the cooling speed of VTRC is higher [21,22]. The rapid cooling speed of the alloy during the TRC process is beneficial for reducing segregation, achieving higher uniformity and expanded solid solubility, refining the microstructure characteristics [23]. It enables better utilization to be made of a variety of transition elements that have limited solid solubilities in magnesium alloy, to improve mechanical and chemical properties [24]. As mentioned above, heat transfer VTRC is more effective method for continuous production of magnesium alloy sheet than HTRC, and the VTRC process enables achieve a wide range of variable casting speed [25], as a result, the processability and application performance of the products are improved.

However, studies of the rapid casting speed of Mg alloy sheets produced by rapid solidification technology using TRC are rare. In this study, based on the VTRC process, the Mg-RE (RE—rare-earth elements) sheets were produced on a vertical twin roll caster and then were annealed. The microstructure of the Mg-RE sheets was investigated. The corrosion behavior properties were studied. Through this study, we expect to the Mg-RE sheet with a special organizational structure to be potential biodegradable material. Meanwhile, it was implanted into the femur of rat to explore its prospect as biological transplantation material.

2. Materials and Methods

Ingots of Mg-RE alloy were prepared by induction melting the mixture of industrial AZ31, Mg-10%La and Mg-20%Ce (wt%) master alloys in an induction furnace at 993 K for 30 min under the protection of high-purity argon. The chemical compositions of the ingots were measured by X-ray fluorescence spectrometry, and the results are listed in Table 1. Figure 1 is a schematic diagram of the manufacturing process for magnesium alloy sheets. The Mg-based sheets were prepared by VTRC. The roll, which had a diameter of 300 mm and a width of 100 mm, was made from a copper alloy. Twin roll casting experiments were carried out under casting conditions with a casting speed of 10 m/min and a pouring temperature 973 K. Because the casting produces separation force during the casting process, the metal block was set at the side of the moving roller to form a supporting force in order to minimize the gap between the rollers as much as possible during the casting process. Afterward, the melt of magnesium alloy flowed through a nozzle to a position between running rolls. The initial roll gap was set to 0 mm. An oil tank was set under the roll and the Mg-RE alloy sheet was completely submerged in the tank when the casting process was completed to prevent further grain growth. The final thickness of the sheets between 0.5 mm and 1.1 mm and the width of the strip rang is 25 mm to 50 mm.

From the Mg-RE master ingots and the as-extruded Mg-RE sheets, samples with a dimension of $10 \times 10 \times 1$ mm^3 were firstly grounded with SiC papers to1200 grid, and then by diamond pastes down to #0.25 μm grade. The microstructures of the polished surfaces were observed using a field emission scanning electron microscope (FE-SEM, JMS-6301, Tokyo, Japan) and the elements distribution maps were observed by electron probe micro-analysis (EPMA, JXA-8530F, Tokyo, Japan). The alloy phases were obtained by an X-ray diffractometer (XRD, D/Max 2500 PC, Tokyo, Japan). For TEM (HF-3300, Tokyo, Japan) analysis, the focused ion beam (FIB, JIB-4500, Tokyo, Japan) was used for preparation. During the original work of sample preparation, the alloy sheet was cut into 10 mm^2 square shapes, and the thinned section thickness was about 60 μm. Square metal sheets with an area of 1.5 mm^2 were cut off from the as-cast sheet and stuck together on a Mo grid with some resin glue, and then the sample was further cut with FIB to obtain the ultimate sample thickness of 0.1 μm.

Table 1. Composition and the atomic radius of the Mg-RE alloy (atomic radius difference between Mg (Al) and other elements is symbolized by ARD$_{Mg}$ (ARD$_{Al}$).

Elements	Mg	Al	Si	Mn	Cu	Fe	Zn	La	Ce
at%	95.253	3.460	0.145	0.132	0.040	0.041	0.053	0.297	0.579
wt%	90.770	3.660	0.160	0.284	0.101	0.090	0.136	0.187	0.182
Radius/nm	0.160	0.143	0.134	0.132	0.128	0.126	0.139	0.187	0.182
ARD$_{Mg}$/%	-	10.63	16.25	17.50	20.00	21.25	13.13	16.87	13.75
ARD$_{Al}$/%	11.88	-	6.29	7.69	10.50	11.88	2.78	30.77	27.27

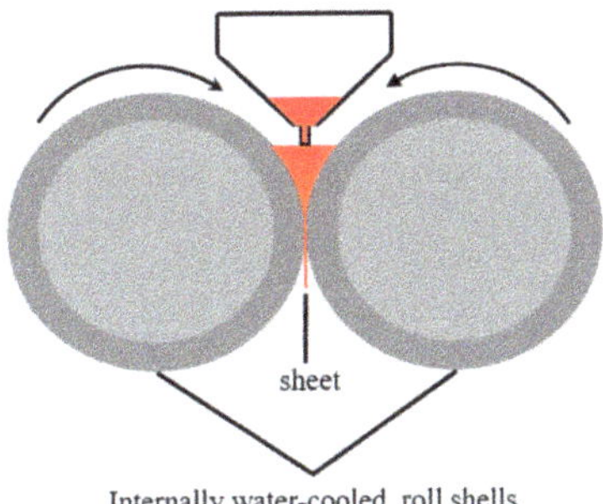

Figure 1. Schematic diagram of the manufacturing process for magnesium alloy sheets.

The corrosion behaviors of the alloys were studied by potentiodynamic polarization (HZ700, Tokyo, Japan) and electrochemical impedance spectroscopy (Modulab XM, Tokyo, Japan), using a three-electrode cell comprising an auxiliary electrode of platinum counter, a reference electrode of Ag/AgCl electrode, and a working of the samples. The exposure area was 1 cm^2. After immersion for 0.5 h, electrochemical impedance tests were operated at the open circuit potential with signal amplitude of 10 mV over a frequency varying between 100,000 Hz and 0.01 Hz. After immersion for 1 h, potentiodynamic polarization was performed at a scanning rate of 1 mV/s. All electrochemical tests were conducted in PBS (phosphate buffer saline) solution at 310 K. A triplicate electrochemical test was carried out to ensure the reproducibility of the results. For immersion tests, for each sample, only one side of 1 cm^2 was exposed. Then, the samples were immersed in PBS solution at 310 K for 30 days. After immersion, the section corrosion morphologies were observed by SEM. The concentration of metal ions in the solutions were analyzed by Inductively Coupled Plasma (ICP) (ICPS-7000 Sequential Plasma Spectrometer, Tokyo, Japan).

All animal experiments were approved by the University of Saitama Institute of Technology Animal Care and Use Committee (Grant NO. 2019-5). The project recognition date is 20 May 2019. 5 white rats (12 weeks of age, 0.36 ± 0.02 kg) from Tokyo University Institute of medicine were used in this study. Rats were anesthetized with isofluane and the right thighs were shaving and disinfected. The skin and muscle of thighs were carefully retracted to expose the femurs. Single screw fixation was used in this implantation experiment. Sheets were 22 × 4.8 mm^2 with a thickness of 1 mm. Screws were 5 mm in length, with an outer shaft diameter of 1.75 mm and shaft inner diameter of 1 mm. Animals were monitored daily for general behavior, movement, and food and water intake. High resolution microCT (R.mCT2, Rigaku, Tokyo, Japan) was used to assess Mg-RE sheets degradation and new bone formation.

3. Results and Discussion

3.1. Microstructure Characteristic

Figure 2 reveals the surface morphologies and EDS analyses of the as-cast Mg-RE alloy ingot. It can be seen from Figure 2a that acicular intermetallic compounds crystallize in the Mg-RE alloy ingots. Figure 2b presents the combined EDS results, and the acicular compound crystallization in the red rectangle contains higher contents of Ce elements. Figure 3a presents the microstructure of the Mg-RE alloy sheet at a thickness of ~1.1 mm and a width of ~50 mm obtained at the casting speed of 30 m/min, in which the microstructure of the Mg-RE alloy sheet is characterized by dendrites of fine grains and a closely spaced secondary dendrite axis. Apart from that, as shown by the red rectangular area in Figure 3a, it can be seen that there are portions where no appreciable crystalline features can be observed. Figure 3c shows the XRD patterns of the studied alloys. It can be concluded that the as-cast Mg-RE alloy sheets (Figure 3a) consisted mainly of α-Mg, La-Al and Ce-Al, and it is worth noting that a broad peak appears at the angle of 20°~30°, indicating that the sample may contain both crystalline and amorphous phases.

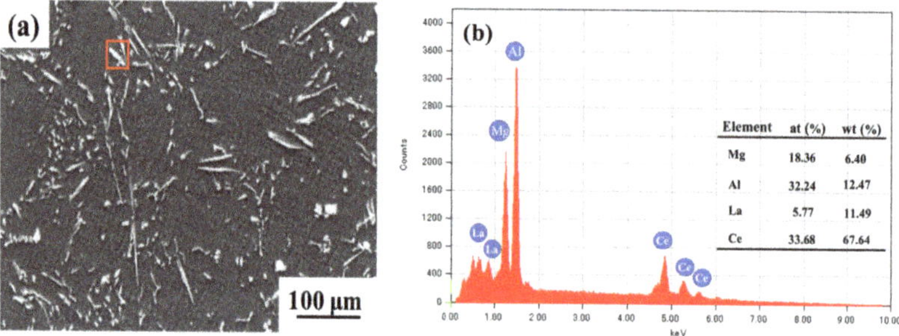

Figure 2. SEM morphologies (**a**) and the precipitation by EDS analysis (**b**) of as-cast Mg-RE alloy ingot.

The crystallite size of each detected phase in Figure 3c could be calculated using the Scherrer equation, which is expressed by $D_{hkl} = K\lambda/B_{hkl} \cos\theta$ [26], where D_{hkl} is the grain size perpendicular to the lattice planes, *hkl* are the Miller index of the planes being analyzed, K is a constant numerical factor called the crystallite-shape factor, λ is the wavelengths of the X-rays, B_{hkl} is the width of the X-ray diffraction peak in radians and θ is the Bragg angle. The calculated results are listed in Table 2. It turns out that the grain sizes of the detected phase are very fine. However, the X-ray tube on the line focus side is unsuitable for analyzing such a specific area without crystals, as shown in Figure 3a, as the line focus range is 0.1~0.2 mm wide and 8~12 mm long [27]. To solve these problems, the structure of the areas without crystals was analyzed by means of micro area X-ray diffraction. In the current operation, a collimator that was 0.03 mm in diameter was situated at the point focus side of the X-ray tube. Therefore, very small specific areas could be analyzed without reflecting unnecessary regional structural information. Figure 3d displays the μ-XRD of the areas without crystals of the Mg-RE sheet. The amorphous structure was determined by a peak that does not correspond to any sharp crystalline peak. As shown in Figure 3b and Table 2, very fine grains and dendrites with closely spaced secondary dendrite axes can be found around a large amorphous region.

Figure 4 shows the solute element distribution of Mg-RE with TRC. It can be seen that Mg elements are evenly distributed along the matrix, while Al/Ce/La are concentrated in amorphous areas. In addition, Al, Ce, La element segregation exists between crystal phase and the amorphous region. Meanwhile, the grain boundary is prone to segregation, because the relative atomic radius difference (ARD$_{Mg}$) between Mg and other elements is more than 10% [28], as shown in Table 1. It can also be found that Al, Ce and La elements are enriched in the amorphous phase region, which may be related to the fact that the alloy is prone to producing very stable Al-RE compounds under the solidification condition of low cooling rate. Figure 5 presents the values of enthalpy of mixing ($\Delta H^{mix}_{[AB]}$) calculated by Miedma's model for atomic pairs between major elements of Mg-RE sheet samples, in which the enthalpy of mixing between Mg-Al, Mg-La and Mg-Ce are −2 KJ/mol, −7 KJ/mol and −7 KJ/mol, respectively, while the enthalpy of mixing between Al-La and Al-Ce are −38 KJ/mol, which is greater than that between Mg and other major elements [29]. The design of Mg-RE alloy conforms to the three rules summarized by Inoue et al. [30] for the glass forming ability (GFA) of alloys: first, a multi-component system consisting of more than three major elements; second, the difference in atomic size between major elements is large (greater than 10%), and in line with the relationship of large, medium and small; third, the mixed heat between the main elements is a suitable negative value. In other words, Mg-RE alloy has good glass-forming ability.

Table 2. Crystallite sizes of the Mg-RE alloy sheet, calculated by Scherrer equation.

(hkl)	(100)	(002)	(101)	(102)	(110)	(103)
D_{hkl}/nm	5.1	14.2	7.8	10.6	16.6	17.9

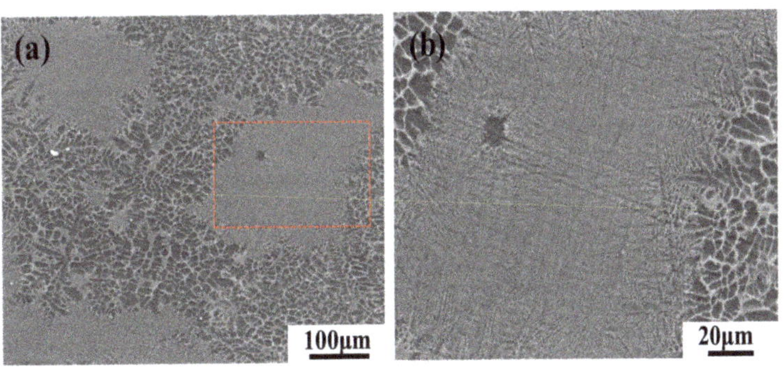

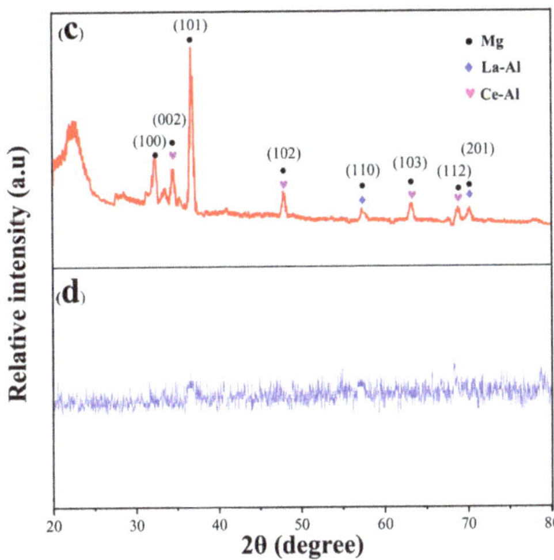

Figure 3. The SEM micrographs and X-ray diffractometry (XRD) patterns of as-cast Mg-RE sheet: (**a**,**c**) SEM micrographs and XRD of the Mg-RE sheets. (**b**,**d**) SEM micrograph and μ-XRD of local amorphous region.

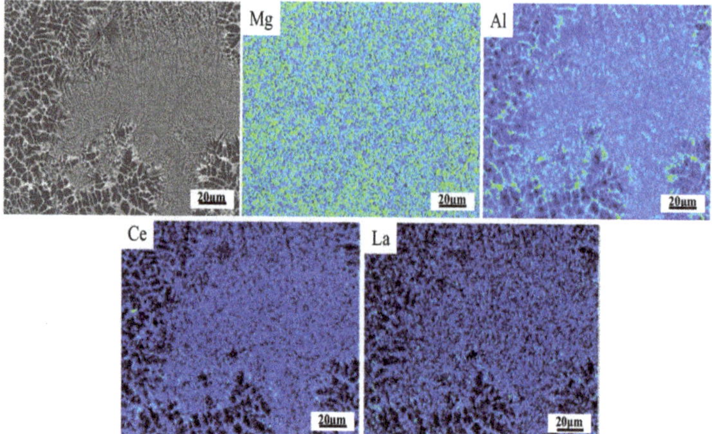

Figure 4. The major element distribution of the Mg-RE alloy sheet with TRC samples.

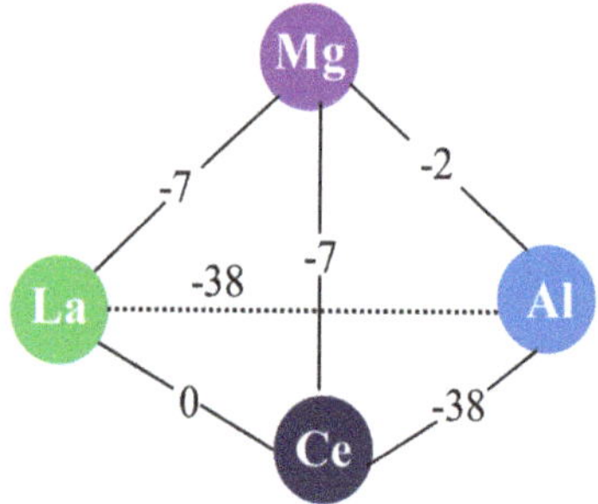

Figure 5. The values of $\Delta H^{mix}_{[AB]}$ (KJ/mol) calculated by Miedma's model for atomic pairs between major elements of Mg-RE sheet samples.

To ascertain the amorphous phase structure, many initial specimens were prepared using the FIB technique, and TEM observation was further performed. Figure 6c shows that the first step of TEM sample preparation is to find the amorphous phase on the surface of the sample in the SEM diagram, and then cut the specific phase area with FIB technology which make the cutting area length was 30~60 μm. Figure 6a presents the cross section of the sample is sliced by FIB technique. Since part of the sample was cut with FIB on the surface, it is easy to find the section position to be cut. The sample was cut into steps, originally, and a slice thickness of 0.1 μm was ultimately prepared, as shown in Figure 6b. Figure 6d presents the TEM image and selected area diffraction pattern (SADP) of the amorphous phase, in which amorphous circular halos are not distinctly visible, and a poor crystallinity is shown, although few spots of electron diffraction exist. This may be due to the small size of the amorphous phase in the current TEM sample or the slight oxidation of the Mg-RE alloy sheet after cutting by FIB technology.

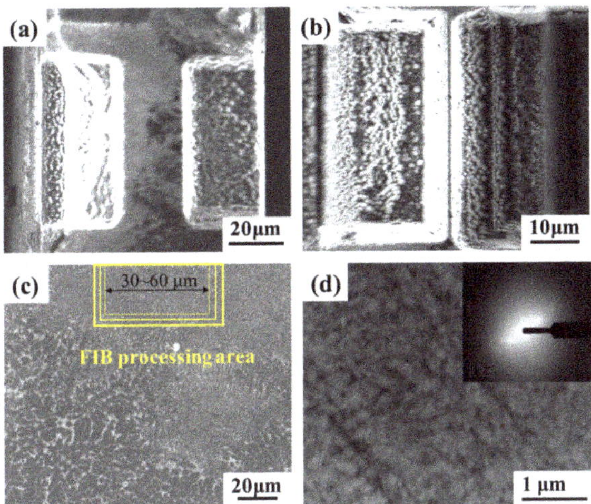

Figure 6. (**a**,**b**) TEM specimen by FIB technique of Mg-RE sheet with TRC; (**c**) SEM and FIB processing area on the Mg-RE sheet surface; (**d**) TEM image and high-resolution morphology.

3.2. Electrochemical Measurements

Figure 7 shows the electrochemical behaviors of the prepared Mg-RE alloy ingot sample and the sheet with TRC sample in PBS solution at 310 K. In addition, the fitting results are summarized in Table 3. As stated above, after the two-roll casting process, the Mg-RE alloy sheet presented a more positive potential (-1.08 $V_{Ag/AgCl}$), with comparable potentials to the Mg-RE alloy ingot (-1.37 $V_{Ag/AgCl}$). Based on electrochemical theory, in the process of electrode reaction, the ions in the solution were mainly in charge of conveying the transformation to the surface of the electrode. In the cathode area, the Mg-RE alloy dissolves into metal cation. Because the metal cation ion concentration is too high, the charge exchange process cannot be carried out as soon as possible. The cloud of ions blocks the electrodes' ability to charge, which is called polarization resistance (Rp). In general, the larger the Rp of the metal materials, the larger the ion cloud on the electrode surface, thus preventing charge exchange. The corrosion potential (E_{corr}) of the samples is mainly determined by the relative size between anode and cathode reaction rates, which reflects the reaction trend [31]. Furthermore, the corrosion current density (I_{corr}) shows a decreasing trend: Mg-RE sheet with TRC sample (1.51×10^{-4} µA) < Mg-RE ingot sample (1.74×10^{-3} µA). Lower I_{corr} indicates better corrosion resistance. Therefore, it is demonstrated that the Mg-RE sheet sample with TRC possesses a higher corrosion potential, a smaller current density, and a better corrosion resistance. Due to the influence of part of the amorphous phase being formed in the Mg-RE alloy after rapid cooling solidification by two-roll casting, the corrosion resistance was enhanced.

The electrochemical impedance data were determined for the corrosion potential in PBS and presented in Nyquist plots (Figure 7b, and Bode plots (Figure 7c). The equivalent circuit for electrochemical impedance is shown in Figure 7b. *Rs*, *Rt*, and *Rf* represent the solution resistance between the reference electrode and the working alloy sample, the resistance of charge transfer and the resistance of the corrosion product layer on the surface of the sample, respectively. Additionally, CPE1 and CPE2 illustrate the capacitance of the corrosion product layers and charge separation at the positions where hydrogen evolution increases. In the Nyquist plots shown in Figure 7b, the magnitude of the radius curvature has different values, showing a decreasing trend: Mg-RE ingot sample < Mg-RE sheet with TRC sample, which is also illustrated by the decreasing impedance modulus trend of the Mg-RE alloys in the curves in Figure 7c. It is well known that the size of the Nyquist curve is an important parameter that reflects corrosion resistance. That is to say, better corrosion resistance and behavior of the metal and alloy matrix is related to higher |Z| modulus at lower frequency, which is inversely proportional to the corrosion rate of the alloy. The Bode phase plot is shown in Figure 6d, and it can be found that the phase angles corresponding to high frequency are in a decreasing order as follows: Mg-RE sheet with TRC sample > Mg-RE ingot sample, which could be attributed to the protective properties of the surface film layers.

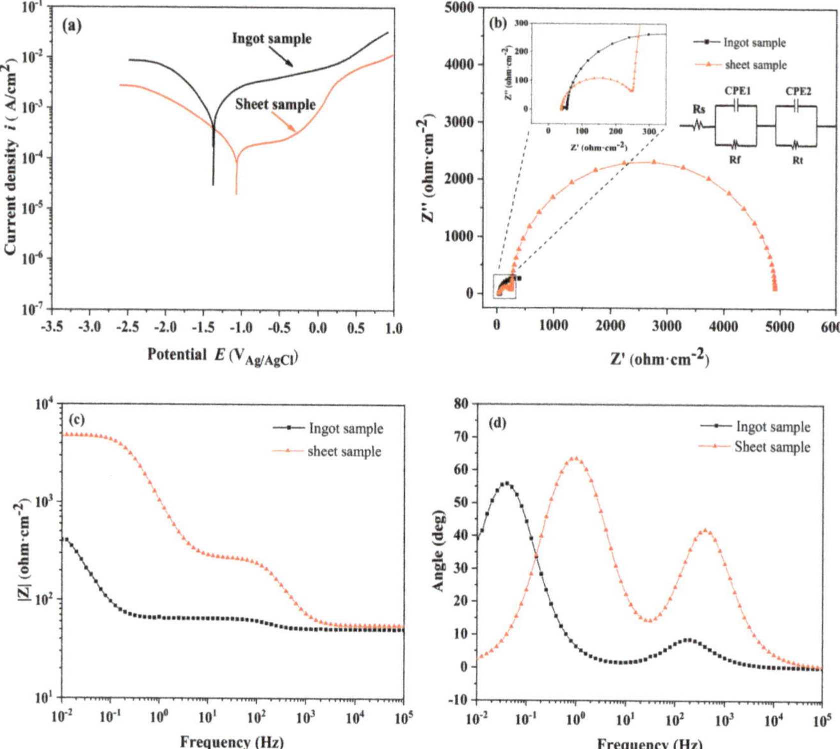

Figure 7. Electrochemical behaviors of the Mg-RE alloy ingot sample and the sheet with TRC sample in PBS solution: (**a**) polarization curves; (**b**) equivalent circuit and Nyquist plots of the real part Z′ vs. the imaginary part Z″; (**c**) Bode plots of |Z| vs. frequency; and (**d**) Bode plots of phase angle vs. frequency.

Table 3. Result of the electrochemical polarization tests in PBS solution.

Sample	E_{corr} (V)	I_{corr} (µA)	Rs (Ω·cm^{-2})	Rf (Ω·cm^{-2})	CPE1 (F)	CPE2 (F)	Rt (Ω·cm^{-2})	Rp (Ω·cm^{-2})
Ingot	−1.37	1.74×10^{-3}	50.73	22.72	7.1×10^{-5}	2.60×10^{-2}	528.37	410.12
Sheet	−1.08	1.51×10^{-4}	56.61	216.18	4.03×10^{-6}	1.57×10^{-4}	4655.12	4920.25

3.3. Immersion Test

Figure 8 presents the cross-sectional SEM micrographs of corrosion morphology of the Mg-RE alloy ingot sample and the sheet with the TRC sample in PBS solution at 310 K for 10 days. It can be observed from Figure 8a that the corrosion layer covered the Mg-RE ingot sample surface, which extended to the inside of the matrix with crack features. In addition, the maximum length of the corrosion cracks extending to the interior of the matrix was 30 µm, and they were distributed unevenly along the cross section of the Mg-RE ingot sample. The improved corrosion-resistance properties of the Mg-RE alloy with TRC are reflected in Figure 8b, in that the corrosion layer is thinner and more uniform. The results of the energy spectrum analysis indicate that the content of elements in different location of Mg-RE alloy ingot is dissimilar, which is mainly connected with the corrosion behavior of the alloy in PBS solution. The mechanism of corrosion analysis is as described below.

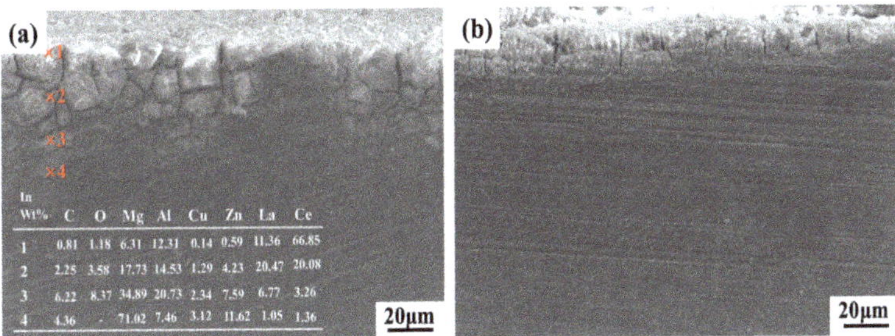

In Wt%	C	O	Mg	Al	Cu	Zn	La	Ce
1	0.81	1.18	6.31	12.31	0.14	0.59	11.36	66.85
2	2.25	3.58	17.73	14.53	1.29	4.23	20.47	20.08
3	6.22	8.37	34.89	20.73	2.34	7.59	6.77	3.26
4	4.36	-	71.02	7.46	3.12	11.62	1.05	1.36

Figure 8. Cross-sectional SEM micrographs of corrosion morphology of (**a**) the Mg-RE alloy ingot sample and (**b**) the sheet with TRC sample in the PBS solution at 310 K for 10 days.

The multiple layers of Mg-RE alloy ingot primarily consisted of C, O, Mg, Al, Ce and other elements during the immersion test in PBS solution. Mg transformed into the stable Mg^{2+} ion in the initial stages. Meanwhile, the cathodic reaction occurred on account of the galvanic corrosion between the matrix and the secondary phase, accompanied by the hydrogen evolution [32]. The cathode reaction formed a heterogeneous thin porous layer which was predominantly magnesium hydrate on the surface of the Mg-RE alloy. This prevented contact of the solution and the substrate, resulting in a decrease in the corrosion rate [33]. Moreover, the chloride ions in the PBS solution were able to easily penetrate the membrane and react with the magnesium hydrate compounds. Therefore, magnesium hydrate compounds were converted into magnesium chloride compounds, which are more likely to dissolve into magnesium ions and chloride ions [34]. The dissolution of the compound leads to a decrease in the magnesium hydrated compounds around the protective layer, leading to further dissolution of the matrix.

Magnesium hydrated films constituted on the surface of magnesium are generally referred to as crystals. Previous studies have found that the composition and structure of magnesium and magnesium alloy surface films can be changed using a rapid solidification process. The conversion of magnesium hydrated films from a crystal form into an amorphous film structure improves the corrosion resistance. Amorphous films have better protection than crystalline films, and films without grain boundaries are better able to resist ion motion than crystalline films [35]. The Mg-RE alloy sheet obtained by the TRC process has a unique structure due to its crystal structure surrounding the amorphous structure, which may be the reason for the improvement of the corrosion resistance of the magnesium hydrated film.

Figure 9 presents the Mg, Al, Ce and La metallic ion concentrations of the solution for (a) the Mg-RE alloy ingot sample and (b) the sheet with the TRC sample at day 10, day 20 and day 30 under immersion testing in PBS solution at a temperature of 310 K. Specifically, both alloys showed a general trend of increasing Mg ion concentration in the PBS solution with increasing immersion time from day 10, day 20 and day 30. In contrast, for Al ion concentration, both alloys showed a decreasing trend. Generally, the Mg-RE alloy ingot sample showed greater average Mg and Al ion concentration in the PBS solution at each time point—day 10, day 20 and day 30. From day 10 to day 20 after immersion, the magnesium ion concentration for the sheet sample was obviously less than the ingot sample after 10, 20 and 30 days. This may be caused by the rapid corrosion of alloys in the initial corrosion stage and then the beginning of passivation to slow down the corrosion. The decrease of aluminum in the corrosive solution may have resulted from the formation of corrosion products on the surface of the sample.

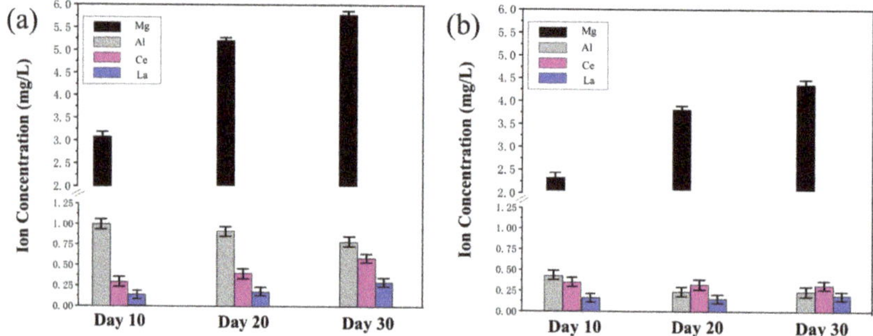

Figure 9. Metallic ion concentrations of the PBS solution at 310 K with (**a**) Mg-RE alloy ingot sample and (**b**) sheet with TRC sample at day 10, day 20 and day 30 under immersion testing.

Figure 9a,b also shows that throughout the entire 30 day culture period, the Mg-RE with TRC sample group showed significantly lower release rates of Ce and La ions than the Mg-RE alloy ingot sample, and the group showed a general trend of increasing Ce and La ion concentration with increasing immersion time, but no significant change for the Mg-RE with TRC sample group. Overall, the content of metallic ions concentrations in the corrosion solution of the Mg-RE alloy sheet produced by rapid solidification TRC method less than that of Mg-RE alloy ingot sample within the same amount of immersion time. Which is probably induced by the special microstructure formed under rapid cooling and exhibits an improved corrosion resistance.

3.4. In Vivo Implantation

The Mg-RE sheet with TRC samples was successfully inserted into the subcutaneous tissue of the rat thigh and attached to the femur with an AZ31 screw during surgery, despite the slight mismatches between the Mg-RE sheet and femur owing to manual surgical placement and the fact that there is only one screw to fix the femur and alloy sheet. Slight skin swelling was observed in the hind limbs of the rats during the visual examination of the rats during the study period. The bulge disappeared after a few days.

Figure 10 reveals the surface volume of the degrading Mg-RE alloy sheet and bone growth changes, depicted using MicroCT 3D rendering after various implantation times. As can be seen from Figure 10a,c, two weeks after surgery, no new bone production was found around the Mg-RE alloy sheet and femur, and the screw between them was clearly visible. There are no obvious corrosion pits on the surface of magnesium alloy sheet. Figure 10b,d, six weeks after the surgery, new bone was formed between the Mg-RE alloy sheet and the femur. Figure 10e presents on the cross section of the bone–alloy joint, the new bone is generated and surrounds the original femur. The sheets are attached to the femur by new bone, which appears to be relatively stable in living organisms. In the red area shown in Figure 10, corrosion pits began to appear on the surface of magnesium alloy sheet after six weeks of surgery. This indicates that the films with improved corrosion resistance of Mg-RE alloy sheet began to degrade in vivo after 6 weeks.

Mg-RE sheet degradation and femoral change were assessed by volume quantification in MicroCT 3D rendering separately as shown in Figure 11. Before surgery, the volume of self-made Mg-RE alloy bone sheet was 94.79 mm^3. The Mg-RE sheet corroded, resulting in 9.32 ± 0.86 mm^3 volume loss after 6 weeks. With regard to the changes in the femur, the volume of the femur did not change significantly during the two weeks following the surgery. In fact, for some time after the surgery, the rat lost significantly more weight, the reason for this being that it took some time for the body to adapt the a foreign body. After about five days, the rats returned to their normal mental state and began to eat normally. By 6 weeks, as new bone had formed between the sheet and the femur, a noticeable change in bone volume was observed from 2 to 6 weeks. The rat showed no abnormal physical or physiological responses during subsequent breeding.

Uniquely, magnesium alloy application has the potential to enhance bone formation. Several previous studies have shown that the potential performance, presenting increased bone mass, mineral apposition, and bone mineral density around magnesium alloy implants in bone [36–39]. In our present study, new bone appears in the area where the femur contacts the alloy sheet, which confirmed our findings that the Mg-RE alloy with TRC degradation can lead to the promotion of bone formation. The new bone formation is not common in absorbable polymers or permanent metal devices, thus emphasizing the unique advantages of magnesium fixtures. In this way, the degraded fixture will be gradually replaced in the future.

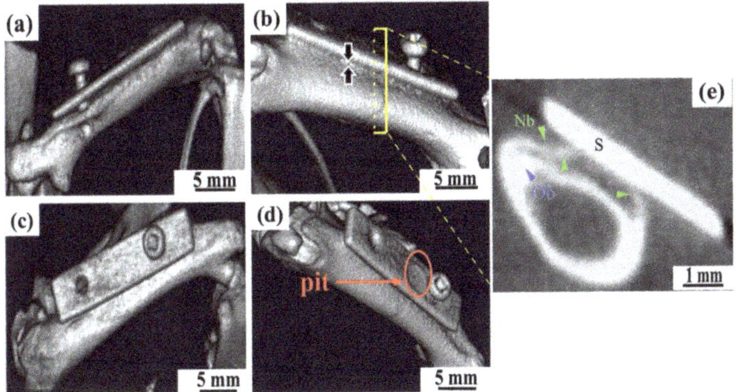

Figure 10. Determination of the volume of the degrading Mg-RE alloy sheet and bone growth changes, depicted in MicroCT 3D rendering after implantation times of 2 weeks (**a**,**c**) and 6 weeks (**b**,**d**). A cross section slice of femoral-sheet contact area observed after 6 weeks (**e**) shows that the new bone (Nb is represented in green) was formed between the Mg-RE alloy sheet (S) and the original bone (Ob is represented in blue).

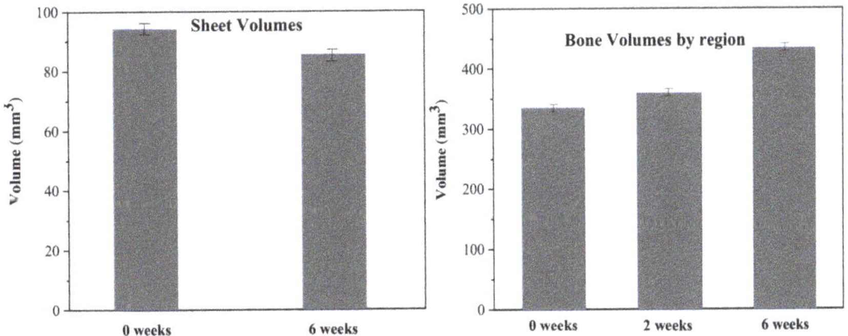

Figure 11. Mg-RE sheet degradation after 6 weeks was estimated by the volume of quantification. This shows a sheet volume loss of 9.32 ± 0.86 mm^3 after 6 weeks. MicroCT 3D rendering after 2 weeks shows there was no obvious change in bone growth around the degrading Mg-RE sheet. By 6 weeks, as new bone formed between the sheet and the bone, a notable change in bone volume was observed from 2 to 6 weeks.

4. Conclusions

(1)   The Mg-RE alloy sheet was prepared using a vertical-type twin-roll caster method. Its microscopic characterization experiments show that the crystal structure is crystalline phase containing amorphous phase.

(2)   EPMA experiments show that Al, La and Ce elements are enriched in the amorphous phase region and the grain boundary region. However, Mg is evenly distributed throughout the microscopic region. This shows that segregation is more likely to affect Al, La and Ce elements.

(3)   Electrochemical tests and immersion test results revealed that Mg-RE sheet with TRC has a better corrosion resistance than master alloy, and a uniform corrosion layer on the surface.

(4)   In vivo, as an implant material, the tests show that Mg-RE alloys sheets were safe with respect to rat physical fitness and induced new bone formation; thus, they were promising for utilization as implant materials in the future.

Author Contributions: H.W. (Haijian Wang) and D.J. conceived and designed the experiments; H.W. (Haijian Wang) and H.W. (Haiwei Wang) performed the experiments; H.W. (Haijian Wang) and D.J. analyzed the data; and H.W. (Haijian Wang) wrote the paper.

Funding: This research received no external funding.

Acknowledgments: This research receives support from the High-Tech Research Center and Nano-technology Project at Saitama Institute of Technology. All animal experimental protocols were conducted in compliance with the Guiding Principles for the Care and Use of Animals in the Field of Physiological Sciences approved by the council of the Physiological Society of Japan, and were permitted by the Animal Institutional Review Board of Saitama Institute of Technology in accordance with the guidelines of the U.S. National Institutes of Health.

Conflicts of Interest: The authors declare no conflict of interest.

References

1.   Green, A.L.; Ma, E. Bulk Metallic Glasses: At the Cutting Edge of Metals Research. *MRS Bull.* **2007**, *32*, 611–619.

2.   Chen, Q.Z.; Thouas, G.A. Metallic implant biomaterials. *Mater. Sci. R* **2015**, *87*, 1–57. [CrossRef]

3.   Wise, D.L.; Trantolo, D.J.; Altobelli, D.E.; Yaszemsk, M.J.; Grasser, J.D. *Human Biomaterials Application*; Humana Press: Totowa, NJ, USA, 1996.

4.   Yaszemsk, M.J.; Trantolo, D.J.; Lewandrowski, K.U.; Hasirci, V.; Altobelli, D.E.; Wise, D.L. *Biomaterials in Orthopedics*; Marcel Dekker Inc.: New York, NY, USA, 2004.

5.   Matusiewicz, H. Potential release of in vivo trace metals from metallic medical implants in the human body: From ions to nanoparticles—a systematic analytical review. *Acta Biomater.* **2014**, *10*, 2379–2403. [CrossRef] [PubMed]

6.   Heary, R.F.; Parvathreddy, N.; Sampath, S.; Agarwal, N. Elastic modulus in the se-lection of interbody implants. *J. Spine Surg.* **2017**, *3*, 163–167. [CrossRef] [PubMed]

7.   Andreas, A.K.; Jörg, F.L.; Florian, H.D.T. Rapid Solidification and Bulk Metallic Glasses—Processing and Properties. In *Materials Processing Handbook*; CRC Press: Boca Raton, FL, USA, 2007; pp. 17-1–17-44.

8.   Masumoto, T.; Maddin, R. The mechanical properties of palladium 20 a/o silicon alloy quenched from the liquid state. *Acta Metall.* **1971**, *19*, 725–741. [CrossRef]

9.   Liebermann, H.; Graham, C. Production of amorphous alloy ribbons and effects of apparatus parameters on ribbon dimensions. *IEEE Trans. Magn.* **1976**, *12*, 921–923. [CrossRef]

10.   Narasimhan, M.C. Continuous casting method for metallic strips. U.S. Patent 4,142,571, 22 October 1979.

11.   Inoue, A.; Nakamura, T.; Sugita, T.; Zhang, T.; Masumoto, T. Bulky La-Al-TM (TM=Transition Metal) Amorphous Alloys with High Tensile Strength Produced by a High-Pressure Die Casting Method. *Mater. Trans. JIM* **1993**, *34*, 351–358. [CrossRef]

12.   Barekar, N.S.; Dhindaw, B.K. Twin-roll casting of Aluminum alloys—An overview. *Mater. Manuf. Process.* **2014**, *29*, 651–661. [CrossRef]

13. Chen, H.S.; Miller, C.E. A Rapid Quenching Technique for the Preparation of Thin Uniform Films of Amorphous Solids. *Rev. Sci. Instrum.* **1970**, *41*, 1237–1238. [CrossRef]

14. East, D.R.; Kellam, M.; Gibson, M.A.; Seeber, A.; Liang, D.; Nie, J.F. Amorphous magnesium sheet produced by twin roll casting. *Mater. Sci. Forum* **2010**, *654*, 1078–1081. [CrossRef]

15. Lee, J.G.; Park, S.S.; Lee, S.B.; Chung, H.T.; Kim, N.J. Sheet fabrication of bulk amorphous alloys by twin-roll strip casting. *Scr. Mater.* **2005**, *53*, 693–697. [CrossRef]

16. Oh, Y.S.; Lee, H.; Lee, J.G.; Kim, N.J. Twin-Roll Strip Casting of Iron-Base Amorphous Alloys. *Mater. Trans.* **2007**, *48*, 1584–1588. [CrossRef]

17. Urata, A.; Nishiyama, N.; Amiya, K.; Inoue, A. Continuous casting of thick Fe-base glassy plates by twin-roller melt-spinning. *Mater. Sci. Eng. A* **2007**, *449*, 269–272. [CrossRef]

18. Suzuki, T.; Anthony, A.M. Rapid quenching on the binary systems of high temperature oxides. *Mater. Res. Bull.* **1974**, *9*, 745–753. [CrossRef]

19. Lee, J.G.; Lee, H.; Oh, Y.S.; Lee, S.; Kim, N.J. Continuous fabrication of bulk amorphous alloy sheets by twin-roll strip casting. *Intermetallics* **2006**, *14*, 987–993. [CrossRef]

20. Hofmann, D.C.; Roberts, S.N.; Johnson, W.L. Twin Roll Sheet Casting of Bulk Metallic Glasses and Composites in an Inert Environment. U.S. Patent 20130025746 A1, 31 January 2013.

21. Ding, P.-D.; Pan, F.-S.; Jiang, B.; Wang, J.; Li, H.-L.; Wu, J.-C.; Xu, Y.-W.; Wen, Y. Twin-roll strip casting of magnesium alloys in China. *Trans. Nonferrous Metals Soc. China* **2008**, *18*, s7–s11. [CrossRef]

22. Hu, X.; Ju, D.; Zhao, H. Thermal flow simulation of twin-roll casting magnesium alloy. *J. Shanghai Jiaotong Univ.* **2012**, *17*, 479–483. [CrossRef]

23. Luo, L.T.; Gong, X.B.; Li, J.Z.; Kang, S.B.; Cho, J.H. Microstructure and Mechanical Properties of Severely Deformed Mg-4.5Al-1.0Zn Alloy Processed by Asymmetric Rolling on Ingot and Twin Roll Cast Strip. *Mater. Res.* **2016**, *19*, 207–214. [CrossRef]

24. Park, S.S.; Park, W.J.; Kim, C.H.; You, B.S.; Kim, N.J. The Twin-Roll Casting of Magnesium Alloys. *JOM* **2009**, *61*, 14–18. [CrossRef]

25. Pei, Z.P.; Ju, D.Y.; Li, X. Simulation of critical cooling rate and process conditions for metallic glasses in vertical type twin-roll casting. *Trans. Nonferrous Met. Soc. China* **2017**, *27*, 2406–2414. [CrossRef]

26. Holzwarth, U.; Gibson, N. The Scherrer equation versus the 'Debye-Scherrer equation'. *Nat. Nanotechnol.* **2011**, *6*, 534. [CrossRef] [PubMed]

27. Pecharsky, V.K.; Zavalij, P.Y. Properties, Sources, and Detection of Radiation. In *Fundamentals of Powder Diffraction and Structural Characterization of Materials*; Springer: Boston, MA, USA, 2009; pp. 107–132.

28. Jung, I.-H.; Sanjari, M.; Kim, J.; Yue, S. Role of RE in the deformation and recrystallization of Mg alloy and a new alloy design concept for Mg–RE alloys. *Scr. Mater.* **2015**, *102*, 1–6. [CrossRef]

29. Takeuchi, A.; Inoue, A. Calculations of mixing enthalpy and mismatch entropy for ternary amorphous alloy. *Mater. Trans. JIM* **2000**, *41*, 1372–1378. [CrossRef]

30. Inoue, A. Stabilization of metallic supercooled liquid and bulk amorphous alloys. *Acta Mater.* **2000**, *48*, 279–306. [CrossRef]

31. Anawati, A.; Asoh, H.; Ono, S. Effects of alloying element Ca on the corrosion behavior and bioactivity of anodic films formed on AM60 Mg alloys. *Materials* **2017**, *10*, 11. [CrossRef] [PubMed]

32. Song, Y.; Han, E.; Shan, D.; Yim, C.D.; You, B.S. Effect of hydrogen on the corrosion behavior of the Mg–xZn alloys. *Corros. Sci.* **2012**, *60*, 238–245.

33. Zhang, B.; Hou, Y.; Wang, X.; Wang, Y.; Geng, L. Effects of solidification cooling rate on the corrosion resistance of Mg–Zn–Ca alloy. *Mater. Sci. Eng. C* **2011**, *31*, 1667–1673. [CrossRef]

34. Bakhsheshi-Rad, R.; Abdul-Kadir, M.R.; Idris, M.H.; Farahany, S. Relationship between the corrosion behavior and the thermal characteristics and microstructure of Mg–0.5Ca–xZn alloys. *Corros. Sci.* **2012**, *64*, 184–197.

35. Makar, G.L.; Kruger, J. Corrosion of magnesium. *Intern. Mater. Rev.* **1993**, *38*, 138–153. [CrossRef]

36. Witte, F.; Kaese, V.; Haferkamp, H.; Switzer, E.; Meyer-Lindenberg, A.; Wirth, C.J.; Windhagen, H. In vivo corrosion of four magnesium alloys and the associated bone response. *Biomaterials* **2005**, *26*, 3557–3563. [CrossRef]

37. Janning, C.; Willbold, E.; Vogt, C.; Nellesen, J.; Meyer-Lindenberg, A.; Windhagen, H.; Windhagen, F.; Thorey, F.W. Magnesium hydroxide temporarily enhancing osteoblast activity and decreasing the osteoclast number in peri-implant bone remodelling. *Acta Biomater.* **2010**, *6*, 1861–1868. [CrossRef] [PubMed]

38. Yang, J.X.; Cui, F.Z.; Lee, I.S.; Zhang, Y.; Yin, Q.S.; Xia, H.; Yang, S.X. In vivo biocompatibility and degradation behavior of Mg alloy coated by calcium phosphate in a rabbit model. *J. Biomater. Appl.* **2012**, *27*, 153–164. [CrossRef] [PubMed]
39. Yoshizawa, S.; Brown, A.; Barchowsky, A.; Sfeir, C. Magnesium ion stimulation of bone marrow stromal cells enhances osteogenic activity, simulating the effect of magnesium alloy degradation. *Acta Biomater.* **2014**, *10*, 2834–2842. [CrossRef] [PubMed]

Communication

Effect of Hydrogen Charging on Pop-in Behavior of a Zr-Based Metallic Glass

Lin Tian [1,2,*], Dominik Tönnies [1], Moritz Hirsbrunner [1], Tim Sievert [1], Zhiwei Shan [2] and Cynthia A. Volkert [1,*]

[1] Institute of Materials Physics, University of Göttingen, 37077 Göttingen, Germany; dominik.toennies@phys.uni-goettingen.de (D.T.); moritz.hirsbrunner@stud.uni-goettingen.de (M.H.); tim.sievert@stud.uni-goettingen.de (T.S.)

[2] Center for Advancing Materials Performance from the Nanoscale (CAMP-Nano) & Hysitron Applied Research Center in China (HARCC), State Key Laboratory for Mechanical Behavior of Materials, Xi'an Jiaotong University, Xi'an 710049, China; zwshan@mail.xjtu.edu.cn

[*] Correspondence: ltian@phys.uni-goettingen.de (L.T.); volkert@ump.gwdg.de (C.A.V.); Tel.: +49-551-39-25002 (L.T. & C.A.V.)

Received: 12 November 2019; Accepted: 18 December 2019; Published: 22 December 2019

Abstract: In this work, structural and mechanical properties of hydrogen-charged metallic glass are studied to evaluate the effect of hydrogen on early plasticity. Hydrogen is introduced into samples of a Zr-based (Vit 105) metallic glass using electrochemical charging. Nanoindentation tests reveal a clear increase in modulus and hardness as well as in the load of the first pop-in with increasing hydrogen content. At the same time, the probability of a pop-in occurring decreases, indicating that hydrogen hinders the onset of plastic instabilities while allowing local homogeneous deformation. The hydrogen-induced stiffening and hardening is rationalized by hydrogen stabilization of shear transformation zones (STZs) in the amorphous structure, while the improved ductility is attributed to the change in the spatial correlation of the STZs.

Keywords: metallic glass; hydrogen; indentation; pop-in; plasticity

---

1. Introduction

Metallic glasses (MGs) or amorphous metals are materials composed of metal components but without crystalline structure [1]. In addition to applications as high-performance structural materials [2], the unique structure of MGs makes them promising candidate materials in many other fields. As one of the important applications, MGs can be used as storage and separation media for hydrogen gas (H_2), which is a well-known clean energy source. Zr-based MGs for instance have been observed to absorb large amounts of hydrogen [3] with a content even comparable to the best crystalline hydrogen-storage materials [4] due to their large number of interstitial-like sites and favorable hydrogen–metal chemistry. The high solubility and moderate diffusivity of hydrogen [5] in the amorphous structure also increases H_2 permeability in MGs [6,7]. With high H_2 permeability, high strength, and corrosion resistance, MG films can be used as H_2 separation membranes in hydrogen selective devices [6,8]. As a critical part of the device, MGs must survive mechanical pressure in the gaseous H_2 environment. The performance and stability of MGs in H_2 have significant impact on device safety, and therefore the effects of hydrogen on the mechanical properties of MGs have received increasing attention.

Although the effect of hydrogen in MGs has been studied by several groups in recent years, debates remain on questions such as whether hydrogen is a detrimental [7] or beneficial [9,10] element for MGs, or whether doping hydrogen increases [11] or decreases [12] the strength of MGs. Like in crystalline materials, hydrogen can induce embrittlement in MGs [7,13–15]. In contrast, positive effects of hydrogen on plasticity have also been found in some alloys. For instance, in Zr-based

MGs [9,16–18] which have large negative enthalpies of mixing with hydrogen [19], the hydrogen addition increases both glass-forming ability and ductility, opening a new approach to make both stable and ductile MGs. It has been proposed that hydrogen addition in MGs can introduce local structural heterogeneities which prevent single, catastrophic shear bands and promote the formation of multiple shear bands [16,18,20]. A recent work on elastic heterogeneity actually shows that the spatial correlation of the heterogeneities plays an important role in tuning ductility of MGs [21]. Since hydrogen addition is potentially useful in MG alloy design, the underlying mechanisms of how hydrogen addition improves ductility and stability of MGs deserve further investigation.

In this work, the effect of solid-solution hydrogen on pop-in behavior of an MG is studied. A pop-in during spherical indentation results from an abrupt shear event along a shear band trajectory [22] which represents early plasticity and is a smaller version of the shear bands observed in bulk MGs [22,23]. The results are interpreted in terms of the hydrogen-induced changes to the MG structure and potential energy landscape, and we discuss how hydrogen can be used to control shear band formation. A Zr-based MG of similar composition to those used in several studies in the literature [17,24] is used.

2. Materials and Methods

A Zr-based Vit 105 MG ($Zr_{52.5}Cu_{17.9}Ni_{14.6}Al_{10}Ti_5$) plate with a thickness of 2 mm was produced by suction casting into a Cu-mold. The as-cast samples were cut into thin samples by electrical discharge machining. On each of the samples, the as-cut surface was mechanically polished while the as-cast surface was kept intact for indentation tests. The final sample had an area of ~15 mm^2 and a thickness of 1 mm.

In contrast to previous studies [11], where hydrogen was added to the alloy by processing the melt in a H_2-containing gas, hydrogen charging of samples was performed after quenching using a cathodic charging technique in a 0.5 mol/L sulfuric acid solution. The samples were charged at a current density ~10 mA/cm^2 (lower than the critical current density for the formation of hydride [25]) for various charging times in order to control the dissolved hydrogen content (c_H). Hydrogen contents up to $c_H > 0.5$ H/M (hydrogen-to-metal ratio) were investigated, with most studies focusing on $c_H \leq 0.29$ H/M where no specimen cracking was observed. The structure of the samples before and after hydrogen charging was investigated with X-ray diffraction (Cu-Kα line).

Nanoindentation tests were conducted at room temperature using a Nano Indenter G200 (formerly MTS; now KLA-Tencor, Milpitas, CA, USA). Indentation with a Berkovich indenter in the standard XP indentation head was performed on the samples to measure hardness H and Young's modulus E using the continuous stiffness measurement technique [26]. Reduced Young's modulus E_r is calculated using the slope of the unloading curve and precalibrated contact area of the indenter. The modulus of the sample E_s can be derived from the equation:

$$1/E_r = \left(1 - v_i^2\right)/E_i + \left(1 - v_s^2\right)/E_s, \tag{1}$$

where Young's modulus of the tip $E_i = 1140$ GPa. Poisson's ratios of the diamond tip and the sample are $v_i^2 = 0.07$ and $v_s^2 = 0.37$, respectively. The indentation depth reaches 2 μm in order to reduce the effect of surface roughness.

A spherical diamond tip with a diameter of $R = 650$ nm was used in the DCM-1 indenter head. The tip radius was calibrated on a fused silica reference sample. Arrays of 12 × 12 indents which covered an area of 33 × 33 μm^2 were measured with a constant loading rate of 0.1 mN/s to a maximum load of 1 mN corresponding to a displacement of ~50 nm. Displacement bursts, which are commonly referred to as "pop-ins", were visible in the load-displacement curves and automatically identified from displacement-rate data using a MATLAB script (R2013b, MathWorks, Natick, MA, USA). See details in [27]. In order to measure c_H, the average hydrogen content of the sample, melt extraction was carried out on the samples after indentation tests using a Hydrogen analyzer (G8 Galileo, Bruker, Billerica, MA, USA). The hydrogen analyzer was calibrated by a commercial standard hydrogen containing sample (501-529, Leco, St Joseph, MI, USA). All the indentation tests and melt extraction analysis were performed within 72 h after the samples were charged with hydrogen.

3. Results

X-ray diffraction spectra were recorded and compared for each sample before and after hydrogen charging. The result of X-ray diffraction only reveals a small shift of the broad maxima characteristic of the nearest neighbor distances. In a sample charged to $c_H > 0.5$ H/M (the highest hydrogen content in this work) the position of the broad maxima is shifted to lower 2θ values by 0.5 degrees relative to the as-cast material (see Figure 1a). The lower diffraction angle indicates that on average the hydrogen-charged sample has larger atomic spacing than the uncharged sample. No evidence of hydride formation was found in the samples.

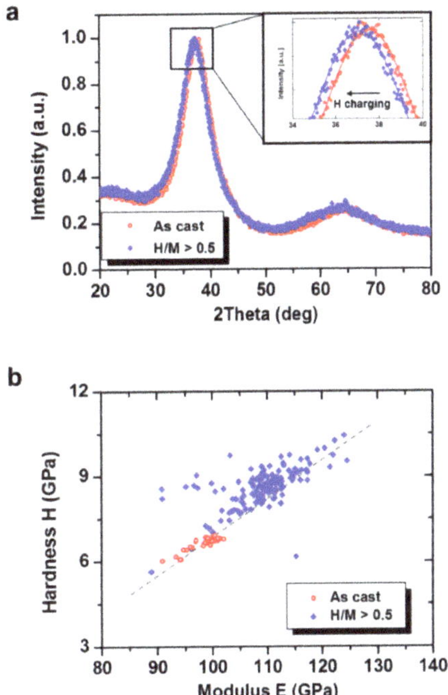

Figure 1. (**a**) X-ray diffraction spectra of Vit 105 metallic glass (MG) samples with and without hydrogen charging. The hydrogen content is defined as the hydrogen-to-metal ratio. (**b**) Hardness and modulus (Berkovich indentation) of samples with and without hydrogen charging. The dashed line is to guide the eye.

Modulus and hardness measured by Berkovich indentation on samples with and without hydrogen charging for $c_H > 0.5$ H/M are plotted in Figure 1b. The average modulus and hardness of the as-cast sample were 98 ± 3 and 6.6 ± 0.2 GPa, respectively. Although there is considerable scatter in the values, both the modulus and the hardness clearly increase as a result of hydrogen charging, leading to average values of 110 ± 7 and 8.7 ± 0.8 GPa, respectively. The modulus and hardness values in both the uncharged and charged states are linearly correlated with each other, suggesting that both properties are determined by the same features of the structure and short-range chemical ordering. Meanwhile, the wider distribution of modulus and hardness values in the charged state indicates larger variations in local structure (more structural heterogeneities) due to hydrogen incorporation on the micrometer length scale of the Berkovich indents. Some of the data points do not show the linear correlation, in that the modulus is decreased without a significant change in hardness. We attribute this to the formation of surface relaxation or surface cracks which are observed in some regions of the highly charged specimens [28,29].

Mechanical properties of the bulk sample were measured by Berkovich indentation while local plasticity was studied by spherical indentation which is sensitive to early plasticity events in MGs [27]. The first pop-in in the load-displacement curve is attributed to the onset of plasticity and the corresponding maximum shear stress along the shear band trajectory defines the yield stress [22]. Samples were prepared for spherical indentation tests with c_H-values of 0, 0.11, 0.13, and 0.29 H/M. Before indentation tests, the sample surfaces were checked with optical microscopy to make sure there were no cracks. Representative load-displacement curves of indentations from the four different samples are plotted in Figure 2a. Pop-ins can be seen in the data from samples with c_H-values 0, 0.11, and 0.13 H/M. The first pop-in is attributed to the initiation of detectable plasticity [27] and is marked by a solid arrow in each curve. In contrast, the sample with c_H = 0.29 H/M does not show any pop-ins. In order to further analyze the pop-ins, displacement rates were calculated for each curve (Figure 2b) where pop-ins with a displacement larger than a threshold value can be easily identified. Compared to other samples, displacement rate curves for the sample with c_H = 0.29 H/M are very smooth. It is worth noting that there is residual plastic deformation in the indents of the 0.29 H/M sample even though no pop-ins are observed. The data point indicated by the arrow in Figure 2a demonstrates that deformation of the material under the indenter is not fully recoverable. Therefore, the deformation during indentation is already beyond the elastic region but is not able to initiate a pop-in. A statistical evaluation shows that 100% of the indents in the uncharged c_H = 0 H/M sample have pop-ins. This percentage reduces to about 80% in the c_H-value 0.11 and 0.13 H/M samples. In the c_H = 0.29 H/M sample, there are no pop-ins at all in all indentations demonstrating that local homogeneous deformation is occurring. This result is summarized in Figure 2c.

The strength of the sample is evaluated using the maximum shear stress when the first pop-in occurs. Analysis of elastic stress distribution based on Hertzian contact theory [30] reveals that the maximum stress locates underneath the center of the indenter at a depth of half the contact radius [31]. The maximum shear stress is approximated by:

$$\tau_{max} \approx AP_{pop-in}/\left(\pi \times R \times h_{pop-in}\right), \tag{2}$$

with P_{pop-in} and h_{pop-in} being the load and displacement at the onset of the pop-in, respectively, diameter of the spherical diamond tip R = 650 nm, and a pre-factor A = 0.4413 [27]. Cumulative distributions of the maximum shear stresses at the first pop-ins for the c_H-value 0, 0.11, and 0.13 H/M samples are plotted in Figure 3. The pop-in stresses vary by as much as 2 GPa for a given sample, indicating significant structural heterogeneity on a length scale between the size of the indented volume (ca. 1 μm) and the size of the indent array (ca. 35 μm). The average maximum shear stress for uncharged sample 1 is about 2.9 GPa. A clear shift of maximum shear stresses to higher values is shown in hydrogen-charged samples 2 and 3 compared to sample 1. The average maximum shear stress for the charged sample 3 is about 3.2 GPa, 10% higher than that of sample 1. This result showing hydrogen-induced hardening effect is qualitatively consistent with modulus and hardness measurement with Berkovich indentation, as shown in Figure 1b. It can also be noticed that the shape of the distribution curves for samples 2 and 3 is different from that of sample 1. The curve of sample 2 is stretched while the curve of sample 3 is compressed even though they have similar c_H. This difference suggests considerable structural heterogeneity in the hydrogen-charged MG over length scales much larger than the array size of ca. 35 μm. There is no cumulative distribution curve for sample 4 because no pop-ins were observed for any of the 144 indentation tests.

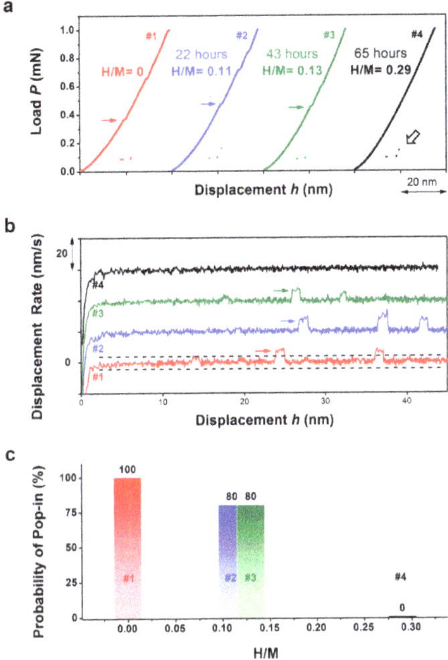

Figure 2. (**a**) Typical load displacement curves for indents in samples with different hydrogen content (c_H). The curves are shifted for the sake of clarity. (**b**) Displacement rate as a function of displacement for each curve in (a). The first pop-ins are marked with arrows. The curves are shifted for the sake of clarity. (**c**) The percentage of indents that have no pop-in out of 144 tests in each sample.

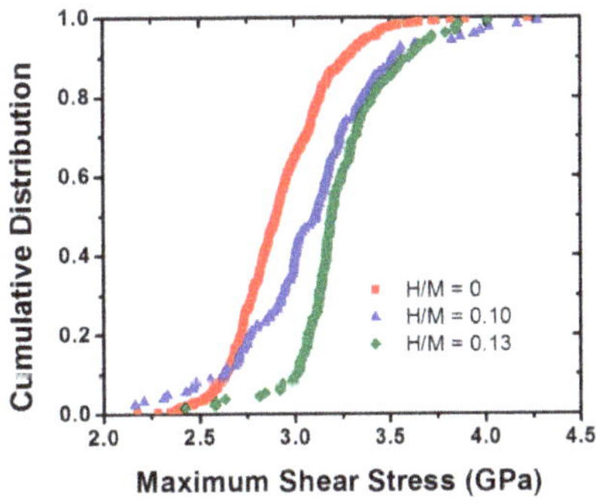

Figure 3. Cumulative distribution of the first pop-ins against the maximum shear stress of samples with different c_H.

4. Discussion

The deformation behavior of MGs is often discussed in terms of a distribution of activation energies for atomic rearrangements [32]. Underlying this distribution of activation energies is the potential energy landscape of the disordered structure [33]. When atomic rearrangements occur, either due to deformation or structural relaxation, they change the glass structure by shifting potential and activation energies. Similarly, hydrogen solid-solution charging of an MG results in local atomic rearrangements and a c_H-dependent increase of the interatomic distances [28,29,34], thereby also locally shifting potential and activation energies. In the beginning of hydrogen charging, the favored or so-called trapping sites for hydrogen are those with high local potential energies, such as sites with large free volume [35], or geometrically unfavored motifs (GUMs) [36]. Hydrogen also has been found to fill interstitial sites of loosely packed Zr-rich tetrahedral-like sites (t-sites) [3,34,37,38]. These trapping sites are stabilized by the hydrogen [39], lowering the local potential energy and generally resulting in an increase in the local activation energies for atomic rearrangements, thereby restricting deformation [11].

The study here reveals a clear increase in both modulus and hardness due to hydrogen charging. The linear relation between modulus and hardness suggests that the underlying mechanisms controlling stiffness and deformation by atomic rearrangement depend on the same aspects of the atomic structure. A similar correlation between the stiffness and barrier height for local shear rearrangements has been thoroughly investigated and discussed for metallic glass-forming liquids [33]. Furthermore, the fact that the modulus and hardness have the same relation in the uncharged and charged states indicates that the structural changes introduced by hydrogen do not produce fundamentally different mechanical responses than the structural heterogeneities already present in the uncharged disordered structure. The simplest explanation for our observations is that the hydrogen stabilizes regions of low shear modulus, thus increasing the stiffness and hindering atomic rearrangements.

Both increases and decreases of modulus and hardness have been observed in the literature due to hydrogen charging, depending on the alloy composition. MGs with a high Zr composition consistently show increases in mechanical properties on charging, likely because the Zr-rich clusters are made more stable by the inclusion of hydrogen [11,18,40] due to the large negative mixing enthalpy. The incorporation of hydrogen in these clusters presumably makes them more resistant to deformation, increasing the modulus and hardness, as well as the pop-in stress of the MG. On the contrary, hydrogen incorporated into a stable icosahedron cluster may raise the potential energy and make the cluster less stable to atomic rearrangements. As a result, the MG may appear to be softened by hydrogen addition [20].

Beyond the increase in hardness and modulus, the most noticeable effect of hydrogen charging in this study is the replacement of pop-ins by stable or homogeneous flow. The improved ductility and malleability of MGs induced by hydrogen has usually been attributed to an increase in structural heterogeneities in the literature, although an exact mechanism has not been discussed [16,18,20]. There is no doubt that the hydrogen addition changes local environments in the MGs, thereby changing the potential energy distribution which likely accounts for changes in hardness and modulus. In addition to the change in the potential energy distribution, the spatial distribution of sites with different potential energy should be considered. Our recent work suggests an increase of correlation length of medium-range order in a $Zr_{51}Cu_{49}$ MG during hydrogen loading [41]. The spatial correlation of the heterogeneities may play a decisive role in the ductility of MGs. In a simulation study by Wang et al. [21], the mechanical properties of samples with the same potential energy distribution but different correlation lengths for the local shear modulus are compared. To first order, the correlation length simply describes the size and spacing of potential STZs (elastically soft sites) in the MG. Their results reveal that when the correlation length increases from 0.5 to 5 nm, a transition in deformation mechanism from "stress-dictated" shear band nucleation and growth to a "structural-dictated" strain percolation occurs [21]. This is associated with a decrease in the tendency for strain localization and an increase in the plastic strain to failure. The basic understanding of this observation is that more widely spaced STZs cannot easily interact through their stress fields, thus hindering the formation of shear bands through collective excitation. Building on the model of Wang et al., we propose that hydrogen stabilization of STZs decreases their density, thereby increasing the correlation length and hindering strain localization and pop-in instabilities. We further suggest that the high mobility of the hydrogen may allow it to change sites in response to changing strain fields, thus inhibiting propagation of excitations between STZs and hindering pop-in formation. If hydrogen is able to move in response to strain fields introduced by other hydrogen atoms, some form of self-organization of the hydrogen is expected to be active.

Potential sites that can be filled with hydrogen are limited in MGs since the maximum solubility of hydrogen is low. When the sample is overcharged with hydrogen [28,29], cracks presumably form due to hydrogen-induced expansion and the possible motion of hydrogen in the resultant strain fields. Precise control of c_H will be required when using hydrogen to optimize the mechanical behavior of MGs.

Heterogeneities at various length scales have been discussed here, ranging from the flexibility volumes, free volume defects, STZs, and GUMs at the nanometer scale up to the observed variations in modulus, hardness, and pop-in stresses at the micrometer scale. Whether the hydrogen-induced changes in nanometer scale heterogeneities can result in the observed micrometer-scale variations will depend on the distribution of spatial correlations as well as the ability of hydrogen to move in response to the hydrogen-induced strain fields. Confirmation of the nanometer-scale heterogeneity changes as well as evidence for possible self-organization due to hydrogen charging remain elusive due to technical limitations. However, state-of-the-art methods such as fluctuation electron microscopy [42,43] may open the possibility to track the evolution of nanoscale heterogeneities during hydrogen charging.

5. Conclusions

Our results show a clear effect of hydrogen on the mechanical behavior of a Vit 105 MG. With increasing hydrogen content up to 0.29 H/M, the load of the first pop-in increases, consistent with the observed increase in hardness and modulus, while the occurrence of pop-ins decreases. This results in a transition from pop-in or shear banding to homogeneous deformation at a hydrogen content of 0.29 H/M. The behavior is interpreted in terms of hydrogen stabilization of STZs in the amorphous structure due to the negative enthalpy of mixing of hydrogen with Zr. Thus, charging of MGs with negative heats of mixing with low concentration of hydrogen may offer a robust method to increase modulus and hardness and prevent catastrophic shear band formation in MGs.

Author Contributions: Conceptualization, L.T., D.T. and C.A.V.; Methodology and Formal Analysis, D.T., M.H. and T.S.; Writing—Review and Editing, L.T., D.T. and C.A.V.; Funding Acquisition, L.T., Z.S. and C.A.V. All authors have read and agreed to the published version of the manuscript.

Funding: L.T. acknowledges the Natural Science Foundation of China (Grant 51501144), the China Postdoctoral Science Foundation (2015M580842), and the Alexander von Humboldt Foundation for financial support. C.A.V. and D.T. acknowledge financial support by the German Research Foundation (DFG) through an individual grant (VO 928/9-1). Z.W.S. acknowledges support by the National Key Research and Development Program of China (No. 2017YFB0702001) and Natural Science Foundation of China (51231005 and 51621063).

Acknowledgments: We would also like to thank Feng Jiang and Mingcan Li for providing the samples.

Conflicts of Interest: The authors declare no conflict of interest.

References

1. Klement, W.; Willens, R.H.; Duwez, P. Non-crystalline Structure in Solidified Gold-Silicon Alloys. *Nature* **1960**, *187*, 869–870. [CrossRef]
2. Inoue, A.; Takeuchi, A. Recent Development and Applications of Bulk Glassy Alloys. *Int. J. Appl. Glass Sci.* **2010**, *1*, 273–295. [CrossRef]
3. Bankmann, J.; Pundt, A.; Kirchheim, R. Hydrogen loading behaviour of multi-component amorphous alloys: Model and experiment. *J. Alloys Compd.* **2003**, *356–357*, 566–569. [CrossRef]
4. Zander, D.; Tal-Gutelmacher, E.; Jastrow, L.; Köster, U.; Eliezer, D. Hydrogenation of Pd-coated Zr-Cu-Ni-Al metallic glasses and quasicrystals. *J. Alloys Compd.* **2003**, *356–357*, 654–657. [CrossRef]
5. Wang, Y.I.; Suh, J.Y.; Lee, Y.S.; Shim, J.H.; Fleury, E.; Cho, Y.W.; Koh, S.U. Direct measurement of hydrogen diffusivity through Pd-coated Ni-based amorphous metallic membranes. *J. Membr. Sci.* **2013**, *436*, 195–201. [CrossRef]

6. Yamaura, S.; Sakurai, M.; Hasegawa, M.; Wakoh, K.; Shimpo, Y.; Nishida, M.; Kimura, H.; Matsubara, E.; Inoue, A. Hydrogen permeation and structural features of melt-spun Ni-Nb-Zr amorphous alloys. *Acta Mater.* **2005**, *53*, 3703–3711. [CrossRef]

7. Paglieri, S.N.; Pal, N.K.; Dolan, M.D.; Kim, S.M.; Chien, W.M.; Lamb, J.; Chandra, D.; Hubbard, K.M.; Moore, D.P. Hydrogen permeability, thermal stability and hydrogen embrittlement of Ni-Nb-Zr and Ni-Nb-Ta-Zr amorphous alloy membranes. *J. Membr. Sci.* **2011**, *378*, 42–50. [CrossRef]

8. Dolan, M.D.; Dave, N.C.; Ilyushechkin, A.Y.; Morpeth, L.D.; McLennan, K.G. Composition and operation of hydrogen-selective amorphous alloy membranes. *J. Membr. Sci.* **2006**, *285*, 30–55. [CrossRef]

9. Dong, F.; Su, Y.; Luo, L.; Wang, L.; Wang, S.; Guo, J.; Fu, H. Enhanced plasticity in Zr-based bulk metallic glasses by hydrogen. *Int. J. Hydrog. Energy* **2012**, *37*, 14697–14701. [CrossRef]

10. Su, Y.; Dong, F.; Luo, L.; Guo, J.; Han, B.; Li, Z.; Wang, B.; Fu, H. Bulk metallic glass formation: The positive effect of hydrogen. *J. Non-Cryst. Solids* **2012**, *358*, 2606–2611. [CrossRef]

11. Zhao, Y.; Choi, I.C.; Seok, M.Y.; Kim, M.H.; Kim, D.H.; Ramamurty, U.; Suh, J.Y.; Jang, J.i. Effect of hydrogen on the yielding behavior and shear transformation zone volume in metallic glass ribbons. *Acta Mater.* **2014**, *78*, 213–221. [CrossRef]

12. Dandana, W.; Yousfi, M.; Hajlaoui, K.; Gamaoun, F.; Yavari, A. Thermal stability and hydrogen-induced softening in $Zr_{57}Al_{10}Cu_{15.4}Ni_{12.6}Nb_5$ metallic glass. *J. Non-Cryst. Solids* **2017**, *456*, 138–142. [CrossRef]

13. Schroeder, H.W.; Koster, U. Hydrogen embrittlement of metallic glasses. *J. Non-Cryst. Solids* **1983**, *56*, 213–218. [CrossRef]

14. Lin, J.J.; Perng, T.P. Embrittlement of amorphous $Fe_{40}Ni_{38}Mo_4B_{18}$ alloy by electrolytic hydrogen. *Metall. Mater. Trans. A-Phys. Metall. Mater. Sci.* **1995**, *26*, 197–201. [CrossRef]

15. Jayalakshmi, S.; Fleury, E. Hydrogen embrittlement in metallic amorphous alloys: An overview. *J. ASTM Int.* **2010**, *7*, 1–23.

16. Dong, F.; Lu, S.; Zhang, Y.; Luo, L.; Su, Y.; Wang, B.; Huang, H.; Xiang, Q.; Yuan, X.; Zuo, X. Effect of hydrogen addition on the mechanical properties of a bulk metallic glass. *J. Alloys Compd.* **2017**, *695*, 3183–3190. [CrossRef]

17. Granata, D.; Fischer, E.; Löffler, J.F. Hydrogen microalloying as a viable strategy for enhancing the glass-forming ability of Zr-based bulk metallic glasses. *Scr. Mater.* **2015**, *103*, 53–56. [CrossRef]

18. Granata, D.; Fischer, E.; Löffler, J.F. Effectiveness of hydrogen microalloying in bulk metallic glass design. *Acta Mater.* **2015**, *99*, 415–421. [CrossRef]

19. Yamanaka, S.; Tanaka, T.; Miyake, M. Effect of oxygen on hydrogen solubility in zirconium. *J. Nucl. Mater.* **1989**, *167*, 231–237. [CrossRef]

20. Zhao, Y.; Choi, I.C.; Seok, M.Y.; Ramamurty, U.; Suh, J.Y.; Jang, J.i. Hydrogen-induced hardening and softening of Ni-Nb-Zr amorphous alloys: Dependence on the Zr content. *Scr. Mater.* **2014**, *93*, 56–59. [CrossRef]

21. Wang, N.; Ding, J.; Yan, F.; Asta, M.; Ritchie, R.O.; Li, L. Spatial correlation of elastic heterogeneity tunes the deformation behavior of metallic glasses. *npj Comput. Mater.* **2018**, *4*, 19. [CrossRef]

22. Packard, C.; Schuh, C. Initiation of shear bands near a stress concentration in metallic glass. *Acta Mater.* **2007**, *55*, 5348–5358. [CrossRef]

23. Greer, A.L.; Cheng, Y.Q.; Ma, E. Shear bands in metallic glasses. *Mater. Sci. Eng. R Rep.* **2013**, *74*, 71–132. [CrossRef]

24. Dong, F.; He, M.; Zhang, Y.; Luo, L.; Su, Y.; Wang, B.; Huang, H.; Xiang, Q.; Yuan, X.; Zuo, X.; et al. New insights into melt hydrogenation effects on glass-forming ability in a Zr-based bulk metallic glass. *J. Non-Cryst. Solids* **2018**, *481*, 170–175. [CrossRef]

25. Gostin, P.F.; Eigel, D.; Grell, D.; Uhlemann, M.; Kerscher, E.; Eckert, J.; Gebert, A. Stress-corrosion interactions in Zr-based bulk metallic glasses. *Metals* **2015**, *5*, 1262–1278. [CrossRef]

26. Oliver, W.C.; Pharr, G.M. An improved technique for determining hardness and elastic modulus using load and displacement sensing indentation experiments. *J. Mater. Res.* **1992**, *7*, 1564–1583. [CrossRef]

27. Tönnies, D.; Samwer, K.; Derlet, P.M.; Volkert, C.A.; Maaß, R. Rate-dependent shear-band initiation in a metallic glass. *Appl. Phys. Lett.* **2015**, *106*, 171907. [CrossRef]

28. Ismail, N.; Uhlemann, M.; Gebert, A.; Eckert, J. Hydrogenation and its effect on the crystallisation behaviour of $Zr_{55}Cu_{30}Al_{10}Ni_5$ metallic glass. *J. Alloys Compd.* **2000**, *298*, 146–152. [CrossRef]

29. Jayalakshmi, S.; Fleury, E.; Leey, D.Y.; Chang, H.J.; Kim, D.H. Hydrogenation of $Ti_{50}Zr_{25}Co_{25}$ amorphous ribbons and its effect on their structural and mechanical properties. *Philos. Mag. Lett.* **2008**, *88*, 303–315. [CrossRef]

30. Johnson, K.L. *Contact Mechanics*; Cambridge University Press: Cambridge, UK, 1987.

31. Bei, H.; Lu, Z.; George, E. Theoretical strength and the onset of plasticity in bulk metallic glasses investigated by nanoindentation with a spherical indenter. *Phys. Rev. Lett.* **2004**, *93*, 125504. [CrossRef]

32. Hufnagel, T.C.; Schuh, C.A.; Falk, M.L. Deformation of metallic glasses: Recent developments in theory, simulations, and experiments. *Acta Mater.* **2016**, *109*, 375–393. [CrossRef]

33. Johnson, W.L.; Demetriou, M.D.; Harmon, J.S.; Lind, M.L.; Samwer, K. Rheology and Ultrasonic Properties of Metallic Glass-Forming Liquids: A Potential Energy Landscape Perspective. *MRS Bull.* **2011**, *32*, 644–650. [CrossRef]

34. Samwer, K.; Johnson, W. Structure of glassy early-transition-metal-late-transition-metal hydrides. *Phys. Rev. B* **1983**, *28*, 2907. [CrossRef]

35. Spaepen, F. A microscopic mechanism for steady state inhomogeneous flow in metallic glasses. *Acta Metall.* **1977**, *25*, 407–415. [CrossRef]

36. Ding, J.; Patinet, S.; Falk, M.L.; Cheng, Y.; Ma, E. Soft spots and their structural signature in a metallic glass. *Proc. Natl. Acad. Sci. USA* **2014**, *111*, 14052–14056. [CrossRef] [PubMed]

37. Harris, J.; Curtin, W.; Tenhover, M. Universal features of hydrogen absorption in amorphous transition-metal alloys. *Phys. Rev. B* **1987**, *36*, 5784. [CrossRef]

38. Chuang, A.C.-P.; Liu, Y.; Udovic, T.J.; Liaw, P.K.; Yu, G.-P.; Huang, J.-H. Inelastic neutron scattering study of the hydrogenated $(Zr_{55}Cu_{30}Ni_5Al_{10})_{99}Y_1$ bulk metallic glass. *Phys. Rev. B* **2011**, *83*, 174206. [CrossRef]

39. Suh, D.; Asoka-Kumar, P.; Dauskardt, R.H. The effects of hydrogen on viscoelastic relaxation in Zr-Ti-Ni-Cu-Be bulk metallic glasses: Implications for hydrogen embrittlement. *Acta Mater.* **2002**, *50*, 537–551. [CrossRef]

40. Dong, F.; He, M.; Zhang, Y.; Luo, L.; Su, Y.; Wang, B.; Huang, H.; Xiang, Q.; Yuan, X.; Zuo, X.; et al. Effects of hydrogen on the nanomechanical properties of a bulk metallic glass during nanoindentation. *Int. J. Hydrogen Energy* **2017**, *42*, 25436–25445. [CrossRef]

41. Tian, L.; Yang, Y.-Q.; Zhao, X.-A.; Wang, Z.-J.; Xie, D.-G.; Tönnies, D.; Roddatis, V.; Volkert, C.; Shan, Z.-W.; Univerisity of Göttingen, Göttingen, Germany. Unpublished work. 2019.

42. Gibson, J.M.; Treacy, M.M.J.; Voyles, P.M.; Jin, H.C.; Abelson, J.R. Structural disorder induced in hydrogenated amorphous silicon by light soaking. *Appl. Phys. Lett.* **1998**, *73*, 3093–3095. [CrossRef]

43. Voyles, P.M.; Gibson, J.M.; Treacy, M.M.J. Fluctuation microscopy: A probe of atomic correlations in disordered materials. *J. Electron Microsc.* **2000**, *49*, 259–266. [CrossRef] [PubMed]

MDPI

St. Alban-Anlage 66

4052 Basel

Switzerland

Tel. +41 61 683 77 34

Fax +41 61 302 89 18

www.mdpi.com

Metals Editorial Office

E-mail: metals@mdpi.com

www.mdpi.com/journal/metals